Sweet, Reinforced and Fortified Wines

DEDICATED TO OUR FRIEND AND SCIENTIST, BERNARD
Bernard Donèche
1950–2012

Sweet, Reinforced and Fortified Wines

Grape Biochemistry, Technology and Vinification

Edited by

Fabio Mencarelli
DIBAF, University of Tuscia
Viterbo, Italy

Pietro Tonutti
Institute of Life Science
Scuola Superiore Sant'Anna
Pisa, Italy

WILEY-BLACKWELL
A John Wiley & Sons, Ltd., Publication

Wiley-Blackwell is an imprint of John Wiley & Sons, formed by the merger of Wiley's global Scientific, Technical and Medical business with Blackwell Publishing.

Registered office: John Wiley & Sons, Ltd, The Atrium, Southern Gate, Chichester, West Sussex, PO19 8SQ, UK

Editorial offices: 9600 Garsington Road, Oxford, OX4 2DQ, UK
The Atrium, Southern Gate, Chichester, West Sussex, PO19 8SQ, UK
111 River Street, Hoboken, NJ 07030-5774, USA

For details of our global editorial offices, for customer services and for information about how to apply for permission to reuse the copyright material in this book please see our website at www.wiley.com/wiley-blackwell

Library of Congress Cataloging-in-Publication Data

Sweet, reinforced, and fortified wines : grape biochemistry, technology, and vinification /
[edited by] Fabio Mencarelli, Pietro Tonutti.
pages cm
Includes bibliographical references and index.
ISBN 978-0-470-67224-2 (cloth)
1. Wine and wine making–Chemistry. 2. Grapes. 3. Fortified wines. I. Mencarelli, F. (Fabio)
II. Tonutti, P. (Pietro)
TP548.S96 2013
663′.2–dc23

2012047725

A catalogue record for this book is available from the British Library.

Wiley also publishes its books in a variety of electronic formats. Some content that appears in print may not be available in electronic books.

We would like to thank Gill Whitley for her linguistic input.
Cover design by www.hisandhersdesign.co.uk

Set in 10/12pt Times by Aptara® Inc., New Delhi, India

First Impression 2013

Contents

List of Contributors

Daniele Accordini
Cantina Valpolicella Negrar
Negrar, Verona, Italy
Email: daniele.accordini@cantinanegrar.it

Andrea Bellincontro
DIBAF, University of Tuscia
Viterbo, Italy
Email: bellin@unitus.it

Claudio Bonghi
Department of Agronomy, Food, Natural Resources, Animals and Environment
University of Padova
Legnaro, Padova, Italy
Email: claudio.bonghi@unipd.it

Gabriele Chilosi
DIBAF, University of Tuscia
Viterbo, Italy
Email: chilosi@unitus.it

Kleopatra Chira
Faculty of Oenology, Université Bordeaux Segalen
Institut des Sciences de la Vigne et du Vin
Villenave d'Ornon, France
Email: cleochira@yahoo.com

Danilo Corradini
National Research Council (CNR) Institute of Chemical Methodologies
Montelibretti, Rome, Italy
Email: danilo.corradini@imc.cnr.it

Renzo Cotarella
Marchesi Antinori srl
Firenze, Italy
Email: renzo.cotarella@antinori.it

Riccardo Cotarella
Falesco Winery
Montecchio, Terni, Italy
Email: riccardo.cotarella@falesco.it

Bernard Donèche (deceased)
Faculty of Oenology, Université Bordeaux Segalen
Institut des Sciences de la Vigne et du Vin
Villenave d'Ornon, France

Claudio D'Onofrio
Department of Agriculture, Food and Environment
University of Pisa
Pisa, Italy
Email: c.donofrio@agr.unipi.it

Daniele Eberle
Terredavino SpA
Barolo, Cuneo, Italy
Email: eberle@terredavino.it

Osvaldo Failla
Deptartment of Agricultural and Environmental Sciences
University of Milan
Milan, Italy
Email: osvaldo.failla@unimi.it

Teresa García-Martínez
Department of Microbiology
University of Córdoba
Córdoba, Spain
Email: mi1gamaj@uco.es

Juan C. García-Mauricio
Department of Microbiology
University of Córdoba
Córdoba, Spain
Email: mi1gamaj@uco.es

Vincenzo Gerbi
DIVAPRA – Food Technology Sector
University of Torino
Grugliasco, Torino, Italy
Email: vincenzo.gerbi@unito.it

Tim Hogg
Escola Superior de Biotecnologia
Universidade Católica Portuguesa
Porto, Portugal
Email: thogg@porto.ucp.pt

Zoltán Kerényi
Corvinus University of Budapest
Institute Vitivinicole Research Station Kecskemét
Kecskemét, Hungary
Email: zoltan.kerenyi@uni-corvinus.hu

Alberto Mattiacci
Department of Communication and Social Research
University of Rome, La Sapienza
Rome, Italy
Email: alberto.mattiacci@uniroma1.it

Fabio Mencarelli
DIBAF, University of Tuscia
Viterbo, Italy
Email: mencarel@unitus.it

Silvia Mencarelli
Wine public relations consultant
Pisa, Italy
Email: silviaelettramencarelli@gmail.com

Luigi Moio
Department of Food Science
University of Naples Federico II
Portici, Naples, Italy
Email: moio@unina.it

Juan J. Moreno-Vigara
Department of Agricultural Chemistry
University of Córdoba
Córdoba, Spain
Email: qe1movij@uco.es

Nikolin Musabelliu
Winemaker
Toronto, Ontario, Canada
Email: nikolinmusabelliu@yahoo.com

Isabella Nicoletti
National Research Council (CNR) Institute of Chemical Methodologies
Montelibretti, Rome, Italy
Email: isabella.nicoletti@imc.cnr.it

Costanza Nosi
Department of Management Sciences
University Roma Tre
Rome, Italy
Email: cnosi@uniroma3.it

Paola Piombino
Department of Food Science
University of Naples Federico II
Portici, Naples, Italy
Email: paola.piombino@unina.it

Luca Rolle
DIVAPRA – Food Technology Sector
University of Torino
Grugliasco, Torino, Italy
Email: luca.rolle@unito.it

Laura Rustioni
Department of Agricultural and Environmental Sciences
University of Milan
Milan, Italy
Email: laura.rustioni@unimi.it

Giancarlo Scalabrelli
Department of Agriculture, Food and Environment
University of Pisa
Pisa, Italy
Email: gscalabrelli@agr.unipi.it

Attilio Scienza
Department of Agricultural and Environmental Sciences
University of Milan
Milan, Italy
Email: attilio.scienza@unimi.it

Pierre Louis Teissedre
Faculty of Oenology, Université Bordeaux Segalen
Institut des Sciences de la Vigne et du Vin
Villenave d'Ornon, France
Email: p.teissedre@u-bordeaux2.fr

Pietro Tonutti
Institute of Life Science
Scuola Superiore Sant'Anna
Pisa, Italy
Email: pietro.tonutti@sssup.it

Andrea Vannini
DIBAF, University of Tuscia
Viterbo, Italy
Email: vannini@unitus.it

Andrea Zanfi
Wine journalist
Grosseto, Italy
Email: info@andreazanfi.it

Glossary

Note to the Reader:

A glossary of the main technical, scientific and commercial terms, referring to grapes and wines, used in this book is provided below.

Dehydration: The dynamic process of water loss from the berries occurring on-vine (over-ripening, late harvest, icing, *Botrytis*) or after harvest under more or less controlled environmental conditions.

Drying: The process of intense water loss of the berries after harvest carried out under open-air conditions.

Fortified wines: Wines produced with freshly harvested or dehydrated berries by adding alcohol or spirits.

Raisins wines: A commercial category of wines obtained from more or less dehydrated grapes regardless of the method of dehydration or drying.

Reinforced wines: Dry wines (generally, but not exclusively, red) characterized by higher alcohol and secondary metabolites content produced with partially dehydrated berries (weight loss less than 25% of initial fresh weight).

Straw wines: A commercial category of wines (e.g. *Vin de paille*) obtained from grapes partially dehydrated after harvest on straw (e.g. straw mats) or similar organic material.

Withering: The consequence of the long dehydration process in grape berries.

Introduction

Fabio Mencarelli[1] and Pietro Tonutti[2]

[1] *DIBAF, University of Tuscia, Viterbo, Italy*
[2] *Institute of Life Science, Scuola Superiore Sant'Anna, Pisa, Italy*

The history of wine is a story of sweet wines. The first wines used by human beings were sweet wines; not because they selected these wines above others, but because of the climatic conditions where the grapes were harvested and processed, and the way in which these grapes were processed, which did not permit a complete fermentation. The wine was probably a mixture of alcohol, sugars and acetic acid, and to reduce the off-flavours in the wine, ancient peoples used to add honey and dry aromatic herbs. The Mediterranean basin is the cradle of these sweet wines, which were also defined, erroneously, as dessert wines. The climactic conditions of this basin were favourable for leaving the grapes on the vine to over-ripen, or for picking the bunches and allowing them to dry in the sun and wind. Most of the commercially well-known sweet wines are from the countries bordering the Mediterranean Sea; but not all these wines are sweet, a few of them are dry and full bodied such as Amarone in Italy. For this reason, we like to define these wines not simply as 'dessert wines' or 'sweet wines', but as wines made from grapes undergoing a dehydration process which can occur on-vine or off-vine. In addition, several wines that are traditionally and commercially very important, coming or not from on- or off-vine dehydrated berries, are fortified with alcohol. Thus, the world of this type of wines is very complex, and with this book we have tried to provide readers with the most exhaustive pool of information.

The book is divided into four sections. The opening chapter provides an historical overview and description of the cultural impact of sweet wines through the centuries in the Mediterranean area. The chapters in Part 2 cover the different technical and scientific aspects of the whole production chain, from the field to the cellar. The various solutions and diverse options for managing the vineyard and the dehydration process are thoroughly described. The following chapters specifically focus on the changes, at physiological and physico-chemical levels, characterizing the berries during on-vine and postharvest dehydration, and in relation to noble rot infection. The second section ends with the chapters extensively reporting issues, methods and technical solutions for optimizing vinification in order to achieve the best-quality sweet wines. Part 3 of the book comprises 11 chapters specifically dedicated to well-known reinforced, fortified or sweet wines produced in different areas of the world. By describing the specific characteristics and quality traits, these chapters highlight the cultural diversity and variety of protocols applied in this segment of the wine industry. The final section, Part 4, deals with the sweet wine market and marketing, and not only reports current statistics and figures, but also presents opportunities and strategies for the future.

There are several publications dealing with these special wines in many of the producing countries, but they are just guide books. Thanks to the contribution of a number of different

Sweet, Reinforced and Fortified Wines: Grape Biochemistry, Technology and Vinification, First Edition.
Edited by Fabio Mencarelli and Pietro Tonutti.

authors, well-known specialists (researchers and winemakers), we believe that this book is comprehensive in its descriptions of technical issues, scientific topics and the characteristics of the most important special wines.

Finally, given that these wines are often 'niche' products, not all of them are included in this edition. Any additional information from the readers on other special wines produced around the world will mean that this book can be considered a success, and will be greatly appreciated for future updates.

Viterbo/Pisa, October 2012

Part 1 History

1 Sweet Wines: The Essence of European Civilization

Attilio Scienza

Department of Agricultural and Environmental Sciences, University of Milan, Milan, Italy

Quis non malarum quas amor curas
habet haec inter obliviscitur?
(Among such delights, who cannot fail to forget
the sad cares that passion may bring?)
Horace, *The Book of Epodes*

1.1 HISTORICAL PATH

The aim of this chapter is to cover the broad subject of sweet wines throughout European civilization. A historical path will guide us from their origins to the present date.

The following are the constant elements in the production and trade of sweet wines:

- sweet wines have always been considered as luxury goods and imported by merchants;
- ordinary wines were normally produced and consumed within the local area, whereas sweet wines were generally produced close to commercial areas such as harbours;
- for light wines, climates and soils are key factors in order to obtain certain features; the quality of sweet wines, however, is more influenced by technology in the vineyard (choice of grape variety, late harvest) and in the cellar (concentration and stabilization techniques);
- consumption of sweet wines has always been regarded as fashionable rather than as a complement to food (unlike dry wines); as with all fashions, the consumption of sweet wines has had its own ups and downs, but its importance in the production of wine on a worldwide scale is always high, even though the market offers a full range of alternative drinks.

In *Les Memoires de la Mediterranée* (1999), Braudel claims that it is no great effort to feel at home within the familiar Mediterranean environment, whether in Venice, Provence, Sicily, Malta or Istanbul. There is an endless theme that links these places with their glorious past, through the names of their wines (especially the sweet wines). These places are the borderline between prehistory and our traditional history. We can identify this story with the birth of the first agricultural civilization not just in the Mediterranean region, but throughout Europe: the Fertile Crescent revolution.

Sweet, Reinforced and Fortified Wines: Grape Biochemistry, Technology and Vinification, First Edition.
Edited by Fabio Mencarelli and Pietro Tonutti.

1.2 ORIGINS

Wine is the symbol of the ancient peoples who developed in the Mediterranean region: the cradle of civilization. The first Sumerian evidence in the Fertile Crescent goes back to 3000 BC. The myth of wines spread throughout Aleppo, Ebla, Mori, Ugarit: the tablets of Paleo-Babylonian archives contain names of feasts and banquets where wine played a central role.

The origin of the word wine, '*vine*', in Hittite means 'stick of the Bacchants'; this identifies the sacredness of its use. Throughout the Mediterranean, we have similar words in the various linguistic groups. Even if they do not sound close, they share a common semantic root: *wo-no* in Linear B, *woinos* in Greek, *wo-i-no* in Mycenaean, *g-vino* in Georgian, *yayin* in Hebrew, *vinum-vinum* in Etruscan and *vinum* in Latin.

The Sumerian pictograms indicating the vine, the vineyard and wine are very similar to the ancient signs TIN and GESTIN. They are interpreted as a grape bunch and a pointed amphora. The correct translation of the signs of the ideogram GESTIN-HEA is not 'white wine', as it was believed in the past, but 'vine + sun', which literally means 'dried vine', therefore 'raisin' (McGovern, 2003).

GESTIN SA also means that wine is red, and it is followed by several adjectives such as 'good, sweet, pure, new'.

The adjective –LAL occurs quite frequently too; it means honey. As a matter of fact, honey was frequently added to fermenting wine must, in order to preserve it longer; in Linear B script, the ideogram 'wine' is often modified by associating the word 'honey'. The ideogram GES-IN DURU (interpretation reported by the *Chicago Hittite Dictionary*) means 'fresh grapes', while GESTIN generally indicates wine; if associated with DU-GA it means first quality wine; –VS means second quality and HALLUM is a wine which becomes vinegar. In the Mari archives, an account book of the court mentions the existence of a bitter wine and of a good one: the expression SHA SHATE' BELIYA means '[of the quality] which is drunk by my lord'.

Why was this wine sweet? The clay tablets on which the administrator of the royal warehouses carefully recorded wines constantly show the logogram GESTIN-HAD-AV or GESTIN UD, indicating dried grapes; they were widely used for food rations, as an offer to the gods or as an ingredient in medical preparations.

The red wine SA GESTIN KUB was diluted with water in order to better symbolize blood: this is clearly stated in many Ugaritic prayers of the Near East.

In the Hittite and Thracian traditions, offering sweet wine was a privilege of the king: a precise social symbol of power. In the Hittite tradition, it is through the power of sweet wine that the King-Priest *Ullikummi* is able to capture the snake *Illuyanka*. This representation often occurs in Greek mythology: a further piece of evidence of the assimilation of the oriental culture.

LA'L GESTIN-KU was the sumerogram for a natural sweet wine that could not be consumed in ordinary situations; it was precious, so it had to be offered to the gods (Gennari, 2005).

There is evidence of a preference for the sweet taste of wine also during the Egyptian transition: the jars in the tomb of Tutankhamun contained sweet wines. It is actually during the period of the New Empire that we have the first evidence of the use of heat to concentrate must; this technique was used to produce sweet and alcoholic wines for long storage.

Almost 1500 years BC, when Egypt was ruled by the Hyksos, a Semitic people from Syria and Palestine, sweet wines started to be produced in the town of Avaris, in the Nile delta. This was discovered in the mid 1990s, when archaeologists found a structure for pressing grapes and identified a vineyard called *Kaenkeme*, where a wine '... which was sweeter than honey' was produced. Through the analysis of remains inside the ollas, molecular archaeology confirmed that the wine was red. The red writings (*ostraka*) on the jars stated the production areas, the style of the wine and the addition of resin or terebinth (*Pistacia terebinthus*). The indication 'sweet' (*vip*) is the most frequent one, even though it could also have marked the addition of figs and honey.

Vine and wine are also often indicated by the term *kur*. *Kurum* has the semantic value of red wine, *karanu* in Akkadian, *carenum* in late Latin and *careno* in ancient Italian, with the meaning of cooked must. This shows how the practice of drying grapes and concentrating must through heat had the same importance in the production of sweet wines. These two techniques were often used together. There is a city in the inland of Judea called Lachish where archaeologists found an Iron Age jar (second millennium BC) with ancient Semitic inscriptions like 'wine made with black raisins', 'smoked wine', 'very dark wine'; this confirms the habit of mixing wine from dried grapes with must that was concentrated through direct contact with fire, which gave the wine a smoky-caramel taste.

The terms that we find in Ninurta's Georgics, dating back to the second millennium BC, refer to sweet red wines, with an explicit reference to blood, as they were used in rituals. For these purposes, people chose vines yielding wines that were light in colour and had a yellow rim, so that the wine looked like blood without the addition of water, as required by the ceremonial. This is the reason why in Magna Greece and Sicily we can still find such grape varieties as Frappato, Nerello and Gaglioppo, low in total anthocyanin and high in cyanin, which is responsible for the light yellow nuance in wines, especially those obtained by drying grapes.

In the ancient Greek tradition, men who were responsible for the community agreed that sweet wine was the best way to establish a new relationship with foreign people (*philoxenìe*).

Homeric poems are full of episodes, actual *topoi*, where sweet wine was the unifying element of relationships among men.

In Book VII of *The Odyssey*, among the Phaeacians, Zephyrus blows and dries the ripe grapes in the sun; in Book IX, Ulysses offers Polyphemus the sweet, black wine that was given to him by Maron; in Book XI a sacrifice is offered mixing milk, honey and sweet wine; in Books XII, XIII and XIV there are references to the colour of wine, which is red or looks like fire.

In the Homeric poems there are many descriptions of wine evidencing the poet's attention to the aesthetic side of wine. Colour was almost always red: purple wine (to extinguish the funeral stake), rubicund nectar, rubicund wine, reddish wine, black wine, vermilion liquor, ancient wine, purple nectar, rosé nectar, sweet wine, black nectar (Maron's sweet wine), soft wine, incorruptible wine (sweet, alcoholic?), soft liquor (sweet?), smoky wine (alcoholic), sweet grape's liquid, tasty wine, powerful wine.

1.3 DIFFUSING THE MYTH OF WINE

The wine known as wine *par excellence*, which was traded by the Phoenicians and Greeks throughout the whole of the Mediterranean area, where the symposium ritual was spreading, was almost certainly a sweet wine. It was a luxury good destined for the upper classes and it was also among the most valuable goods for exchange. Only wines with high sugar levels could travel for such trade purposes.

We usually ascribe the improvements of viticulture and enology to the meeting of the oriental and occidental cultures. In fact, new productive varieties were introduced, along with drying techniques to improve the ageing potential of wine. Moreover, new containers were more resistant, easier to produce and to transport: further facilitating trade. The most important innovation was the intuitive introduction of the Greek *emporium*. Wine was transformed from a simple alimentary product to a bargaining chip; moreover, it was associated with the worship of a god who was the protector of viticulture. Such an interpretation, which could seem ideological, is broadly justified by the fact that wine and vine have a huge symbolic importance in the European culture, especially in religion and politics (for ruling and controlling). Possibly, this is the reason why innovation was limited to small daily steps forward by the growers to lighten their work load or to improve the productivity of their plants. Enology underwent even slower changes, and ancient techniques still survive in some parts of the Caucasus, Portugal and Greece.

Production techniques were highly influenced by the development of the local economies and consumption habits. In places characterized by subsistence farming, wine was produced for self-consumption. In such cases, the main features of mixed farming did not change until the arrival of the American diseases. In those places where grapes were grown for wine trade, farming and winemaking techniques underwent several changes. Vines were located along the main trade routes and near harbours, thus they benefited both from proximity to a border (improvement by comparison) and from consumer feedback, demanding new wines made using new techniques. The development of colonies in South America, South Africa and Australia, together with the increasing cost of transport, led to the production of wines and spirits that, through fortification, could both resist long trips and be less bulky to transport. Innovation led to the selection of the most suitable grape varieties for over-ripening (or for noble rot) and to winemaking techniques involving the addition of high levels of sugar and alcohol; as distillation spread, alcohol was used to fortify musts and wines.

Over the centuries, winemakers tried to meet the taste of consumers who, according to the current fashion, wanted wine to be white, red, alcoholic, fortified, sweet, and they adapted viticulture and winemaking according to the market.

The so-called 'permanent evolution' phase started towards the end of the seventeenth century, due to three key factors: knowledge, competition and investment. Producers aimed at improving quality wines for long ageing, sparkling and sweet wines.

The development of chemistry and fermentation allowed the improvement of yeasts and the introduction of sulphur dioxide (SO_2) as an antiseptic and a preservative. The other huge factor was the industrial production of bottles and corks.

1.4 CLIMATE CHANGES, THE DEVELOPMENT OF VITICULTURE AND THE PRODUCTION OF SWEET WINES

During the third millennium BC, the habits of the Sumerians and their relationship with the gods reveal that the earth was threatened by unpredictable violent forces; heavy rainfalls arrived inopportunely, flooding entire villages. A rupestral relief from the third millennium shows the Hittite King pouring sweet wine for the god of the sky, Tarhunta, asking him to save his vineyards. A Hittite tablet about climate changes states '([Observe]) raisins. As they preserve wine in (their) heart, [. . .] (even) you, god of the tempest, preserve prosperity, vigour, long life (and) the joy of the king, of the queen (and) of the children in (your) heart'.

This prayer is justified by the climatic changes that were destroying the vineyards. Drought became so severe in Mesopotamia at that time, that most southern towns were abandoned and new residential and trade centres were established in the northern part of the region, towards Anatolia. New contacts with the inhabitants of the towns of Ebla and Mari, together with commercial relationships with the Mediterranean ports of Cananeia and Ugarit, enhanced the diffusion of vine and the production of sweet wines in Thrace and Greece from the second millennium BC. The serious drought that struck Anatolia, Mycenae and Crete in 1200 BC was only the starting point of a series of famines that pushed the so-called *Peoples of the Sea* to invade the more fertile Mediterranean lands, causing wars and devastation that led to the end of the Egyptian kingdom and pushed the Greeks towards the west, founding colonies in Italy, France and Spain, where they started the production of sweet wines.

There was then another long period characterized by favourable climate conditions for the development of agriculture (especially grain). Around the first century AD, the climate changed once again in northern Europe. In 92 AD, Emperor Domitian had to forbid viticulture in all Roman territories outside Italy, so that grain could be grown in its place, since drought made it impossible to grow enough in Sicily and northern Africa.

Emperor Probus also grappled with climate change. Two hundred years after Domitian, he had to encourage viticulture along the borders by introducing Pannonian grape varieties, since wine supply from the *Venetiae* region was extremely reduced due to the mismanagement of water in those territories, and the grape varieties from Campania and Latium struggled in the cold weather of the Danube and Rhine regions.

Also in Latium, growers struggled to late harvest their grapes to produce sweet wines, and therefore suffered competition from Greece, whose wines were very expensive, but demand was so high that they were sold anyway. The Latin Georgics (Cato, Pliny, Columella) report about an alternative way of producing sweet wines by concentrating must up to a quarter or a half of the original volume. This product was then added to fermenting wine must in various proportions (from 1/4 to 1/30).

Around the year 1000, Europe benefited from a warm climate that extended viticulture towards high latitudes, such as Scotland, and very high altitudes, up to 1200 m above sea level. This phase was called *optimum climaticum* and viticulture, together with olives, expanded within the alpine valleys and spread throughout continental Europe, thanks to religious institutions.

At the end of the fourteenth century, Europe had to face the first negative effects of a climatic phase called the 'Little Ice Age', which, with the big 1709 frost, destroyed most of the continental agriculture. This period officially ended in 1850, with the '*Irish potato famine*'.

Towards the end of the feudal system, the location of European vineyards underwent a drastic change: vines disappeared from England and from the internal alpine valleys. Poor climate conditions led to a lack of grain, and consequent famine. Black Death spread and wines were of such poor quality that they were unable to last until spring without becoming too acid. The nobles and higher clergy were not prepared to tolerate drinking low-quality wines, and they started to look again at eastern Mediterranean production. Thanks to the mercantile activity of the Republic of Venice, wines like Greek Malmsey, Vinsanto and Muscats spread throughout Italy and northern Europe. The huge commercial success of such wines encouraged the production of similar wines in many Italian locations and along the oriental Adriatic coast. This was helped by the Turkish conquest of Crete in 1564, because Venice was deprived of the vineyards yielding most of its Malmsey. However, the merchants in Venice were anxious not to lose such a precious market, so they encouraged the production of sweet and aromatic wines from any suitable grape variety in different locations, and traded all these wines under the name of Malmsey. This is the first example in Europe of a famous varietal name being used for other varieties that had no genetic relationship. They were distinguished by their location of origin and by the peculiar features of the grape bunch. Their only common point was to yield a sweet, aromatic and alcoholic wine. This is also the first example of a grape variety named after a wine: usually, it happens the other way round. French and Italian agronomists between the sixteenth and eighteenth centuries suggested several techniques to improve late harvest even in unfavourable climatic conditions (such as stacking up whole bunches of grapes for several days to drain the must) or winemaking (like adding concentrated must).

This climatic change did not strike the southern regions of Italy, which became the most advocated areas for the production of sweet wines.

Large quantities of wine produced in the area of Santa Severa left the Tyrrhenian coast from the harbour of Tropea. They were called Greek wines, and reached the markets of northern Europe, competing with the (*sweet* and *round*) Malmsey from Venice or the so-called *strong* wines from southern France (sweet, aromatic, alcoholic).

Calabria gained a key position in the trade of sweet and often aromatic wines in many northern Italian towns, both for their quality and for their competitive price compared to the wines from Greece (Candia) and Lebanon (Tyre). In fact, the latter were imported by the Venetian merchants and suffered from the so-called 'revolution of freight charges', which made it unprofitable to trade low-quality wines. As a consequence, low-quality wines lost their market share and many marginal production areas (in the mountains, far from harbours and cities) that had risen up during favourable climate conditions – when there was a demand for wine as it was perceived as a status symbol – were abandoned.

1.5 THE SWEET WINES OF CLASSICAL GREECE: GRAPE VARIETIES, TECHNIQUE AND PRODUCTION AREAS

In ancient times, people were particularly keen on sweet foods, even though at that time they were not easy to find, which is why sweet wines were considered the highest pleasure. Through the addition of dried grapes or by cooking must, producers could make sweet drinks by adding calcareous powder (obtained by crushing shells), or sea water. This lowered acidity, making the wine sweeter and less aggressive. The wines obtained by late-harvested grapes were called *saprias oinos*, while *melilotes* was a concentrated must with the addition of honey and salt. In spite of being very expensive, honey was the most popular sweetener, but a similar effect was obtained by concentrating must through heat. Adding sea water to honey (making a mixture called *thalassomeli*) was common in the preparation of the most famous sweet wines.

Discorides is the first author to whom we ascribe the distinction between *Pramnios* or *Cretico* wine, obtained by late-harvested grapes, and *Siriaisos* or *Hepsema*, mainly produced in Crete from cooked must.

The first evidence of the production of sweet wines in Greece is found in the poetry of Hesiod, who had worked as a farmer, and addressed himself to the people of his homeland: Boeotia.

In his masterpiece *The Works and Days*, in the section about autumn (609–614), he says, 'But when Orion and Sirius are come into mid-heaven (towards mid-September), and rosy-fingered Dawn sees Arcturus, then cut off all the grape-clusters, Perses, and bring them home. Show them to the sun ten days and ten nights: then cover them over for five, and on the sixth day draw off into vessels the gifts of joyful Dionysus'.

Greek and Byzantine literary sources quote many different grape varieties suitable for making sweet wines from raisins. Some of these varieties called *Xenologos* (the etymology of the name recalls the way they were used in viticulture) were used only for drying. In fact, such wines as *Tharrupia*, *Buconiates*, *Capnios* and *Sirpula* refer respectively to *tharrupio*, the fruit-drying rack made of canes; *buconiates*, the pine cone from where the resin for conservation was extracted; *capnios*, the cooking technique that gave the smoky flavour; and *sirpula*, the operation of twisting the bunch on the plant to make it dry better.

Pramnio (*Iliad*, XI, 631; *Odyssey* X, 225): this grape variety was grown in Izmir, Lesbos and Icarus to produce generous sweet wines. The wine was obtained by fermenting a must particularly rich in sugar, as it drained from over-ripe grapes that were stacked up.

Biblino: from Byblos, in Phoenicia. This grape variety was grown in Thrace and Naxos (Sicily). It was the wine for feasts, the most famous among sweet wines (*Idylls* by Theocritus XIV, 4). Hesiod (*Works and Days*, 609–614) forecast its late harvest. It gave a black wine, intense, for long ageing; it is the wine Ulysses offered to Polyphemus. The grape grown in Chios belongs to the same varietal group.

This is how Hesiod describes it (*Works and Days*, 590–596): 'But at that time let me have a shady rock and wine of Biblis / then also let me drink bright wine, sitting in the shade / from the everflowing spring which pours down unfouled thrice pour an offering of water, but make a fourth libation of wine'.

Phanaios: highly appreciated by Virgil, who called it the king of wines (*Georgics*, II, 98).

Psithia: quoted since the fourth century BC with the name of *anaxandrides* (grapes produced by vines trained on trees). Columella calls it '*species uvae graculae*' and writes about its low productivity and its feature of producing sweet wines after the grapes dry. Virgil highlights its aptitude for the production of sweet good-quality cooked wines and the particular taste of grapes (*Georgics*, II, 93). He calls it *psizio*, and another wine, which was black in colour, he calls *melampsizio*.

Leucothrakia: a sweet white from Thrace, made from dry grapes and bound to the fame of Santorini.

Methymnacea: from Lesbos. Virgil calls its wine *rex vinorum*.

Sticula: appreciated by Pliny (XIV, 9, 11) as a long-lasting grape, but also *dulci vino apta*.

Samia: from the island of Samos, quoted by many Greek and Byzantine authors, it produced the sweet wine *anthosmias* through a peculiar technique: the juice, drained from over-ripe grapes, was mixed with must obtained partially by pressing and partially by boiling. Ash and lime were added too.

Mandes: from the Falcidia region, it produced a wine from over-ripe grapes called 'god's pee'.

Winds played an important role in the choice of the most suitable locations for the production of sweet wines from over-ripe grapes. Athenaeus (I, 26c) stated that sweet high-quality Falernian wine was obtained when, during the final phase of maturation, a southern wind blew (called 'autumn wind' in the Vallese). Martial (III, 58, 6f) quotes several locations uphill where very dry weather conditions allowed harvesting in November, during the first snows. These grapes were called *seras uvas*. Also sweet wines from Macedon were obtained by drying grapes cultivated on the south-facing slopes of the mountains (*sta prossilia*) at altitudes of about 600–700 m above sea level; such grapes were able to dry not because of the heat, but because of the long, dry autumns. Usually, a mix of white and black grapes was used, and the best *listò krasi* (Vinsanto) was called *èliaumènon*.

1.6 THE GREEK SYMPOSIUM AND THE ETRUSCAN CONTINUERS

When one thinks of Classical Greece, images of vascular paintings can quickly come to mind. A deeper analysis of such representations highlights the exclusive function of containing wine that such pottery had, between the sixth and fourth centuries BC (Figure 1.1).

The literary sources and the images on the various pieces of pottery reveal how such consumption was the central element of socialization in the various Hellenic communities. Wine was the gathering point of a specific conviviality, which was different from banquets (*deipnon*), where wine was simply complementary to food.

Such activity was called *symposion*, and it is conceptually close to a private feast, a sort of *drinking-party*. It was also very different from a Roman *convivium*, more similar to a *deipnon*. The Romans called the symposium a *comissatio* or *compotatio*, a word of Greek origin.

The Greek word symposium actually means 'drinking together' (*syn-potein*). It was a social event with social drinking at its centre. Everything followed set rules: type, theme and manner of toasts. The main prescriptions though concerned the ritual of mixing wine and water in the *crater*, because every mixture had a different effect on the guests according to the different phases of their inebriation – which all guests had to reach and which was one of the main values of a symposium: the equality of the prefix *syn* which constitutes the word 'symposium' (Catoni, 2010).

In his *Convivial Questions*, Plutarch says that *Lathikedès* is the perfect wine–water mixture in the proportions of about 3 : 2.

Athenaeus, in the *Deipnosophists* (XI, 462c–f) (a banquet of philosophers), in the elegy of Xenophon, describes the celebration of a symposium: 'For now the floor and all men's

Figure 1.1 An image of a symposium from a Greek Attic red-figure cup attributed to the painter Douris (*ca* 500 to 460 BC). (For a colour version of this figure, please see the colour plate section.)

hands are clean, and all the cups, and since the feasters' brows are wreathed with garlands, while the slaves around bring fragrant perfume in well-suited dishes; and in the middle stands the joyful bowl, and wine's at hand, which never deserts the guests who know its worth, in earthen jars well kept, well flavoured, fragrant with the sweet fresh flowers; and in the midst the frankincense sends forth its holy perfume; and the water's cold, and sweet, and pure'.

The ceremonial of the symposium was strictly connected to the myth of Dionysus: the god who taught men how to grow vines, produce wines and especially in what proportions wine had to be mixed with water. The constant practice of late harvest and of drying grapes in Thrace, homeland of Dionysus, with the consequent high alcohol volume, were the original reasons why it was forbidden to drink pure wine in Greece: it was considered a dangerous drink, similar to a drug. The wine for the symposium was sweet and alcoholic, because only a concentrated wine could be mixed with water. Moreover, it was popular to mix wine with grated cheese, usually goat cheese (this reminds us of the present-day habit of drinking sweet wines with strong cheese), and this could be done only with sweet wines, or wines that were sweetened through the addition of honey, as we can read in *The Iliad* (IX, 638): 'In this (cup) the woman, as fair as a goddess, mixed them a mess with Pramnian wine; she grated goat's milk cheese into it with a bronze grater, threw in a handful of white barley-meal'.

This drink called *kykèon* (from *kikòn*, to mix) is the same as that prepared by the slave Ecamede for Nestor and the injured Machaon, and that Circe offered to the Achaeans before transforming them into pigs (*Odyssey*, XI).

Outside Greece, only the Etruscans and a few communities around the areas of Apulia and Lucania can be considered as continuers of the symposium tradition. This is shown by the many grave relics of the various ages. A few variations were introduced, like the game *kottabos* and the presence of women.

In Taranto, one of the Magna Greece cities where the tradition of the symposium was particularly strong, wine was described as 'light, lacking in violence, fresh, sweet and easy to digest'. Near the agora of Sibari, in Metaponto, a big *kantharos* was found. Around its edge was the following inscription: 'to dispense nectar for the mortals to drink'. This is evidence of the diffusion of the symposium in Magna Greece during an early phase of colonialism. The preparation techniques of sweet wines were also adopted in the colonies; the most ancient trace in Italic is a wall painting on a *guttus* from the second half of the fifth century BC from Sicily; it is interpreted as *vinum defructum*, cooked wine.

The wine that was mainly used in these Magna Greece cities came from Lagaria, an Ionic town founded by Epeo, the legendary builder of the Trojan horse. Strabo, in the second half of the first century BC, calls such wine sweet (*glykùs*) and smooth (*apalòs*).

Another important occasion for drinking sweet wines in Classical Greece was when people went to sanctuaries to consult the oracles. Along the route there were many places where pilgrims could stop and rest. Such hostels usually offered sweet wines (*glukus*). The verb *glukizen* means 'to offer sweet wines' and *glukimus* 'place where sweet wine and food are served'. The words *glukus and glukimos* often appear in inscriptions about banquets, meaning a moment just before dinner when an aperitif was served.

1.7 PRODUCTION AND CONSUMPTION OF SWEET WINES IN ANCIENT ROME

In ancient Rome wine was a way to communicate; it was synonymous with feasting, conviviality and pleasure, during great banquets as well as at the tables of the poor. Wine was always served; this is evidenced by literary citations, still-life wall paintings, sculptures, Bacchic low reliefs, and gravestones representing tools for cultivation, for building barrels, and trade scenes. For the Romans, good wine was a synonym of old wine. Sweet wines kept better than dry ones, even though interrupting fermentation was really hard at that time.

The Latin *Georgics* often quote wine but without describing its organoleptic features. They are precise about the places of origin (Falerno, Retico, Massico, etc.) and about the grape varieties (*aminee*, *elvole*, *rodie*, *dattili*, *eugenie*, *graecula*, etc). Pliny praises the wine from Lebanon, which 'had the perfume of incense': it was a sweet wine produced with raisins and slowly cooked must.

Martial, who is ruthlessly ironical about Rome in the first century, does not forget to mention wine.

There are 90 *topoi* where wine is precisely described, amongst which we have the wines sweetened with honey that were harvested in Crasso, and those with the addition of resin from Vienne.

But which wines did the Romans actually drink? Wines for the upper classes were generally aged (*vinum vetus*, *sequentis gustus*), while among those drunk by the farmers were *posca*, a mixture of water and vinegar, and *lora*, a very light wine obtained by rinsing pomace with water after all the must had been pressed out (this was still done in Italian rural areas until about the second world war). The latter was the drink for the slaves.

Another rather poor wine product was *vappa*, obtained by refermenting a wine with residual sugar during the summer.

Columella stated (XII, 6, 9, 2) that the best wines were those which kept without any special treatment, while the lowest quality wines were those harvested either from vines that were far too young, or from unfavourable *terroir*; these needed the addition of concentrated must or resin.

Columella, just like Hesiod, provides us with precise instructions about how to make *passum* (from sun-dried grapes) and the so-called second *passum*; these are the most detailed enological indications from the period.

The technique described by Pliny (*Naturalis Historia*, 8, 23) is particularly interesting, since it differs from those of both Hesiod and Columella. For some famous wines, such as those from Cos, Lesbos and Clazomene, he introduces, before fermentation, maceration of the sun-dried grapes in a mixture of sea water and cooked must which had to boil for one night and one day. This prevented the action of acetic bacteria and slowed down yeasts, in order to obtain a wine higher in sugar.

The best mix though was of old Falernian wine, which had a bitter character, with Greek honey (*vinum melle conditio*).

The diffusion of must enrichment techniques (by adding honey or by heating the must itself) was also the result of Roman military expansion throughout Europe, which carried the wine culture with it.

The use of *defructum* and *sapa* (boiled concentrated must) and of *mulsum* (sweetening through addition of honey) helped to improve low-quality wines for export.

Cooked must or wine had two names, depending on the degree of concentration: *defructum* (from *defeuere*, which means to chill after boiling; although this is misleading because it could also mean that heat was the result of the fermenting process or of fire heating), and *sapa*, which had a much higher concentration, from two-thirds to one-third of the initial volume. During the heating process, especially if the containers were made of copper, wine often acquired a bitter, unpleasant taste; therefore certain aromatic substances were added, such as iris or Greek hay. Pliny preferred a lead container, Columella agreed, but it had to be treated with oil. Cato shows no preferences. Two days after fermentation started, *defructum* was poured into the *dolia* in the proportion of 1 : 48, in order to increase the alcohol content.

There were two types of *defructum*, a simple one with chalk and salt, which preserved and stabilized wines, and a spicy one (mainly used to produce a medicine), to which resin was added together with spikenard, cinnamon, iris, saffron.

Cato (*De Agri Cultura* XXIV, 1) gives us a recipe for Greek wine, 'gather carefully well-ripened Apician grapes, and add to the *culleus* of must two *quadrantals* of old sea-water, or *a modius* of pure salt. If the latter is used, suspend it in a basket and let it dissolve in the must. If you wish to make a straw-coloured wine, take equal parts of yellow and Apician wine and add a thirtieth part of old boiled wine. Add a thirtieth part of concentrated must to any kind of blended wine'. The author also describes how sweet wine was made in Cos and Chios: over-ripe grapes were put in sea water for three days before pressing.

Sometimes, to enhance the drying process on the plant, the bunch was twisted (Palladius, XII, 22) or the branch was cut (Pliny, XIV, 89).

Apicius, a famous cookery expert in ancient Rome, gives his own recipe for producing a sweet wine to use as an ingredient for very elaborated dishes (*De re coquinaria*, I, I,1): mix 15 parts of honey with about one litre of wine into a bronze jar and let them cook together. When it starts to boil, remove from fire and let it cool down. Repeat this three times. Add four ounces of grained pepper, three *scrupoli* (24 parts of an ounce) of resin, a *drama* of spikenard and saffron, five toasted date stones and date pulp soaked in wine. Add 18 *sestori* of diluted wine, heated on the fire.

In Italy, the affirmation of quality wines linked to their place of origin (especially in Campania, with Falerno) started late, around the second century AD.

A quality wine was usually a 5- to 25-year-old white wine, sweet or fortified. Horace is the only author who, towards the end of the Republic and the beginning of the Empire, senses the change in the aristocracy's tastes for styles of wine. In his *Satires* and *Odes*, Greek wines are hardly mentioned, with the exception of the wine from Chios. We have many wines from Campania and Latium, such as Cecubo, Massico and Falerno. This is very important, because it shows that Roman taste was shifting from sweet to dry wines. Moreover, wine was no longer drunk for ritual purposes; it became a complement to food. Late harvests were gradually abandoned to produce less sugary and more acid wines.

Sweet wines were still produced during the expansion of the Empire, and reached places far away from their production areas; this is demonstrated by the many findings of coeval jars, particularly those of the Dressel series, which take the place of Greek and Etruscan jars.

We have an interesting report by Diodorus Siculus, who describes the Gauls as heavy drinkers who did not mix wine with water. They made the fortune of the Roman merchants, as they would pay the highest prices to have the sweet Roman wine. In Europe, the sweet taste of wine became a synonym for quality. Pliny the Elder reminds us that the Gauls quickly picked up the art of winemaking from the Roman legionaries, although they used different varieties (*allobrogica*, *biturica*, *raetica*, *carbunica*, etc). Moreover, they produced wines similar to the Greek and Latin *passum*: sweet wines that could be kept in barrels of circled wood (a novelty at that time).

Jars are still used in the trade of wines from Provence which, during the second to third centuries AD were counterfeited using Italian names, such as *amienum*, a wine from a Greek vine which was popular in Campania, *passum*, from dried grapes, and *mulsum*, a wine with the addition of honey.

Columella was born in Cadiz, in Andalusia, called Betica at that time. There they produced heated concentrated musts similar to *defructum* and *sapa*. They were used to improve low-quality wines from Rome, instead of using honey, which was more expensive. The phenomenon became so big, that some merchants put the statement '*sine defrito*' on their jars, to underline that the wine was of high quality.

During the Imperial Age, the Greek merchants brought to Rome wine from the oriental Mediterranean, but they no longer used Cos or Rhodes jars (as in the past), as they were a synonym for low-quality, counterfeited wines. Dressel 2/4 were used instead, and they were produced where the wine was made. They contained a *passum* wine, like the one from Crete. It was the best known; it had an aroma of Muscat, as it was obtained from late-harvested bunches that were twisted on the plant, in order to enhance the loss of water from the grapes without spoiling the aroma characteristics.

Consumption of wine in ancient Rome was prohibited to women. A husband could control this by kissing his wife. Wine was associated with betrayal; it was not Catulus' 'basia' (*Da mihi basia mille, deinde centum* ...), the love kiss, but the 'osculu' (from *os*, mouth) to control the breath. The only alcoholic drinks that were permitted were sweet ones. Actually, these were not *vina* (wines obtained by fermenting grapes) but *dulcia*: a mix of cooked must and water, often with the addition of myrrh, as Pliny makes clear (*Naturalis Historia*, XIV, 83): '*medium inter dulcis vinumque est quod graeci aigleucos vocant hoc est* sempre *mustum*'. Women were allowed to drink sweet wines that tasted more like honey and milk drinks (during the 'Parilia' a mixture of honey and milk was drunk) and these drinks were concocted as opposed to wine. This prohibition disappeared towards the second century BC.

1.8 THE ECONOMIC AND SOCIAL ROLE OF SWEET WINES IN THE MIDDLE AGES

Between the fourteenth and sixteenth centuries, people were particularly concerned to highlight the lifestyle differences between the social groups. Foods, but particularly wines, were used by the upper classes as a status symbol; every social class had its own style of wine.

These included the sweet wines from Cyprus, the so-called Greek wines coming from the different routes of the crusades (traded by the Venetians, who at the same time were transporting the soldiers during the Fourth Crusade), or wines like Grenache, Garnacha, or Vernacce, which is both a grape variety and a Mediterranean style of wine. All these were sweet, fortified, very alcoholic and were drunk only by the upper classes.

This is a period when white wines become more fashionable; the bread for nobles had to be white, made from wheat flower, whilst that for farmers and slaves was black, made from rye. In the same way, black wine was for the 'working classes' while the upper classes, who did not need to work, would drink whites and *clarets*. This was also due to the effect of the Little Ice Age, when red berries struggled to ripen and the white and amber 'Greek wines' arrived in Europe.

White wines were dominant during the Middle Ages, and they were an imitation of those coming from Greece.

Cooked medieval wines were 'strained, smelly, adulterated', obtained by cooking wine or must in order to make them last longer (a real problem at that time). Some Byzantine writings provide us with precise instructions on how this was done. The most popular technique was to cook must, reducing it by a quarter; wine would then last up to 3–4 years. If reduction was higher than that, up to three-quarters, the result was called *sapa*; it was no longer a wine, but a food dressing.

All these wines were still called by their Latin names (*defructum, passum, carenum, sapa*), although vernacular terms started to appeared, such as *defervendo* instead of *defructum, defraudato*, meaning adulterated, fake, not natural; *sapa* was called *vin sireo* (*sireum*) and *passum* became *vin passo* and was a name for any sweet wine. Each cellar, especially in central Italy, had a room for heating wines; the average capacity was of about 3 hL: '*pulcra de ramine apta ad coquendum vinum*'.

Given their high alcohol content and pleasant flavour, which tempted people to exceed, doctors and medieval moralists advised people to drink only little quantities on particular occasions, such as feasts and weddings, or for medical reasons.

Wines were then classified in opposite categories: on one side the sour/bitter wines low in alcohol, and on the other side the sweet and alcoholic ones.

Medieval and Renaissance cookery, which were popular at all levels of society, focused not so much on food matching but rather on the contrast between 'cold' and 'warm' foods and drinks, in order to reach a 'temperate' balance. Sweet wines were considered 'warm'.

In the Middle Ages, what 'tasted good' was considered to be healthy: a good wine would certainly have a good effect. Hence, a wine was chosen depending on its flavour, which was a guarantee of health.

'Sweet' was the most popular 'warm' flavour, and it was often contrasted to sour (like vinegar), considered cold (Bartolomeo Anglicus, *De proprietatibus rerum*, 1230–1240). People who were recovering from an illness were advised to drink sweet, clear, red wine. It had to be drunk in moderation, as it had good nutrients.

According to the cold–warm theory, there were two other elements to be considered in the choice of the style of a wine: the season and the geographical location of consumption. Pope Paul III Farnese was very scrupulous in following such prescription (as his cellar man Sante Lancerio reports in 1549): cold wines (low in alcohol) were consumed in the summer and sweet, alcoholic (warm) wines were chosen in winter. Sweet wines though were not recommended for popes, since they were associated with lust.

This distinction in the consumption of wine becomes an actual literary *topos* in Florence, during the second half of the fourteenth century; authors, to highlight moral or comical situations, emphasized the warming properties of sweet wines.

In these stories we find Malmsey, Romania, Vinsanto, Vernaccia of Coniglia, and they were consumed on solemn occasions, such as imperial visits or weddings (even the poor ones, as sweet wine was a symbol of sexuality, as in Boccaccio's *The Decameron*). For the latter reason, pilgrims were discouraged from drinking them along their journey to the Holy Land: they were easy to find on the way, but they stimulated lust. This is also the reason why they were always included in recipes for helping conception, together with pork, pepper and ginger.

The carnival songs of the *clerici vagantes* during the late Middle Ages praised sweet wines, as we read in this line by Morando from Padua: '*Vinum dulce, gloriosum/pingue facit ed carsosum/atque pectus aferit* . . . '.

Ippocrasso was a special medieval wine, typically on the tables of the upper classes up to the seventeenth century, when it was substituted by brandy. Because of its warming and digestive properties, it was served at the end of the meal with desserts. It was made with high quality red wine, spices and sugar.

The popularity of sweet wines encouraged market growth and many wines were traded under the old popular Roman names. In Lombardy there was a sweet wine called Falerno ('*vinum phalernum suavissimum, et pulcrum* . . . '), obtained by mixing an old wine with sweet must and fresh grapes.

1.9 VINSANTO: A METAPHOR FOR MEDIEVAL SWEET WINES

Vinsanto was the most popular and copied wine of the Italian Middle Ages. While Malmsey was the favourite among young men from the upper classes, Vinsanto was popular among the high clergy. The origin of the name is unknown: it could come from *xantos*, yellow, because of its colour, or from its liturgical use in the past Byzantine age; or because of the period when the dried grapes were pressed: during Holy Week for the *Vino Santo* from the Trento area.

Santo could also refer to 'pure wine', as it was produced without any addition or correction. It is reported that the Greek patriarch Bessarion, during the council of Florence in 1349 said, 'this pure wine must come from Xantos', referring to the famous wines from dried grapes produced in Thrace, hence the name 'santo'.

A more charming etymology though links the name to the fact that in the beginning of the Oriental Churches, 'saint' was used to define someone who was different, and this could also be said of a wine produced from dried grapes: it was rare, and therefore different from other wines. The Hebrew *quades*, saint, from which the word *jahweh*, god, originated comes from the root *qd(d)*, meaning 'to split', 'to be different', and it is connected to the Akkadian *quaddusu*, 'to be shiny', as a distinctive sign of holiness (Scienza, 2006).

A more prosaic origin of the name could be found on the island of Santorini, in the Cyclades, known as the Venetian Stone Fleet as they were on the route to Crete and Constantinople. During the Classical Age it was called Thera; after the First Crusade in 1204 it was renamed Santorini because of the church of St Irene or St Erini. Along with Malmsey from Crete, Ariousios from Chios and Commandaria from Cyprus, Santorini's Vinsanto was generally called Greek wine and was placed within the category of sweet wines able to travel long distances.

The first descriptions of the organoleptic features of this wine are rather late compared to its fame, as it is reported by Baron J. Pitton de Tournefort towards the end of 1700: the wine was used only by the Orthodox Church of the Black Sea and of southern Russia.

1.10 THE 'REVOLUTION' OF DRINKS AND THE DEVELOPMENT OF SWEET WINES IN CONTINENTAL AND ATLANTIC EUROPE

With the so-called 'revolution' of drinks at the end of the seventeenth century, the consumption of wines, spirits and the diffusion of non-alcoholic drinks like coffee, tea and chocolate underwent a deep transformation, leading to the recession of some of the main wine-producing areas in France, such as Bordeaux, and the rise of others like Cahors, which produced a wine richer in colour compared to *claret*, which was losing popularity among English drinkers.

At the same time, viticulture began to develop in some areas that produced poor wines, but had a vocation for spirits (Cognac and Armagnac), and which benefited from their location close to Atlantic harbours on the routes to the northern countries (Unwin, 1993).

The real novelty though was the creation of new wines such as Champagne and Port. On one side, consumers were asking for new products; on the other, merchants were trying to improve conservation techniques for wines that had to travel in increasing quantity, in order to satisfy a growing market. From the fifteenth century, there was a series of experiments to improve wine resistance: bigger barrels were better than small ones to prevent the wine from becoming vinegar, adding cooked wine helped as well. The wines produced in Jerez start to be called *sherris-sack.* Such wines had better oxygen stability and developed a sort of yeast (flor yeast) veil on the surface, which prevented wine from alterations. In the Porto region, because of custom's restrictions imposed by the French government in 1700, the English developed the production of wines suitable for transport and bottling, with a high sugar level, and by adding spirit to fermenting must.

The innovation of bottles improved wine ageing and resistance to transport, and also made it possible to stock wine and control its evolution over time. It also meant that merchants could manage their stocks better, and did not have to lower prices when wine began to lose stability.

Once again, the evolution of a product coincided with availability of the financial resources: bottled wines became a status symbol. The development of glassmaking in Murano made available different tools for serving wine: consumers could actually appreciate the appearance of what they were drinking, which was not possible with pewter, tin or silver cups. This might seem obvious to us, but it was a huge cultural innovation at that time; this is evident in Veronese's painting *Wedding at Cana* in 1563. Wines, especially sweet ones, began to be served in transparent jugs called 'Venice-style jugs'. Therefore, wines had to be clear, and enological techniques had to improve.

The sweet wines from the Atlantic coast start to be produced towards the end of the seventeenth century due to a strong Dutch demand in the northern markets. The Thirty Years War (1618–1648) between Catholics and Protestants was ending, and the middle classes were going through the so-called 'revolution' of drinks.

It was a period of great innovations in the world of wines, such as the first adoption of a brand by a single winery (Château Haut-Brion) in England, the first botrytized wines called *rotten wines* or *vins pourris*, the first fortification of must with spirit and the production of the first thick bottle.

In 1648, Holland became an independent country (under the Treaty of Westphalia), and its economy was mainly based on trade. The Dutch founded colonies in America (New Amsterdam), South Africa, India, and they were very competitive, especially against English ships.

They were restricted from sailing in the Mediterranean, which was controlled by the powerful Venetian Fleet and their Gascon allies. The market of sweet wines though was too important; therefore they started to look at alternative production areas. Bordeaux was their choice, as it was close to the area of Charente, where they purchased spirits, and sweet wines were not produced there as the environment was not suitable for the accumulation of sugar in the berries. The Dutch encouraged the growers to change their farming techniques and grape varieties, in order to obtain sweeter musts through late harvests. Moreover, they taught them about the use of sulphur dioxide (SO_2) in fermentation (*allumettes hollandaise* or *sulfur wicks*), which the Dutch used to sanitize barrels and clarify wines before bottling. They used various filtration techniques and cold stabilization to get rid of the yeasts (for instance, they invented the so-called Dutch filter), but because of difficulties in controlling fermentation, wines came out in all sorts of styles (off-dry, sweet, etc.) and they were not always appreciated by the consumers. The addition of cane sugar and spirit helped to standardize the taste, in order to meet the demands of different markets. Because of the war with France, the Dutch and English did not restrict their focus to the area of Sauternes but widened their interests towards the Atlantic coasts of Spain and Portugal. During the Middle Ages, these wines were not so popular; they did not travel well because they were low in alcohol and tannins. The main innovation of the Dutch was the introduction of fortification through spirits. This enabled the production of botrytized wines also in Germany (the Rhine region) from 1687 and Anjou, from the Chenin variety, from 1782.

1.10.1 Sauternes wines

Starting in 1774 in the Dordogne Valley, the Dutch forced the growers to late harvest the grapes, in order to allow noble rot (or *pourriture noble*) to attack the berries and improve sugar concentration. The action of the fungus, enhanced by favourable weather conditions (hot sun during the day, humidity at night and soils with little nitrogen and water) also produced particular flavours that were highly appreciated. Moreover, it produced an antibiotic called botryticine, a yeast inhibitor that, in combination with SO_2, allowed the wine to remain sweet. Three grape varieties were chosen: Semillon, Sauvignon and Muscadelle (or Muscat fou) both for their aroma characteristics and because the skin of the berries was ideal for noble rot.

1.10.2 Malaga wines

Wines from Malaga began to attract the interest of Dutch, English and German merchants towards the end of the seventeenth century, as competition against Venetian leadership in the trade of eastern Mediterranean sweet wines. This wine was traded for the first time in 1618 under the name of Mountain wine, and the most appreciated style was *Pedro Ximen puro y legitimo*. Grapes dried in the sun for a few days (*asoleo*), were then crushed and pressed with a so-called 'roman' press. Next the must was poured into big earthenware jars (*tinajas*) and 2–3% of concentrated must was added. After fermentation, wine was transferred to wooden barrels and underwent egg-white fining. These wines were often sweetened through the addition of *vino tierno*, a wine from grapes grown on the hottest slopes and harvested in August, dried for one week, pressed and fermented. The wine was generous, strong, clear, sweet and excellent.

1.10.3 Wines from the Canary Islands and Madeira

The wines from the Canaries, even though the Spanish people planted them at the end of the fifteenth century with Malmsey from Crete, also owe their success to the English who, during the Seven Years War, bought their wine and called it 'fake Madeira'. The real commercial success on the English market, dominated by Venetian Malmseys, came during the sixteenth and seventeenth centuries. Its fortune declined over the following century though, due to the expansion of Port and Madeira.

1.10.4 Wines from Catalonia

Due to the development of the harbour of Vernazza in Liguria, the production of Vernaccia spread in Catalonia. From the fifteenth century, this style of wine is named after its birthplace, and became popular in various European regions. The main variety is Garnacha, or Grenache (Garnatxa) and the production technique involved fermentation of over-ripe grapes to obtain very sweet and alcoholic wines, capable of ageing and resistant to transport.

1.10.5 Wines from Tokaj

Wines from Tokaj were very successful at the Austrian Court during the eighteenth century, and this enhanced their fame. The earliest origins of viticulture in this territory are ascribed to Italian growers who planted the first vineyards on the volcanic and limestone hills of the Bodrog and Tiska valleys.

Tokaji wine is mentioned for the first time at the end of the fifteenth century, and it was the first European wine to be produced by botrytized grapes when Hungary was under Turkish rule.

An anecdote implies that it was not a deliberate choice: because the Turks attacked Tokaj, the harvest was delayed and the grapes were affected by noble rot. In 1641, Hungary was still under the Ottoman domain when the first law was issued to rule the Tokaj wine. This law dictated yields and sugar levels, in order to enable the natural production of sweet wine. In 1720 we have the first laws regarding harvesting a single vineyard several times, according to the levels of noble rot on the berries.

1.11 DEMOCRATIZATION OF WINE IN THE SEVENTEENTH CENTURY EUROPE: DIFFUSION OF FORTIFIED WINES

Few people know that Venice lost the northern Europe markets for sweet wines just before the creation of 'great wines' from blends of bulk wine coming from different places. Merchants misjudged the role that 'wine spirit' could have played in the creation of fortified wines or mute musts (which would have been far cheaper compared to the ones imported from the Mediterranean). Moreover, Venice made a second big mistake: on the English market, merchants tried to compete with Port by imitating absinth, which was fashionable in France. Paradoxically, switching from Malmsey to Vermouth turned out to be a big mistake: the consumer perceived the latter as an adulteration of wine. For over a hundred years, the market fell into the hands of the Dutch, who had an extremely powerful fleet. Later, leadership passed to the English merchants, who by their protectionist policies forced French wines (such as the fortified Muscats from the Midi) to be replaced by Spanish and Portuguese wines. At the same time, the progress of chemistry and microbiology led to the improvement of enological techniques. Meanwhile the market was shifting towards clarets and drinks other than wine, such as tea, coffee and chocolate.

1.11.1 The wines from Porto

Port has been known on the English market since the fifteenth century. The Dutch innovations led to the improvement of biological stability, while wine started to be fortified and transported in barrels around the seventeenth century; although it was only towards the end of the eighteenth century that Port started to be commercialized in bottles. During this period, production areas started to be regulated by law, through a selection of the most recommended sites. The 1820 vintage was particularly good, and the wine remained sweeter. Consumers particularly appreciated this style; this is why the addition of spirit was shifted back to the middle of fermentation. This is how the production of modern Port was codified: a quarter of spirit in four parts of must, added when fermentation is at three-quarters of the process.

1.11.2 Fortified wines from Andalusia

The origin of *Jerez* wine (sherries) is similar to that of Port; it goes back to the seventeenth century, through the action of the Dutch and English, during the wars against France. It was an ordinary white wine with the addition of alcohol for travelling. Towards the end of 1700, protection rules were issued concerning the choice of grape varieties, soils (according to limestone content), winemaking and ageing practices, in order to obtain different styles with very different sugar levels.

1.12 WHEN SWEET WINE IS A REVOLUTION: THE VENETIAN DREAM OF PRODUCING AN ETERNAL WINE

Between the twelfth and eighteenth centuries, Venice was the principal Italian centre for European trade and consumption of wine. Fortified wines were particularly popular, together with Malmseys aimed at a particular fragment of the market: consumers who wanted wines to drink away from meals or with marzipan desserts or biscuits (this ritual traditionally opened each official dinner of the Republic).

According to Greek-Byzantine ritual, which influenced most of the Venetian food habits, such fortified wines were normally consumed after the addition of water (from one third to a half) according to the concentration of sugar and the presence of resin (if they came from Cyprus). The wines came from several locations in the Mediterranean: the Aegean Islands (Santorini, Samos) and the Ionian islands (Corfu, Zakinthos, Cefalonia), as we learn from Customs' documentation where import duties were registered.

Competition among wine merchants from the different Italian maritime republics was very strong. It concerned sweet, aromatic and alcoholic wines produced in the oriental part of the Mediterranean (Malmsey, Vinsanto), and mainly traded by Venice and partly by Genoa (Vernaccia ligure) on the Anglo-Saxon market. The republics also traded Atlantic and German botrytized wines on the rich northern European markets. The Dutch, who were restricted from sailing the Mediterranean, had previously diffused such wines.

Since the fourteenth century, merchants had been seeking innovations in order to exploit the market of sweet alcoholic wines. The aim was not to improve the intrinsic qualities of a product; progress was instead led by the need to be able to adapt to unexpected circumstances that might force a change in choices. These choices were not only represented by winemaking techniques to improve longevity, but they also implied a move towards selecting early and cold-resistant grape varieties and growing areas.

So what were these circumstances at the end of 1300? The climate was changing and Europe was about to face a long, challenging period in this respect. Towards the end of 1700, European agriculture – which had not changed since the Roman Empire – became dramatically different. The quality of wines dropped because grapes struggled to ripen, and therefore the upper clergy and European aristocracy started to look at the sweet and alcoholic wines of the eastern Mediterranean, thanks to the clever intuition of the Venetian merchants who had occupied all the Greek and Turkish harbours while transporting the crusaders to the Holy Land during the Fourth Crusade.

The Venetians did not simply bring the wines into Venice and then ship them straight to northern Europe, though; they mastered the art of blending, in order to meet the tastes and budgets of the different markets. There was an expression '*alla moda di Venezia*' which referred to the practice of blending wines of different vintages, in order to obtain a standard product from wines of different origins and winemaking techniques. This was done in no other place in Europe.

The Turkish conquest of Crete in 1564 forced the Venetians to shift production of Malmsey to closer locations, along the two sides of the Adriatic fjord. The innovation was not so much in the location of vines, but more in an attempt to imitate the style of a wine that was successful on the market. This practice is a constant in the history of all great wines.

The need for a wine that tasted like Malmsey and Vinsanto for the Venice trade, encouraged the production of such wines throughout central and northern Italy; through the use of the widest range of grape varieties, all sharing the common features of a thick skin and a cluster that did not fear withering. However, climate conditions during the Little Ice Age made it more difficult to dry the grapes outdoors, and so they were put in warehouses or hung for 3–4 months, according to the style of wine. This is how the Tuscan Vinsanto was obtained, and also the Vino Santo from Trent, the white and red Recioto, Amarone, Torcolato and many others, which are the expression of the different local cultures.

In the western part of the Mediterranean, which was not on the Venetian trade routes, people started to produce wines similar to Malmsey or Vinsanto from different grape varieties and from the most varied winemaking techniques, which had little to do with drying grapes (*Botrytis*, cooked must, fortification, etc.).

1.13 WINE APPELLATIONS: A WAY TO FIGHT FRAUD AND PRIVILEGE

Towards the end of the eighteenth century, the diffusion of wine (especially sweet and fortified wine) led to several problems connected to the addition of chemicals or vegetal extracts in order to cover faults or to make it difficult to state the place of origin. In fact, it became popular to name a style of wine after the place where it was originally produced. This was very common for German champagne or Australian sherry. There is an interesting passage in the third chapter of Marx' *Capital*. The author says that for fine wines, income is generated by a monopoly (connected to the rarity of the product or its origin) and by the juridical boundary of a territory, which enables the producers within a given area to manage the offer and price of a wine.

Even though Port was the first wine to be legally protected in 1675, the French were the first to conceive a way of controlling and protecting quality and origin. They focused not so much on delimiting a given location, but instead they stressed the fact that only a given *terroir* was able to impart a certain intrinsic quality that could not be added by any winemaking technique (an acquired quality). This distinction is fundamental for applying defence strategies during production and transformation.

In order to safeguard an intrinsic quality, a law can impose certain vineyard features (plantation density, varieties, training systems, yields, etc.); while the acquired quality is protected against sophistication (addition of water, control of dry extract, acetic acid, etc.). The first appellation was Chablis, in 1901. In Germany, addition of sugar was a main issue; this is why classification was not based on the production areas, but rather on the organoleptic features (sugar levels) of the finished product. The first law was instituted in 1879.

1.14 CONCLUSIONS: THE HERITAGE

The political and economical history of Greek Italy changed completely after the second Punic War: Magna Greece was 'romanized', with very few changes in wine production.

Vineyards and varieties in southern Italy remained the same throughout the whole of the Middle Ages until the arrival of Phylloxera. Several native varieties introduced by the Greeks, such as Murgentino, Eugenia and Aminee spread throughout Campania, Latium and Etruria (Pliny, *Naturalis Historia*, XIV, 273).

A few winemaking techniques have survived to this century. Pliny (XIV, 82) and Columella (XII, 39, 1) describe how on several Mediterranean islands such as Santorini, Samos and Pantelleria, grapes were dried in the sun for about three weeks, then macerated with wine that had been produced a month earlier, in order to add sugar and aroma and improve stability.

The Carthaginian author Magon reports about a similar technique, more suitable for the hot climates of northern Africa, which are similar to Pantelleria's climate. Grapes had to be picked at a very late ripening stage, the best bunches had to be selected and grapes had to dry in the sun under a cane protection (against sun during the day, and humidity at night) becoming raisins. Raisins were then soaked in fresh must and were crushed together. The operation was repeated, and the two parts were fermented together for 20–30 days. In the Greek tradition which has passed down to us, preparation of sweet wines had to follow several rules: while there was no distinction about the colour of the grapes (Attiri, Assyrtico) used separately or mixed with Mavro Tragano or Mandilaria for the preparation of Santorini's wines, grapes were classified according to their aromatic features. Muscat grapes from Samos or Malmsey were never mixed with neutral grapes.

Hesiod (227, 607) talks of a wine produced in Syracuse where Muscat, Calabrese (which was considered a Red Muscat with a violet bouquet) and Vernaccia were blended in proportions of one third. This wine is still produced only from White Muscat and in the nineteenth century it was associated with the legendary wine 'Pollio' (from Pollio Argivo, a tyrant of Syracuse).

The Calabrian Greco di Bianco is produced from grapes that are dried according to a 2500-year-old technique: the same recipe that Hesiod gives for the wine of Taso, according to the *Geoponics*.

A very concentrated juice, with a honey texture, obtained by slow cooking called *carenum* can be recognized in the cooked wines from the Italian Marche region.

Protropos is a wine obtained by fermentation in ceramic containers, which are left in the sun for 40 days at the hottest period of the year. This way, the ageing process was accelerated and the wine underwent a sort of pasteurization. This technique is still used to produce the *rancio* wines in Roussillon, such as Banyuls and Maury.

Peculiar climatic conditions, besides the cultivation of Italian varieties, allowed the preservation of ancient Roman techniques for the production of sweet wines also in France and Switzerland.

There is a special wind phenomenon in the area of Gaillac, the so-called 'marin', a warm autumn wind that dries grapes in the canton of Wallis, and the 'foehn', a southern dry wind that allows Amigne to dry on the plant. Amigne is a reminder of the Roman occupation, as the name recalls the Aminee; the Arvine sounds similar to Elvola (Varro, I, 25; Pliny XIV, 46) and Rèze has a link with the Raetica that we find in Pliny (XIV, 80).

Over the last thousand years, Vinsanto and Malmsey have sadly become nothing more than generic sweet wines. This is due to communication and production strategies going back to the Venetian policy: an early example of globalization of the sweet wine market, where a place of origin (such as Santorini) was simply a generic word for a style of wine with several features that the consumer could easily recognize. There was no real connection to the actual growing area. Many sweet wines from different regions in the world could end up the same way, gathered within an international taste where sensory descriptors are more about the drying techniques than the actual *terroir* features. In order to protect the fragile identity of sweet wines made from dried grapes, which are the true heritage of the very first viticulture, it is important to enhance the oldest grape varieties and to go back to the peculiarities of the enological tradition. Moreover, it is important to maintain the relationship of these varieties with their places of origin and with the peoples who invented them.

REFERENCES

Braudel, F. 1999. *Memorie del Mediterraneo* [Italian translation], Bompiani Ed., Rome.
Catoni, M.L. 2010. *Bere vino puro. Immagini del simposio*, Feltrinelli Ed., Milan.
Chicago Hittite Dictionary. The Oriental Institute of the University of Chicago, Chicago, USA. Available at: http://ochre.lib.uchicago.edu/eCHD/ (last accessed 28 November 2012).
Gennari, G. 2005. *Homo Bibens*, Guiglia Ed., Modena.
McGovern, E.P. 2003. *Ancient Wine: The Search for the Origins of Viniculture*, Princeton University Press, Princeton, USA.
Scienza A. 2006. *Atlante dei vini passiti italiani*, Edizioni Gribaudo srl, Cuneo.
Unwin, T. 1993. *Storia del vino. Geografie,culture e miti dall'antichità ai nostri giorni*, Donzelli Editore srl, Rome.

Part 2 Vineyard Management, On-vine and Postharvest Grape Dehydration, Vinification

2 Management of the Vineyard

Osvaldo Failla,[1] Laura Rustioni[1] and Giancarlo Scalabrelli[2]

[1] *Department of Agricultural and Environmental Sciences, University of Milan, Milan, Italy*
[2] *Department of Agriculture, Food and Environment, University of Pisa, Pisa, Italy*

2.1 INTRODUCTION

There is general agreement among viticulturists, both technicians and growers, on the ideal bunch characteristics for grapes that are grown to produce wines through dehydration. Bunches should not have a dense structure; berries should not be large in size and should have a thick skin. Fruit composition should be rich in sugar, acids and secondary metabolites. Grapes must be healthy and free from symptoms of disease or pests, as well as have low susceptibility to rot decay during ripening and postharvest treatments. To achieve this qualitative profile, proper vineyard management is essential and achieved by making appropriate choices including consideration of genetic, environmental and cultural options.

2.2 GENETIC CHOICES

Genetic factors include the choice of cultivar, clone and rootstock. Even if the choice of grape cultivar should be the dominant factor in planning a vineyard suitable for dehydrated grapes, this does not always happen in the viticulture and wine industry of today. Ideally, growers should select a cultivar within the existing assortment, taking into account the grape's morphological and physiological traits, or by planning specific breeding programmes to obtain new cultivars for this specific enological purpose. The reality is, at least in the context of the Old World enological industry, that regulation of wine production and the general marketing strategy take a conservative approach to the choice of grape variety. As a result, each production area, following its local tradition, grows very few cultivars for dehydration, even if they do not fully meet the requirements of the cultivar ideotype (Table 2.1).

There are two north-eastern Italian varieties that deserve a special mention in this respect: Picolit and Moscato rosa (Rosenmuskateller, Muscat rose). They both have female flowers and, in order to achieve successful fruit set, they need pollen from hermaphrodite varieties planted in the same vineyard. They generally produce loose bunches with large seeded berries, frequently mixed with small seedless parthenocarpic berries.

Within a specific cultivar, a certain level of intra-varietal variability may be detected and selected to obtain clonal lines characterized by phenotypes more suitable for producing dehydrated grapes. Bunch compactness, susceptibility to grey mould (*Botrytis cinerea* Pers.),

Sweet, Reinforced and Fortified Wines: Grape Biochemistry, Technology and Vinification, First Edition.
Edited by Fabio Mencarelli and Pietro Tonutti.

Table 2.1 The main Italian varieties committed to the production of dehydration grapes, their corresponding wines and the essential traits of the fruit.

Grapes	Wines	Bunch density	Skin feature	Sensibility to grey rot
Albana	Albana di Romagna Passito	From loose to dense	Thin but hard	Medium
Albarola	Sciacchetrà Cinque Terre	Dense	Thin	High
Aleatico	Aleatico dell'Elba, Aleatico di Puglia, Aleatico di Gradoli	From loose to medium dense	Medium thin	Medium
Ansonica	Ansonica dell'Elba	Medium loose	Thick	Medium
Bervedino	Vin Santo di Vigoleno	Medium dense	Thick	Medium
Boschera	Torchiato di Fregona	Medium dense	Thick and hard	Medium high
Bosco	Sciacchetrà Cinque Terre	From loose to very loose	Thick	Medium
Corvina	Recioto and Amarone della Valpolicella	Dense	Thick and hard	High
Erbaluce	Erbaluce di Caluso Passito	Dense	Medium thick	High
Garganega	Recioto di Gambellara, Recioto di Soave	Medium loose	Thin but hard	High
Gewürztraminer	Trentino	Dense	Thick and hard	Medium
Grechetto	Orvieto Dolce	Medium loose	Thin but hard	High
Greco	Frascati Cannellino or Frascati Dolce	Dense	Thin and soft	Medium high
Lacrima	Lacrima di Morro d'Alba	Loose	Thick and hard	Low
Malvasia di Candia aromatica	Malvasia Passito dei Colli Piacentini, Vin Santo dei Colli Piacentini	From dense to loose	Thick	Medium
Malvasia bianca lunga	Vin Santo, Carmignano, Chianti Chianti Classico, Montepulciano, Toscano	Dense	Medium thick	High
Malvasia del Lazio	Frascati Cannellino or Frascati Dolce	Dense or medium loose	Thick	Medium
Malvasia di Lipari syn. Greco di Bianco	Malvasia di Lipari	Medium loose	Thin and soft	Low
Marzemino	Refrontolo Passito	Medium dense	Thin but hard	Medium high
Melara	Vin Santo di Vigoleno	Dense	Thick	Medium low
Montepulciano	Controguerra Passito rosso	Dense to medium dense	Thick and hard	Medium low
Moscato bianco	Chambave Moscato Passito, Moscadello di Montalcino, Moscato di Loazzolo, Moscato di Siracusa, Moscato di Trani, Piemonte Moscato Passito	From dense to medium loose	Thin	High
Muscat of Alexandria	Passito di Pantelleria	From medium dense to loose	Thick and hard	High
Moscato di Scanzo	Moscato di Scanzo	Loose	Medium thick	Medium

Table 2.1 (*Continued*)

Grapes	Wines	Bunch density	Skin feature	Sensibility to grey rot
Moscato giallo	Alto Adige Moscato giallo passito, Trentino Moscato giallo	Loose	Thick and hard	High
Moscato rosa	Alto Adige Moscato rosa passito, Trentino Moscato rosa or delle rose	Loose	Thin	Low
Mùller Thurgau	Trentino vendemmia tardiva	Medium dense	Thin	High
Nasco	Nasco di Cagliari	From medium dense to medium loose	Thin and soft	Medium
Nebbiolo	Sforzato Valtellina	Medium dense	Thin but hard	Medium low
Nosiola	Vino Santo	Dense	Thin but hard	Medium
Ortrugo	Vin Santo dei Colli Piacentini	Very dense	Thick and hard	High
Passerina	Controguerra Passito bianco	Medium dense	Thick and hard	Low
Picolit	Picolit	Loose or very loose	Medium thick, hard	Low
Pinot grigio	Malvoisie di Nus	Very dense	Medium thick, hard	High
Primitivo	Primitivo di Manduria	Medium dense	Medium thick	High
Prosecco	Torchiato di Fregona	Loose or medium loose	Thin but hard	Medium low
Rondinella	Recioto and Amarone della Valpolicella	Medium dense	Medium thick	Low
Sagrantino	Sagrantino di Montefalco	Very dense	Thick and hard	Low
Sangiovese	Vin Santo Occhio di Pernice	Medium dense	Medium thick	Low
Trebbiano di Soave syn. Verdicchio	Recioto di Soave	Dense	Medium thick	Medium
Trebbiano romagnolo	Vin Santo dei Colli Piacentini, Vin Santo di Vigoleno	Dense or medium loose	Thin but hard	Medium
Trebbiano toscano B.	Vin Santo: Toscano, del Chianti, del Chianti Classico, di Carmignano, di Montepulciano; Controguerra Passito bianco	From medium dense to medium loose	Medium thick and hard	Low
Verdiso	Torchiato di Fregona	Medium dense	Thin and soft	High
Verduzzo friulano	Verduzzo Passito, Ramandolo	Medium dense	Thick and hard	Low
Vermentino	Sciacchetrà Cinque Terre	Medium dense	Medium thick	Medium
Vernaccia di Oristano B.	Vernaccia di Oristano	Medium dense	Thin	Medium
Vernaccia	Vernaccia di Serrapetrona	Dense	Medium thick	Low
Vespaiola	Torcolato di Breganze	Medium dense	Thick and hard	Medium high

Table 2.2 Example of clones of cv. Corvina classified according to their suitability to produce wine from dehydrated grapes.

Clone	Licence year	Yield potential	Qualitative potential	Sensibility to rot	Aptitude to produce wine from dehydrated grapes
RAUSCEDO 6	1969	Medium high	Medium high	Low	High
ISV-CV 7	1980	Medium high	Medium high	Medium low	High
ISV-CV 48	1980	Medium low	High	Low	High
ISV-CV 78	1980	Medium	Medium high	Medium	Low
ISV-CV 146	1980	Medium high	Medium	Medium	Medium
ISV-CV 13	1991	Medium	Medium	Low	Low

richness and complexity in wine flavour, are the main traits to be considered when evaluating the suitability of a clonal line for dehydrated grape production. Examples of selected clones to achieve these aims are shown in Tables 2.2 and 2.3.

Rootstocks, mainly selected from hybrids of American *Vitis* species, play an essential role in plant protection against the lethal attacks of Phylloxera (*Daktulosphaira vitifoliae* Fitch). They may differ in their ability to adapt the grapevines to soils characterized by different lime levels, water shortage or salinity. In addition, rootstocks may modify vine vigour and yield potential and, to a lesser extent, ripening evolution. The possible effect of rootstocks on bunch compactness seems to be an indirect effect of its vigour on the fruiting biology of

Table 2.3 Example of clones of cv. Nebbiolo from Valtellina area selected for their suitability to produce wine from dehydrated grapes.

Clone	Licence year	Yielding traits	Qualitative potential	Sensibility to rot
CH 12	2003	Medium vigorous and productive clone with medium loose bunches and medium-sized berries with thick skins	High sugar levels, medium acidity, high anthocyanins	Very low
CH 21	2002	Medium vigorous and productive clone with long and loose bunches and medium-sized berries intensively covered in flour	High sugar levels, medium acidity, high anthocyanins	Medium
CH 34	2002	Medium vigorous and productive clone with long, loose or medium bunches and medium-sized berries intensively covered in flour	High sugar levels, medium acidity, high anthocyanins	Medium-low
NBL-MI-2	not yet released	Vigorous and productive clone with large winged medium-dense bunches and medium-sized berries with thick skins	High sugar levels, medium acidity, high anthocyanins	Lower than the average
NBL-MI-3	not yet released	Medium vigorous and productive clone with medium-small bunches and small-sized berries with thick skins	High sugar levels, medium acidity, high anthocyanins	Lower than the average

the grapevines. Similarly, the indirect effects of rootstocks on the fruit composition can be seen in vigour and ripening phenology, while mineral nutrition and particularly potassium uptake are directly related to the rootstock and affect berry juice pH and acidity. Interesting data on the effect of rootstocks on bunch and berry structure and the possible consequences on *Botrytis cinerea* rot have been collected by Ferreira and Marais (1987). By comparing grapes of Chenin Blanc grafted onto five different rootstocks; besides the expected rootstock effect on vine vigour and yield, they were also able to show a significant effect on bunch compactness, berry skin and pedicel strength. The two rootstocks lowest in vigour, 99 Richter (*V. berlandieri* x *V. rupestris*) and Jacquez (*V. aestivalis* x *V. cinerea* x *V. vinifera*), induced the least compact bunches, the firmest berry skins and the tightest berry–pedicel attachment. The bunch compactness proved to be correlated to the *Botrytis* rot.

One of the most consistent effects of rootstocks is related to potassium nutrition. High levels of potassium uptake may induce an extra synthesis of malic acid in the berry before veraison to maintain the cation–anion balance in the vacuole with an increase of juice pH due to the change in malate-to-tartrate ratio (Failla *et al.*, 1990; Ruhl, 1991; Corazzina *et al.*, 1993; Failla *et al.*, 1993; Brancadoro *et al.*, 1995; Terrier and Romieu, 2001). Moreover, during ripening, a high potassium flux into the berry will increase the organic acid salification with a further juice pH increase. For this reason, in potassium-rich soils, rootstocks that result in high potassium uptake should be avoided for grapes to be dehydrated. Such rootstocks include, for example: the cultivars derived from the vigorous and root-knot nematode-resistant *V. champinii*, such as 'Freedom', 'Ramsey', 'Salt Creek' and 'Harmony'; some rootstocks obtained from the drought-resistant and deep-rooting *V. rupestris* crossed with the tolerant to calcareous soils *V. berlandieri*, such as '140 Ruggeri'; as well as rootstocks *V. riparia* x *V. berlandieri*, such as 'SO4' (Garcia *et al.*, 2001; Cousins, 2005; Kodur *et al.*, 2010).

Therefore, in vineyards committed to producing grapes to be dehydrated, according to the possible soil-limiting factors, the least vigorous rootstocks should be selected.

2.2.1 A case study: the Aleatico grape variety

Aleatico is a red wine grape variety mainly cultivated in warm areas of central and southern Italy, including the islands of Elba and Capraia, and also Corsica in France, where dessert wines are produced after partial postharvest berry dehydration.

The most famous wine, 'Aleatico dell'Elba Passito', comes from Elba island and has recently been awarded the Italian 'Denominazione di Origine Controllata e Garantita' (DOCG) quality label. Cultivation in this territory is difficult; because of the slopes, most labour is done manually. Innovations need to be introduced in the vineyards in order to cut labour costs and sell the wine profitably (Scalabrelli *et al.*, 2004). Other dessert wine producing areas are located along the Tuscan coastline, followed by Marche, Lazio and Puglia. Although these regions have a higher incidence of flat landscapes, they share the common feature of soils of scarce fertility (Table 2.4).

The Aleatico grapevine has a medium-small cluster, medium-compact, with a single shoulder. Berries are round, medium-sized, of blue-black colour and with a thick skin. A large intra-variety variability has been detected and several clonal lines have been selected (Table 2.5).

The most important aromatic substances contained in grapes are the terpenic compounds, with considerable levels in the free form and, above all, in the glycosidically bound form (Table 2.6). Compared to other aromatic varieties, Aleatico has a higher amount of free terpenes than Brachetto d'Acqui and slightly lower than Malvasia Nera Aromatica and

Table 2.4 Main producing areas of 'Aleatico' wines in Italy.

Region	Denomination	Variety percentage	Wine type, alcohol (% vol) degree	Grape yield (ton ha^{-1})	Territory	Soil fertility
(a) DOP[a] denomination						
Toscana	Elba Aleatico Passito	100	19 of which 12 developed (dvp)	7	Hilly with slope	Poor
Lazio	Aleatico di Gradoli	95	Fortified and fortified riserva: 17.5 of which 15 dvp; Passito: 16 of which 9 dvp	9	Hilly	Poor
Puglia	Aleatico di Puglia	85	18.5 of which 16 dvp	8	Hilly or flat	Poor to medium
(b) Varietal type of DOP denomination						
Marche	Pergola	85	Passito: 15 of which 12 dvp	9	Hilly	Poor to medium
Puglia	Gioia del Colle	85	Aleatico sweet: 15 of which 13 dvp; Sweet fortified: 18.5 of which 16 dvp	12	Hilly	Poor to medium
Puglia	Salice Salentino	85	Sweet: 15 of which13 dvp; Sweet fortified: 18.5 of which 16 dvp	10	Hilly	Poor to medium
Puglia	Terra d'Otranto	90	Aleatico: 15 of which 13 dvp	9	Hilly	Poor to medium
Toscana	Sovana,	85	Superior and riserva: 12 of which 9.5 dvp	9	Hilly or flat	Poor to medium
Toscana	Val di Cornia	100	Passito: 16 of which 13 dvp	6	Flat	Poor to medium
Umbria	Rosso Orvietano	85	11.5 of which 9.5 dvp	10	Hilly or flat	Poor to medium

[a]Denominazione di Origine Protetta.

Moscato bianco (White Muscat, Muscat blanc). On the other hand, geraniol prevails in Aleatico, while in the Muscat, the level of linalool (typical aroma of this grape) is much higher (Table 2.7). Among the other important aromatic components of Aleatico, are nerol and citronellol (Boselli *et al.*, 2009).

Aleatico grapes are rich in phenolic compounds (~10 g L^{-1}) and the non-flavonoid polyphenolic compounds are highly represented (7.54 g L^{-1}). Cinnamic and benzoic acids are usually present in small concentrations and during fermentation the amount decreases even further because they are easily oxidized (Andrich *et al.*, 2009). The high contribution of seeds in determining the total phenolic content of the berry is a varietal characteristic (Scalabrelli *et al.*, 2004). This fact may give rise to some technological problems as the phenolic maturity at harvest, especially due to seeds, can be critical and the wine could become too astringent and bitter, because of a higher extraction of catechin monomers.

The climate of the areas where Aleatico is grown, especially in the Tuscan coastal region and islands, is usually characterized by high temperature and low rainfall which can induce water stress in the vines.

Table 2.5 Main features of 'Aleatico' released clones in Italy and France and candidate clones selected in Tuscany.

Origin	Clone	Bud fertility	Cluster size	Cluster tightness	Berry size	Yield	Earliness of ripening
Tuscany (Italy)	001 - AL-PA-1	Medium	Small	Medium	Medium	Low	Medium
	002 - VCR 438	Medium	Medium	Medium	Medium	Medium	Medium-late
Lazio (Italy)	003 - ARSIAL-CRA 489	(°)					
	004 - AL-VAL 1	(°)					
Tuscany (Italy)	005 - CRA VIC BC SF 3	Medium	Medium	Low-medium	Medium	Low	Medium
	006 - AL-FI.PM	Medium	Medium	Medium	Medium	Medium	Medium
Corsica (France)	785 - ENTAV 15*	Medium	Medium	Medium	Medium	Medium	Medium-early
	802 - ENTAV 57*	Medium-low	Medium	Medium-high	Medium	Medium	Medium
	859 - ENRAV 18*	Medium	Medium	Medium-high	Medium	Medium	Medium-early
	860 - ENTAV 49*	Low	Medium	Medium-high	Medium	Medium	Medium
	861 - ENTAV 53*	Med-low	Medium	Medium-high	Small-medium	Low-medium	Medium-early
	892 - ENTAV 26*	Low	Medium-low	Medium-high	Small	Low	Medium-early
	893 - ENTAV 37*	Medium	High	Medium-high	Medium	Medium-high	Medium-early
	894 - ENTAV 59*	Medium	Medium	Medium	Small-medium	Medium	Medium-early
Tuscany (Italy)	Ale102* (c.c.)	Medium-high	Low	Low-medium	Medium	Medium	Early
	Ale 119* (c.c.)	Medium	Low	Medium	Medium	Low-medium	Early

(°), data not available; c.c., candidate clones; *, data from Scalabrelli and D'Onofrio (2012).

Table 2.6 Content of free and bound terpenes extracted from 'Aleatico' grapes cultivated on Elba Island.

Molecule	Free terpenes ($\mu g\ L^{-1}$)	Bound terpenes ($\mu g\ L^{-1}$)
trans-furanlinalool oxide	1.4	50.6
cis-furanlinalool oxide	1.0	12.6
linalool	19.7	354.3
nerale	27.4	–
α-terpineol	1.5	47.4
geranial	45.8	–
trans-pyran linalool oxide	8.6	33.0
cis-pyran linalool oxide	4.6	4.1
citronellol	10.6	34.9
nerol	179.5	951.7
geraniol	763.4	1705.4
2,6-dimethyl-3,7-octadien-2,6-diol	29.8	49.4
2,6-dimethyl-7-octadien-2,6-diol	<1	15.6
2,6-dimethyl-1,7-octadien-3,6-diol	12.7	13.4
OH-citronellol	8.3	6.8
8-OH-diidrolinalool	1.7	6.9
OH-nerol	–	43.3
trans-8-OH-linalool	2.4	82.9
OH-geraniol + cis-8-OH-linalool	125.1	380.2
Geranic acid	572.7	535.7
p-ment-1-ene-7,8-diol	–	13.6

Research recently conducted on Elba island and in several locations in the Tuscany and Lazio regions, showed that Aleatico has a high thermal sum requirement (at least 1950–2100 GDDs (growing degree-days)), which is usually reached at the beginning of September in the studied environments. Higher sugar accumulations correspond to higher thermal sums, but are not linearly correlated to the aromatic content (Figure 2.1). The bound terpenoids percentage showed that a curvilinear relationship with the temperature excursion occurred during the month before harvest (Figure 2.2).

The districts of Gradoli and Elba island were characterized by particularly low rainfall and high maximum temperatures during the ripening period. This seems to have affected the aroma content. Besides the climatic parameters, the amount of fruit produced per plant is very important for the aroma; therefore, even in cooler climates such as the Mt Amiata area,

Table 2.7 Content of geraniol and linalool in grapes of several varieties from different regions (data from Boselli *et al.*, 2009).

Variety and origin	Geraniol		Linalool		Total
	$\mu g\ kg^{-1}$	Percentage	$\mu g\ kg^{-1}$	Percentage	$\mu g\ kg^{-1}$
Aleatico (Elba)	772	97.47	20	2.53	792
Brachetto (Acqui)	233	95.10	12	4.90	245
Malvasia aromatica (Piacenza)	264	86.27	42	13.73	306
Moscato bianco (Asti)	67	14.47	396	85.53	463
Moscato bianco (Montalcino)	86	14.65	501	85.35	587

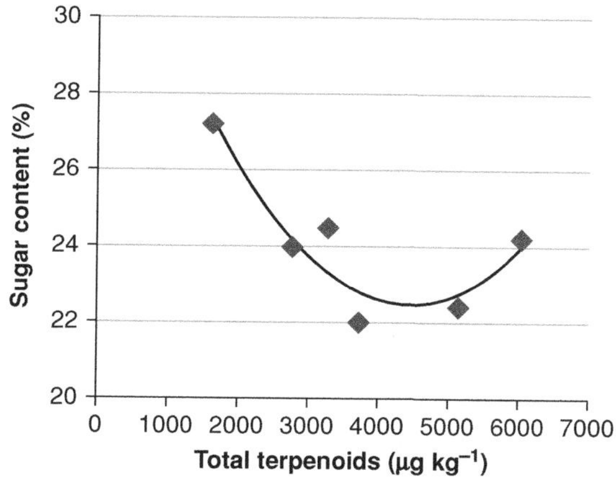

Figure 2.1 Relationship between grape total terpenes and sugar content at harvest of Aleatico grapes cultivated in different locations. Based on data from Boselli *et al.* (2009).

it is possible to reach good terpenic levels by appropriately reducing the yield (Boselli *et al.*, 2009).

Trials conducted on potted plants and in the vineyard suggest that the vines are able to cope with limitation of water availability and that a moderate water deficit can induce a favourable response in grape quality for dessert wine (Scalabrelli *et al.*, 2011; Tuccio *et al.*, 2011). An increase of anthocyanins and sugar content was obtained in non-irrigated vineyards (Figure 2.3). This information could be useful in setting up targeted postharvest dehydration strategies to produce dessert wines (Tuccio, 2011).

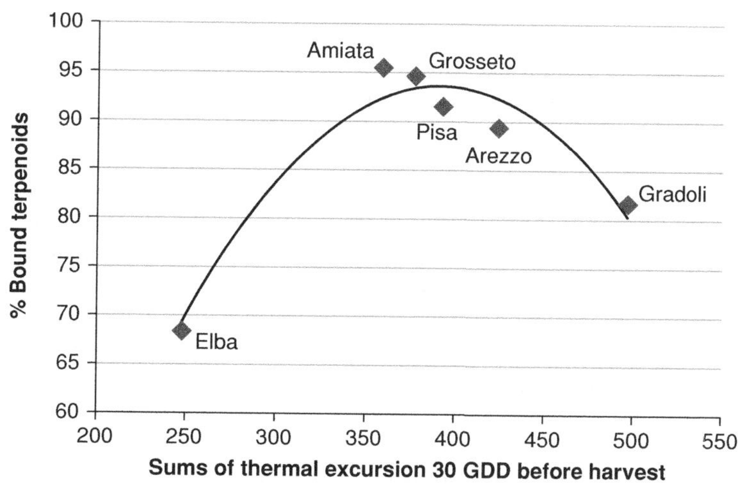

Figure 2.2 Relationship between grape total terpenes and sums of active temperature (>10°C) from the beginning of the cycle to harvest on cv. 'Aleatico' cultivated in different locations. Based on data from Boselli *et al.* (2009).

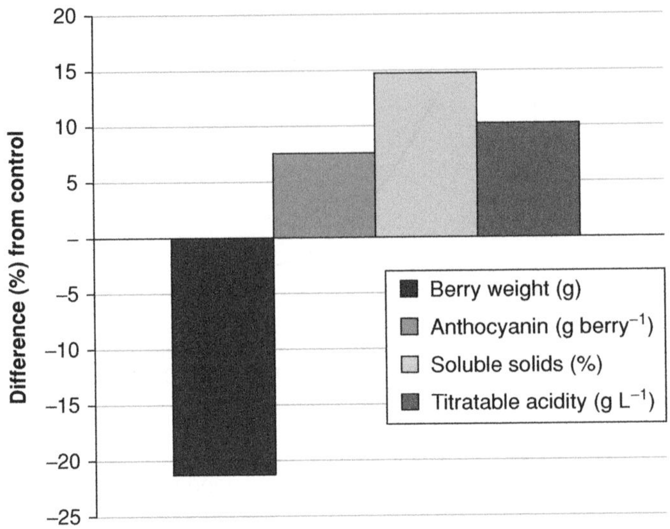

Figure 2.3 Berry characteristics of 'Aleatico' subjected to water stress compared to those from irrigated plots. Based on data from Tuccio (2011).

2.3 VINEYARD DESIGN

Vineyard design includes vine spacing, density and the training system. Here again, the traditional areas for production of grapes aimed at dehydration are characterized by a conservative approach in relation to local vineyard design traditions, and at the same time by trends toward vineyard design aimed at reducing hand labour and increasing mechanization, as well as improving canopy efficiency and yield potential. The move towards a mechanized vineyard system may have a negative effect on the qualitative traits of the grapes for this specific enological process, in comparison to traditional training systems such as 'bush' (Alberello in Italian, Gobelet in French) or 'pergola'. These are two completely different vine architectures with respect to plant size, canopy displacement and pruning method. They have been traditionally adopted in regions that have low and high soil fertility, respectively. Both systems require high manual labour to be properly managed but offer several advantages with regard to the production of grapes to be dehydrated, and in particular the possibility to clearly separate and manage the bunches from the rest of the canopy. This partition is less easy to achieve with more highly mechanized training forms such as 'espalier' (i.e. vertical shoot positioning with trellising), and 'curtain' (i.e. high cordon with downward shoot positioning).

For physiological reasons as well as for grape vigour, which will be discussed later, the canopy of the vineyard committed to the production of grapes to be dehydrated has to be thin and open. This means that it should be composed of few (2–4) spaced leaf layers, obtained with a moderate shoot density (shoot per length of row run). Moreover, bunches should grow without any mechanical constraint such as contact with trellising, canes or other bunches. Adjacent rows should be spaced to avoid shading of the bunch zone by the canopy during most of the day and through the berry growth and ripening periods.

To meet these requirements, in general terms, the vineyard design should consider wider vine spacing and a lower vine density compared to designs adopted in the same area by vineyards producing grapes for conventional processing. Because it is impossible to harvest by machine, among the mechanized training forms (espaliers and curtains), the espalier form appears to have greater advantages with regard to effective canopy management and proper quality grape manipulation.

2.4 VINEYARD AND CANOPY MANAGEMENT

While the general canopy architecture is defined by the training system, the proper fine canopy structure and its physiological performances are assured by the cultural practices included in the canopy management process. Canopy management consists of several operations conducted during the growth period to modify the shoot position and/or its length and leaf area. This improves the canopy microclimate and regulates shoot vigour and the fruit/leaf area ratio. Vines that do not have a good balance between growth and production require intense canopy manipulation (summer pruning). All these cultural cares are extremely important for vineyards aiming to produces grapes to be dehydrated. In particular, within the canopy they have to ensure a correct light and thermal microclimate for berry formation and ripening as well as reducing disease risk by improving air movement and spray penetration. When vineyards are designed to produce grapes for dehydration, excess of shoot vigour should be avoided. Shoot vigour can be regulated by the correct choice of rootstock, planting density and winter pruning, choosing the adequate node number per metre of row. Moreover, soil management and fertilization may contribute to achieving the best vine balance. In temperate climates, green cover crops can be utilized to reduce vine vigour induced by the competition of grass for water and nutrition, which can have a positive influence on grape quality (Jackson and Lombard, 1993; Materazzi and Triolo, 2000). Although this technique can be adopted, there is a frequent need to regulate growth, leaf and cluster exposure to ensure better ventilation and sunlight penetration into the canopy to avoid grape rot.

Shoot thinning is performed to eliminate excess twigs and increase the distance between the shoots and clusters to improve the microclimate of clusters and leaves. Hedge and top pruning are adopted to reduce excess shoot growth and vigour, depending on the climate. In a vertical canopy, shoot positioning is practised and may be integrated with mechanical topping and hedging if required.

As grapes for dehydration are produced in both cool and warm climates, account should be taken of available water content of the soil, which affects shoot growth and yield. In general, in cool and temperate climates, canopy management is oriented toward increasing light exposure, while in hot climates, training systems and vineyard management should ensure cluster protection from over-exposure to direct sunlight.

2.4.1 Vine balance and cluster thinning

It is generally accepted that grape quality is inversely related to yield, although this relationship cannot be widely generalized as it depends on the variety, the conditions and the vine balance. Achieving the best ratio between yield and shoot growth is considered the key factor to maximize yield efficiency without reducing grape quality. Several indexes are commonly used to monitor vine balance, and these include the leaf area-to-yield and yield-to-pruning weight, known as the Ravaz index. In each environmental condition and variety there is a certain range within which these ratios should fall in order to be adequate. For the leaf area-to-yield index, it is considered necessary to have at least 1 m^2 of leaves for each kg of grape produced, even though in some conditions this surface area may be reduced to 0.8

(Stoll *et al.*, 2009). The Ravaz index should range between 5 and 10, although several authors have suggested lower or higher levels according to different situations (Smart and Robinson, 1991; Kliewer and Dokoozlian, 2005).

Viticulturists are looking for the simplest and most rapid way to determine the vine balance, so there are practical observations that may help to understand if the leaf area is adequate for the cluster's weight. In general, the vine needs at least 1 metre of shoot length bearing 10 main leaves and a few laterals. The cane weight is also considered to be a good indicator in relation to cluster weight. Cluster thinning is practised to reduce the total crop due to the imbalance between predicted yield and leaf area. In other cases, cluster thinning is done because the wine's denomination protocols do not allow a specific yield limit per hectare to be exceeded. Reduction of crop load does not always lead to an improvement in grape quality. The effect mainly depends on vine health, soil fertility, the amount of exposed canopy and its efficiency.

Cluster thinning is expensive and must be done by skilled workers. Usually, the number of clusters per vine to be retained is established and the extra ones are cut off. The simplest way to operate is to leave only one cluster per shoot. The right time for cluster thinning is pre-veraison, just before the increase of berry sugar accumulation, a period in which the shoot growth has ceased, which ensures no shoot elongation compensation and a regular increase in berry growth by cell enlargement.

2.4.2 Early leaf removal: a potentially useful practice for grapes to be dehydrated

To obtain the best bunch structure, a fundamental role is played by the genetic characteristics of the grapes. To improve the bunch structure and obtain a less compact bunch, some viticulture techniques can be considered. A fruit zone leaf removal is often done to improve light exposure and air circulation around the clusters, with possible benefits in terms of ripening, pigmentation and prevention of moulds. Early basal leaf removal has recently attracted many researchers for its modulator effect on fruit set, cluster tightness, berry size, yield and quality. To manipulate the bunch structure, the functional relationship between photosynthate availability around bloom time and the yield profile implies that defoliation should be performed just before flowering with the aim of reducing fruit set and consequently provoking looser clusters (Poni *et al.*, 2006; Intrieri *et al.*, 2008). However, the effects of leaf removal on yield are quite variable, depending upon timing, severity and weather conditions, which affect the photosynthesis of the remaining leaves, as well as the status of the vine's carbohydrate reservoir, which may compensate for the lack of direct supply of photosynthates to the flowering clusters during fruit set. The variability in the impact of leaf removal on yield and its components depends on the depressing effects on fruit set and berry growth in the current year. Hence, a positive or negative side effect on bud floral induction and differentiation for the next year's crop (due to the improvement in canopy microclimate or caused by a shortage of photosynthates, respectively) has to be taken into account (Poni *et al.*, 2006). Nevertheless, for a general evaluation of this practice, it should be underlined that early leaf defoliation can replace the costly and time-consuming cluster thinning as a tool for yield control (Poni *et al.*, 2006).

Variable results have been obtained according to timing, intensity, vine vigour, climate conditions and variety. Pre-bloom basal leaf removal reduces fruit set and berry size with a beneficial effect on quality, especially when grapes are tight and berries tend to be too large. The decrease of fruit set percentage is due to the source–sink relationship where the higher competitive mobilization force of the apex subtracts photosynthate to the cluster. Also berry growth during the initial phase may be reduced. If the sink activity of the apex is too strong, a soft topping may be necessary to equilibrate the shoot/cluster balance. Early basal leaf removal increases green berry exposure to direct sunlight, resulting in an increase of phenolic substances (Dixon *et al.*, 2002) whose function is to protect tissue against light stresses. In addition, carotenoids synthesis is also enhanced: they are converted into norisoprenoids during ripening, increasing the fruit potential aroma compounds. A gain of other aromatic substances like terpenoids has also been reported, even though these effects cannot be generalized, because they depend on the cluster microclimate. Excess of cluster exposure to direct sunlight and temperature regimes during ripening can result in a faster degradation or conversion in the free form, which can be lost during the fermentation process (Storchi *et al.*, 2008). In several experiments, early defoliation increased anthocyanin content, as an effect of less dilution when a decrease of berry weight is observed, or by a real increase in accumulation, which is evident when expressing the amount per berry. In hot climates, high temperature due to direct sunlight exposure induced anthocyanin degradation. Hence, the best way to maintain the positive effect of early defoliation should be by protecting the cluster from over-exposure during ripening. It appears that lateral growth, which is related to variety habitus, vigour, topping and canopy direction, can contribute to obtaining a more suitable microclimate.

Although the effect may be limited or non-existent on cluster tightness and berry size, basal leaf removal performed at full bloom, or just after fruit set, still induces a positive effect on the metabolic pattern due to microclimate modification (Scalabrelli *et al.*, 2010). Meanwhile, the source–sink effect will depend on the amount of leaf area remaining on the shoots.

As most of the basal leaf removal trials reported in the literature are done by hand, which helps in understanding the phenomena, we should look at the results obtained in several trials where this operation was mechanized. A survey on this subject indicates that the evolution of machine design and operation has increased mechanical efficiency. In some cases, their use achieves results that are very close to hand operations (Tardaguila *et al.*, 2010; Filippetti *et al.*, 2011). The viticulturist will decide which method to use according to the cost, size of the vineyard, manual labour requirement and convenience. Investigations into the use of anti-transpirants (pinolene) to depress photosynthesis of basal leaves are in progress; these have a similar effect to defoliation on cluster tightness and berry size (Storchi *et al.*, 2008; Intrieri *et al.*, 2012).

2.4.3 Berry epicuticular waxes, ripening and canopy management

An important problem found during berry dehydration is related to the possible growth of moulds. The berry surface is covered by a wax layer. The physicochemical characteristics of these surface layers play an important role in plant resistance to a variety of biotic and abiotic stresses, including those caused by fungal and bacterial pathogens, phytophagous insects, drought, frost, solar radiation, mechanical abrasion, anthropogenic influences, uptake and efficiency of plant growth regulator, mineral nutrients and pesticide sprays (Rosenquist and Morrison, 1989; Jenks and Ashworth, 1999; Müller and Reiderer, 2005). In fruits, and especially in grapes, the research on wax properties is generally focused on their anti-transpirant properties, particularly important in the postharvest manipulations both for storage or drying processes (Mahmutoğlu *et al.*, 1996; Pangavhane *et al.*, 1999; Di Matteo *et al.*, 2000; Doymaz and Pala, 2002; Doymaz, 2006; Muganu *et al.*, 2011), or for protection against pathogens such as *Botrytis cinerea* (Rosenquist and Morrison, 1988, 1989; Percival *et al.*, 1993). Rosenquist and Morrison (1988) studied the formation, shape and modification of the epicuticular waxes of Thompson Seedless grapes. They observed that a few days after anthesis, the pistil epidermis starts to form small, individual, upright platelets, which rapidly increase in both size and number, reaching the highest density during the berry growth lag phase, to then be spread apart as the berry resumes rapid growth after veraison. In addition to an increase in size, the wax platelets also increased in complexity during berry development, starting as small, simple plates with blunt edges, and finishing as overlapped and lace-like plates, terminating in sharply lobed edges.

Muganu *et al.* (2011) showed that extra-canopy bunches had berries with a wider surface covered by plate-like wax in comparison to intra-canopy bunches, even if surprisingly the latter dehydrated more slowly than the exposed bunches. Percival *et al.* (1993), showed how cluster exposure and berry contact can affect the cuticular covering formation and the occurrence of bunch rot (*Botrytis cinerea*). In the contact of surfaces between berries, the wax platelets structure is lost, allowing easier access for the fungus. This is one of the main reasons why grape bunches involved in dehydration processes should not be compact. Moreover, a less compact cluster permits better air movement between the berries, creating a less favourable environment for the development of moulds.

2.4.4 Bunch thermal and light microclimate for grapes to be dehydrated

The thermal condition of grape berries is a crucial driving variable for the biochemical and physiological phenomena involved in berry formation and ripening. These phenomena are key factors for qualitative profiling, both in terms of primary and secondary metabolites (Jackson and Lombard, 1993). Even if the physiological bases of the relationships between temperature and grape metabolism are far from being fully explained, many efforts have been devoted to defining the most appropriate fruit thermal status to achieve the best grape quality according to the desired wine style.

It has been demonstrated that in cool climates, enhancing cluster exposure to sunlight has positive effects on sugar accumulation, terpenoids, anthocyanins and on disease avoidance. The typical aroma of certain varieties may be increased in shaded clusters (i.e. metoxypyrazines) or decreased with over-exposure (Belancic and Agosin, 2007; Falcao *et al.*, 2007; Scheiner *et al.*, 2010). The summer pruning technique requirement may vary according to site and variety.

Canopy manipulation aims at reducing cluster shading but, on the other hand, over-exposure to direct sunlight and high summer temperatures may have negative effects on the anthocyanin content (Bergqvist *et al.*, 2001; Spayd *et al.*, 2002; Tarara *et al.*, 2008).

The relationship between temperature and the grape ripening processes are generally calculated taking into account the average day air temperature or the day/night average temperature regimes. On this base, sugar accumulation would increase linearly as the temperature rises from 10°C to 30–32°C, and then quickly decline above 35°C, while malate levels would decline proportionally to the temperature (Coombe, 1987). More controversial are the experimental and reviewed data with regard to the thermal and light microclimate effects on accumulation of secondary metabolites such as phenols (anthocyanins, flavonols, tannins, phenolic acids and stilbens), aroma compounds (terpenes, C13-norisoprenoids, methoxipyrazines) and precursors (carotenoids). In fact, the final metabolites accumulation is the consequence of the balance between synthesis and degradation processes which are differently affected by temperature and light. Growth represents a diluting versus concentrating factor which has always to be considered for a proper data interpretation (Coombe and McCarthy, 2000; Roby and Matthews, 2003). Anthocyanins accumulation seems to be reduced as the grape temperature exceeds a threshold of between 30 and 35°C, even if a lot of speculation still exists about the role of lower temperatures, in particular during the night (Dokoozlian and Kliewer, 1996; Bergqvist *et al.*, 2001; Downey *et al.*, 2004; Mori *et al.* 2005; Spayd *et al.*, 2002). Mori *et al.* (2007) evidenced a possible major role of high temperature in anthocyanins oxidation compared to an inhibitory effect on their synthesis. Flavonols accumulation appears to be mainly and positively under the control of UV light intensity (Downey *et al.*, 2004; Spayd *et al.*, 2002). Tannins are synthesized during the first growth steps after fruit set and seem to be quite unaffected by the microclimate conditions, so the environmental variables probably only affect their chemical and physical evolution during maturation (Downey *et al.*, 2004). Terpenes synthesis appears to be positively related to light availability but their accumulation would be impaired by heat excess and earlier interruption of synthesis. Carotenoids synthesis has been proved to be stimulated by light intensity as well as by their successive oxidative conversion into C13-norisoprenoids. Methoxypyrazines seem to be more influenced by ripening course and by light environment rather than temperature (Hashizume and Samuta, 1999) especially during the pre-veraison period (Scheiner *et al.*, 2010).

Therefore, canopy management may provoke differences in the berry composition. Several research works have studied the effects of bunch exposure on grape anthocyanins accumulation. Leaf removal changes the bunch microclimate and, in general, as Spayd *et al.* (2002) demonstrated by exposing Merlot grapes to direct sunlight but maintaining temperature conditions comparable to shaded bunches, light may improve anthocyanin accumulation, while high temperature can have a negative effect. It is not only the quantity of anthocyanins in terms of total amount per unit of grapes weight that influences the style and quality of a wine, but also their composition. Some studies have also indicated that bunch exposure changes the anthocyanin profile (Spayd *et al.*, 2002; Downey *et al.*, 2004; Ristic *et al.*, 2007; Rustioni *et al.*, 2011a) and extractability during winemaking (Rustioni *et al.*, 2011b). Not only the pigments, but the berry composition in general is modified by the canopy management. Morrison and Noble (1990) found a Brix increase in leaf removal treatments, and a significant reduction in titratable acidity, malic acid, pH and potassium concentration. On the one hand leaf removal allows grapes to grow in better microclimatic conditions (limiting *Botrytis cinerea* infections), but on the other hand the loss of acidity can be a negative point.

2.4.5 Potential management of juice acidity

As the dehydration process causes an increase in the sugar content, a high acid content is considered favourable to balance the wine flavour. To obtain a good sugar-to-acid ratio, management of the harvesting time plays a central role. In general, the acidic content decreases during ripening, and, together with mould infections, this is one of the main reasons why grapes to be dehydrated are usually harvested early.

Juice acidity is also determined by the degree of salification, mainly by potassium ions, of tartaric and malic acids. The metabolism of the two acids both during their accumulation and decline phase really differ in terms of pathway, timing, and possible environmental regulatory effects.

Tartrate synthesis is very active during the first steps of berry growth, while malate synthesis is successive and lasts also during the berry growth lag phase. During ripening, malate is actively metabolized by respiratory process and at a lower extent by gluconeogenesis pathways, while tartrate is just diluted in the vacuolar sap which increases in sugar and water content. The role of light on tartrate synthesis has been proposed as having both a direct and indirect effect (Terrier and Romieu, 2001). Tartrate is synthesized from ascorbate following two alternative pathways. One, generally active in the grapevine cells, forms tartrate (plus glycolaldehyde) and a second one, active in the idioblast, special cells sited in the berry outer and inner hypodermids, forms at the same time oxalate and tartrate (DeBolt *et al.*, 2004). The role of light intensity on the stimulation of ascorbate synthesis in fruits has been well documented. Recently, this relation has been proved also in grape berry, where a correlation between ascorbate and tartrate synthesis has been shown (Melino *et al.*, 2011). Moreover, in the idioblasts, oxalate and consequently also tartrate could be synthesized also in response to high calcium influx to sequestrate it in the vacuole. Calcium movements within plants are dependent on the xylematic fluxes, which are driven by transpiration. The exposure of the bunches to direct solar radiation at fruit setting and during the first part of berry growth, by increasing transpiration from the fruit could increase at the same time the calcium intake and possibly the oxalate and tartrate synthesis in the berry (Volk *et al.* 2002).

2.4.6 Vineyard management in warm climates

In warm climates, vineyard plantation must be carefully designed according to the site of cultivation by choosing the most resistant rootstocks to water stress. Vines are generally planted at closer distances than in cool climates to ensure the best soil colonization by the root system. Training systems should have a canopy expansion to preserve water loss by transpiration if irrigation is not available. Arid cultivation techniques to cope with water scarcity are developed in warm climates, which consist of autumn soil ploughing to reduce the water loss by evaporation. When the soils are poor in organic matter, winter crop coverage is practised; crops are cut in late spring and incorporated into the soil to enhance the physical soil fertility.

Traditional untrellised training systems such as Alberello have been developed. Vines are trained very close to the soil, spur pruned and have a low bud load. Under these conditions, the vine becomes more tolerant to water restriction, produces lower yields and metabolism is more directed to metabolic concentration. This system, which can be found in southern Italy and Italian islands, requires a high amount of manual labour so viticulturists are interested in trellises, to increase the height of the canopy from the soil and to establish poles and wires to facilitate mechanization. There are many vineyards that have been transformed in this way and also new vineyards have been established adopting espalier instead of Alberello. Observations made in Pantelleria showed that the clusters produced by this system gave better quality grapes compared to Alberello, because they had less berry rot and a higher content of terpenic aroma, like linalool, nerol, geraniol and α-terpineol. Quality was enhanced on espalier, which had the best microclimatic condition at the cluster level, and especially at lower temperatures (Di Lorenzo and Lo Vetere, 2006).

In dry conditions, where irrigation is not available, lateral shoots are removed to avoid excessive transpiration and severe water stress. When irrigation is available regulated deficit

Table 2.8 Relationship between threshold of pre-dawn leaf water potential (PD Ψw) plant function and wine style (data from Deloire *et al.*, 2003).

PD Ψw (MPa)	Vegetative growth	Berry growth	Photosynthesis	Berry biochemical change	Wine style
0 to −0.3	Normal	Normal	Normal	Normal	1
−0.3 to −0.5	Decreased	Normal or slightly reduced	Normal or slightly reduced	Normal or enhanced	2–3–4
−0.6 to −0.9	Decreased or inhibited	Decreased or inhibited	Decreased or inhibited	Decreased or inhibited	4–5
< −0.9	Inhibited	Inhibited	Partial or complete inhibition	Partial or complete inhibition	5–6

Wine style: 1, 'diluted' with high acidity; 2, 'fruity' and balanced; 3, 'fruity'/'tannic'; 4, 'fruity'/'tannic', concentrated and well balanced; 5, 'tannic', concentrated and not always well balanced, high alcohol degree; 6, 'tannic', sour, not balanced high alcohol degree.

irrigation (RDI) is practised. The best way to apply this method is to monitor water potential (leaf: base water potential; stem: midday water potential) in order to give a limited amount of water to induce moderate stress (MS), taking into account that pre-veraison MS has a higher effect on berry size and yield, and post-veraison MS mainly affects berry composition. In red varieties an increase of sugar content and anthocyanins can be observed in concentration and also by the enhanced expression of key genes involved in anthocyanin biosynthesis (Castellarin et al., 2007a, b; Bucchetti *et al.*, 2011).

To maximize irrigation efficiency it is important to choose the volume of distributed water and the frequency, according to the adopted systems. A literature survey about RDI strategy indicates that the use of micro irrigation systems giving a percentage of the total crop evapotranspiration (30–50%) has to be modulated according to the phenological stage (Deloire *et al.*, 2002, 2003; Scalabrelli *et al.*, 2007; Romero *et al.*, 2010) (Table 2.8). Monitoring the water status is important to avoid severe vine stress, the use of the Scholander pressure chamber is recommended: the cost is reasonable and it is easy to use. Practical observation, such as shoot tip growth cessation (Smart and Robinson, 1991), the amplitude of leaf petiole angle, visible leaf turgor, may help in understanding the initial phase of stress and in programming the suitable duration of water restriction, before irrigating again.

2.5 TIME OF HARVEST AND BERRY PHENOLIC MATURITY

Viticulturists have to make the important decision on when to harvest, according to the enological purpose of the grapes. Factors that influence this decision are the earliness of ripening, climatic conditions and technique of dehydration. When the harvested clusters require further manipulation such as being placed in boxes or hung up for transfer to dehydration warehouses, the skins can be damaged which can lead to infection by undesirable fungi and bacteria. The latter can lead to berry rot and to the loss of the cluster's integrity. In this case, the choice of the harvest date is critical for the berry quality; this is why the ripening stage must be carefully monitored in order to pick the clusters for dehydration before the maximum level of ripeness is reached.

The situation is different for grapes that are allowed to remain in the vineyard in order to produce 'late harvest' wines ('vendemmia tardiva'). These grapes are harvested after over-ripening directly on the vine. Berry maturity is assessed by several methods: physical, chemical, by means of laboratory tests and even with non-destructive sensors. A sensorial analysis assessing berry maturity and phenolic maturity in particular, has been proposed by the Cooperative Wine Institute (Institut Coopératif du Vin (ICV)), and recently revised by Scalabrelli (2006) and Scalabrelli *et al.* (2010) proving to be useful for this purpose.

Good phenolic maturity is important for vinification. This is why particular interest is devoted to stimulating grape dehydration before harvest directly in the vineyard. Acceleration of dehydration may be obtained by several methods, which include the raffle torsion, the raffle incision, the cane incision, the shoot cut and the cane cut; the principle behind these techniques is called in French 'double maturation raisonnée' (Cargnello, 1992).

One of these, the 'cane cut' method (in Italian 'recisione del tralcio') has been adopted in experimental production of dessert white wine (Scalabrelli *et al.*, 2008), and the red wines, Raboso Piave (Ziliotto *et al.*, 2012) and Aleatico (Scalabrelli and D'Onofrio, 2012). In these cases the length of grape dehydration that occurs in the vineyard depends on weather conditions (temperature, wind, air humidity); in general 15 days are sufficient to obtain the right over ripening for harvest. In warm and dry sites, it is advisable to monitor frequently the ripening stage to avoid excess dehydration. This method proves to have many advantages when the cane cut is performed before complete berry ripening and softening; it enhances grape health as bee and wasp damage is avoided because berry dehydration is very rapid and these insects can no longer perforate the skin. In addition, a lot of hand-work and time are saved, as it is not necessary to carry out the repeated manual selection of rot-infected and damaged berries during the postharvest dehydration process. Moreover, berry sugars and acidity concentrate, and the phenolic maturity of the different parts of the berry and its overall maturity are enhanced. The typicity of wines obtained from Aleatico grapes using this method is under evaluation, to assess the possibility of introducing the technique in commercial production (Scalabrelli and D'Onofrio, 2012).

ACKNOWLEDGEMENT

Review paper published within the framework of the project PRIN 2008, coordinator Fabio Mencarelli, supported by the Italian Ministry of Education, University and Research.

REFERENCES

Andrich, G., A. Zinnai, F. Venturi, *et al.* 2009. L'evoluzione della frazione polifenolica durante le fasi di appassimento delle uve nella produzione dell'aleatico. *Accademia Italiana della Vite e del Vino*. Global Print, Gorgonzola, Milan, pp. 9–30.

Belancic A. and E. Agosin. 2007. Methoxypyrazines in grapes and wines of *Vitis vinifera* cv. Carmenere. *American Journal of Enology and Viticulture*, 58(4):462–469.

Bergqvist, J., N. Dokoozlian, and N. Ebisuda. 2001. Sunlight exposure and temperature effects on berry growth and composition of Cabernet Sauvignon and Grenache in the Central San Joaquin Valley of California. *American Journal of Enology and Viticulture*, 52:1–7.

Boselli, M., R. Bandinelli, G. Scalabrelli, *et al.* 2009. Caratterizzazione bioagronomica e aromatica del vitigno Aleatico. *Atti Accademia Italiana della Vite e Vino*. Global Print, Gorgonzola, Milan, pp. 31–49.

Brancadoro, L., L. Valenti, and A. Reina. 1995. Rootstock effect on potassium content of grapevine. *Acta Horticulturae*, 383:115–124.

Bucchetti B., M.A. Matthews, L. Falginella, *et al.* 2011. Effect of water deficit on Merlot grape tannins and anthocyanins across four seasons. *Scientia Horticulturae*, 128:297–305.

Cargnello, G. 1992. Premières recherches sur la 'double maturation raisonnée' du raisin au vignoble. *Quaderni Viticoltura Enologia Università di Torino*, pp. 101–114.

Castellarin, S.D., M.A. Matthews, G. Di Gaspero, *et al.* 2007a. Water deficits accelerate ripening and induce changes in gene expression regulating flavonoid biosynthesis in grape berries. *Planta*, 227:101–112.

Castellarin, S.D., A. Pfeiffer, P. Sivilotti, *et al.* 2007b. Transcriptional regulation of anthocyanin biosynthesis in ripening fruit of grapevine under seasonal water deficit. *Plant, Cell and Environment*, 30:1381–1399.

Coombe, B.G. 1987. Influence of temperature on composition and quality of grapes. *Acta Horticulturae*, 206:23–33.

Coombe, B.G. and M.G. McCarthy. 2000. Dynamics of grape berry growth and physiology of ripening. *Australian Journal of Grape and Wine Research*, 6:131–135.

Corazzina, E., O. Failla, A. Scienza, *et al.* 1993. Influence of rootstock and mineral nutrition on must acidity in grapevines. *Proceedings of the IV International Symposium on Grapevine Physiology*, Istituto Agrario San Michele all'Adige, Università di Torino, Italy, 11–15 May, pp. 217–220.

Cousins, P. 2005. Evolution, genetics, and breeding: Viticultural applications of the origins of our rootstocks. In *Proceedings of the Grapevine Rootstocks: Current Use, Research, and Application. 2005 Rootstock Symposium* (eds P. Cousins and R.K. Striegler), Osage Beach, Missouri, 5 February 2005. Mid-America Viticulture and Enology Center, Southwest Missouri State University, pp. 1–7.

DeBolt, S., J. Hardie, S. Tyerman, *et al.* 2004. Composition and synthesis of raphide crystals and druse crystals in berries of *Vitis vinifera* L. cv. Cabernet Sauvignon: Ascorbic acid as precursor for both oxalic and tartaric acids as revealed by radiolabelling studies. *Australian Journal of Grape and Wine Research*, 10:134–142.

Deloire, A., F. Lopez, and A. Carbonneau. 2002. Réponse de la vigne et le terroir: éléments pour une méthode d'étude. *Progrès Agricole et Viticole*, 4:78–86.

Deloire, A., B. Federspiel, H. Ojeda, *et al.* 2003. La vigne et l'eau. *GESCO XIII Jornadas*, 3–8 February.

Di Lorenzo, R. and R. Lo Vetere. 2006. Aromatic compounds in 'Zibibbo' grape bush and VSP trained in Pantelleria island. Preliminary results. *First International Congress on Mountain and Steep Slope Viticulture*, Saint Vincent, Aosta, 17–18 March 2006.

Di Matteo, M., L. Cinquanta, G. Galiero, *et al.* 2000. Effect of a novel physical pretreatment process on the drying kinetics of seedless grapes. *Journal of Food Engineering*, 46:83–89.

Dixon, R.A., L. Achnine, P. Kota, *et al.* 2002. The phenylpropanoid pathway and plant defence, a genomics perspective. *Molecular Plant Pathology*, 3:371–390.

Dokoozlian, N.K. and W.M. Kliewer. 1996. Influence of light on grape berry growth and composition varies during fruit development. *Journal of the American Society for Horticultural Science*, 121:869–874.

Downey, M.O., J.S. Harvey, and S.P. Robinson. 2004. The effect of bunch shading on berry development and flavonoid accumulation in Shiraz grapes. *Australian Journal of Grape and Wine Research*, 10:55–73.

Doymaz, I. 2006. Drying kinetics of black grapes treated with different solutions. *Journal of Food Engineering*, 76:212–217.

Doymaz, I. and M. Pala. 2002. The effects of dipping pretreatments on air-drying rates of the seedless grapes. *Journal of Food Engineering*, 52:413–417.

Failla, O., A. Scienza, A. Galluzzi, *et al.* 1993. Relationships between mineral nutrition and organic acid metabolism in grape berry. *Proceedings of the IV International Symposium on Grapevine Physiology*, Istituto Agrario San Michele all'Adige, Università di Torino, Italy, 11–15 May, pp. 227–230.

Failla, O., A. Scienza, G. Stringari, *et al.* 1990. Potassium partitioning between leaves and clusters: role of rootstock. *Vitis*, Special Issue:187–196.

Falcao, L.D., G. De Revel, M.C. Perello, *et al.* 2007. A survey of seasonal temperatures and vineyard altitude influences on 2-methoxy-3-isobutylpyrazine, C-13-585 norisoprenoids, and the sensory profile of Brazilian Cabernet Sauvignon wines. *Journal of Agricultural and Food Chemistry*, 55:3605–3612.

Ferreira, J.H.S. and P.G. Marais. 1987. Effect of rootstock cultivar, pruning method and crop load on *Botrytis cinerea* rot of *Vitis vinifera* cv. Chenin blanc grapes. *South African Journal for Enology & Viticulture*, 8:41–44.

Filippetti, I., G. Allegro, G. Valentini, *et al.* 2011. Effects of mechanical pre-bloom defoliation on cordon de Royat pruned Sangiovese (*Vitis vinifera* L.) vines. *Journal International des Sciences de la Vigne et du Vin*, 45(1):19–25.

Garcia, M., P. Gallego, C. Daverede, *et al.* 2001. Effect of three rootstocks on grapevine (*Vitis vinifera L.*) cv. Negrette, grown hydroponically. I. Potassium, calcium and magnesium nutrition. *South African Journal for Enology & Viticulture*, 22:101–103.

Hashizume, K. and T. Samuta. 1999. Grape maturity and light exposure affect berry methoxypyrazine concentration. *American Journal of Enology and Viticulture*, 50:194–198.

Intrieri, C., I. Filippetti, G. Allegro, *et al.* 2008. Early defoliation (hand vs mechanical) for improve crop control and grape composition in Sangiovese (*Vitis vinifera* L.). *Australian Journal of Grape and Wine Research*, 14:25–32.

Intrieri, C., I. Filippetti, G. Allegro, *et al.* 2012. Effetto di trattamenti antitraspiranti in pre-fioritura sulla morfologia dei grappoli e sulla composizione dell'uva di Sangiovese. *Italus Hortus*, 3–5(1): 95–101.

Jackson, D.I. and P.B. Lombard. 1993. Environmental and management practices affecting grape composition and wine quality: A review. *American Journal of Enology and Viticulture*, 44:409–430.

Jenks, M.A. and E.N. Ashworth. 1999. Plant epicuticular waxes: function, production, and genetics. *Horticultural Reviews*, 23:1–68.

Kliewer, W.M. and N.K. Dokoozlian. 2005. Leaf area/crop weight ratios of grapevines: Influence on fruit composition and wine quality. *American Journal of Enology and Viticulture*, 56:170–181.

Kodur, S., J.M. Tisdall, C. Tang, *et al.* 2010. Accumulation of potassium in grapevine rootstocks (Vitis) as affected by dry matter partitioning, root traits and transpiration. *Australian Journal of Grape and Wine Research*, 16:273–282.

Mahmutoğlu, T., F. Emìr, and Y.B. Saygi. 1996. Sun/solar drying of differently treated grapes and storage stability of dried grapes. *Journal of Food Engineering*, 29:289–300.

Materazzi, A. and E. Triolo. 2000. Inerbimento del vigneto ed incidenza di Botrytis cinerea su 'Sangiovese'. Il 'Sangiovese'. *Atti del Simposio Internazionale. Firenze, Palazzo dei Congressi*, 15–17 February 2000, ARSIA, Florence, pp. 101–105.

Melino, V.J., M.A. Hayes, K.L. Soole, *et al.* 2011. The role of light in the regulation of ascorbate metabolism during berry development in the cultivated grapevine *Vitis vinifera* L. *Journal of the Science of Food and Agriculture*, 91:1712–1721.

Mori, K., N. Goto-Yamamoto, M. Kitayama, *et al.* 2007. Loss of anthocyanins in red-wine grape under high temperature. *Journal of Experimental Botany*, 58:1935–1945.

Mori, K., H. Saito, N. Goto-Yamamoto, *et al.* 2005. Effects of abscisic acid treatment and night temperatures on anthocyanin composition in Pinot noir grapes. *Vitis*, 44:161–165.

Morrison, J.C. and A.C. Noble. 1990. The effects of leaf and cluster shading on the composition of Cabernet Sauvignon grapes and on fruit and wine sensory properties. *American Journal of Enology and Viticulture*, 41:193–200.

Muganu, M., A. Bellincontro, F.E. Barnaba, *et al.* 2011. Influence of bunch position in the canopy on berry epicuticular wax during ripening and on weight loss during postharvest dehydration. *American Journal of Enology and Viticulture*, 62(1):91–98.

Müller, C. and M. Reiderer. 2005. Plant surface properties in chemical ecology. *Journal of Chemical Ecology*, 31:2621–2651.

Pangavhane, D.R., R.L. Sawhney, and P.N. Sarsavadia. 1999. Effect of various dipping pretreatment on drying kinetics of Thompson seedless grapes. *Journal of Food Engineering*, 39:211–216.

Percival, D.C., J.A. Sullivan, and K.H. Fisher. 1993. Effect of cluster exposure, berry contact and cultivar on cuticular membrane formation and occurrence of bunch rot (*Botrytis cinerea* Pers. Fr.) with 3 *Vitis vinifera* L. cultivars. *Vitis*, 32:87–97.

Poni, S., L. Casalini, F. Bernizzoni, *et al.* 2006. Effects of early defoliation on shoot photosynthesis, yield components, and grape composition. *American Journal of Enology and Viticulture*, 57(4):397–407.

Ristic, R., M.O. Downey, P.G. Iland, *et al.* 2007. Exclusion of sunlight from Shiraz grapes alters wine colour, tannin and sensory properties. *Australian Journal of Grape and Wine Research*, 13:53–65.

Roby, G. and M.A. Matthews. 2003. Relative proportions of seed, skin and flesh, in the ripe berries from Cabernet Sauvignon grapevines grown in a vineyard either well irrigated or under water deficit. *Australian Journal of Grape and Wine Research*, 10:74–82.

Romero, P., J.I. Fernández-Fernández, and A. Martinez-Cutillas. 2010. Physiological thresholds for efficient regulated deficit-irrigation management in winegrapes grown under semiarid conditions. *American Journal of Enology and Viticulture*, 61:300–312.

Rosenquist, J.K. and J.C. Morrison. 1988. The development of the cuticle and epicuticular wax of the grape berry. *Vitis*, 27:63–70.

Rosenquist, J.K. and J.C. Morrison. 1989. Some factors affecting cuticle and wax accumulation on grape berries. *American Journal of Enology and Viticulture*, 40:241–244.
Ruhl, E.H. 1991. Effects of potassium supply on cation uptake and distribution in grafted *Vitis champinii* and *Vitis berlandieri* × *Vitis rupestris* rootstocks. *Australian Journal of Experimental Agriculture*, 31:119–125.
Rustioni, L., M. Rossoni, G. Cola, *et al.* 2011a. Bunch exposure to direct solar radiation increases ortho-diphenol anthocyanins in northern Italy climatic condition. *Journal International des Sciences de la Vigne et du Vin*, 45:85–99.
Rustioni, L., M. Rossoni, M. Calatroni, *et al.* 2011b. Influence of bunch exposure on anthocyanins extractability from grapes skins (*Vitis vinifera* L.). *Vitis*, 50:137–143.
Scalabrelli, G. 2006. Razionale gestione dell'acqua di irrigazione in viticoltura. *I Georgofili. Quaderni*, Accademia dei Georgofili. Polistampa, Firenze, 6:78–118.
Scalabrelli, G. and C. D'Onofrio. 2012. *L'Aleatico dai mille profumi*. Ed. De Batte, Livorno.
Scalabrelli, G., F. Loreti, and G. Ferroni. 2004. La selezione clonale dell'Aleatico dell'Elba: aspetti agronomici e ampelografici. *Atti Convegno 'Miglioramento qualitativo delle produzioni vitivinicole e del materiale di produzione', Quaderno ARSIA*, 1:47–59.
Scalabrelli, G., C. D'Onofrio, G. Ferroni, *et al.* 2010. Effect of leaf removal timing on grape and wine quality of 'Ciliegiolo' grapevine. *Italus Hortus*, 17(suppl. 3):79–83.
Scalabrelli, G., G. Ferroni, A. Zinnai, *et al.* 2008. Sensorial characteristics of wines obtained from from experimental vinification of grapevine variety Malvasia Istriana. *Rivista di Viticoltura e di Enologia*, 2-3-4:261–280.
Scalabrelli, G., D. Remorini, L. Tuccio, *et al.* 2011. Qualità dell'uva e del vino e interventi irrigui. *Informatore Agrario*, 20:38–44.
Scalabrelli, G., E. Saracini, D. Remorini, *et al.* 2007. Changes of leaf phenolic substances in two grapevine varieties (*Vitis vinifera* L.) grown in different water conditions. *Acta Horticulturae*, 754:295–299.
Scheiner, J.J., G.L. Sacks, B. Pan, *et al.* 2010. Impact of severity and timing of basal leaf removal on 3-isobutyl-2-methoxypyrazine concentrations in red wine grapes. *American Journal of Enology and Viticulture*, 61(3):358–364.
Smart, R.E. and M. Robinson. 1991. Sunlight into wine. In *A Handbook for Winegrape Canopy Management*, Winetitles, Adelaide, Australia.
Spayd, S.E., J.M. Tarara, D.L. Mee, *et al.* 2002. Separation of sunlight and temperature effects on the composition of *Vitis vinifera* cv. Merlot berries. *American Journal of Enology and Viticulture*, 53: 171–182.
Stoll, M., M. Schweidweiler, M. Lafontaine, *et al.* 2009. Possibilities to reduce the velocity of berry maturation through various leaf area to fruit ratio modifications in *Vitis vinifera* L. Riesling. *Proceedings of the 16th International GiESCO Symposium*, July 12–15, University of California, Davis, USA, pp. 93–96.
Storchi P., M. Leprini, M. Pieri, *et al.* 2008. Effect of defoliation on grape quality of some Malvasias. *Rivista di Viticoltura ed Enologia*, 2-3-4:135–146.
Tarara, J.M., J. Lee, S.E. Spayd, *et al.* 2008. Berry temperature and solar radiation alter acylation, proportion, and concentration of anthocyanin in Merlot grapes *American Journal of Enology and Viticulture*, 59:235–247.
Tardaguila, V., F. Martinez de Toda, S. Poni, *et al.* 2010. Impact of early leaf removal on yield and fruit and wine composition of *Vitis vinifera* L. Graciano and Carignan. *American Journal of Enology and Viticulture*, 61:372–381.
Terrier, N. and C. Romieu. 2001 Grape berry acidity. In *Molecular Biology and Biotechnology of the Grapevine* (ed. K.A. Roubelakis-Angelakis), Kluwer Academic Publisher, pp. 35–57.
Tuccio, L. 2011. Aleatico grapevine characterization: physiological and molecular responses to different water regimes. PhD Thesis, University of Pisa.
Volk, G.M., V.J.J. Lynch-Holm, T.A. Kostman, *et al.* 2002. The role of druse and raphide calcium oxalate crystals in tissue calcium regulation in *Pistia stratiotes* leaves. *Plant Biology*, 4:34–45.
Ziliotto, F., F.M. Rizzini, S. Cavaleri, *et al.* 2012. Evoluzione dei parametri sensoriali, biochimici e molecolari in uve della cv Raboso Piave sottoposte alla Doppia Maturazione Ragionata e alla Vendemmia Tardiva. *Italus Hortus*, 3–5(1):9–14.

3 Technology and Management of Postharvest Dehydration

Fabio Mencarelli and Andrea Bellincontro

DIBAF, University of Tuscia, Viterbo, Italy

3.1 INTRODUCTION

When a winery decides to start a wine project using the postharvest technique of grape dehydration, the winemaker should have some knowledge of the physiology of grape dehydration and of the fundamentals of water loss. Why? Because 'withering, with the different practices and technologies, produces different grapes and thus different wines' (Di Lorenzo, 2007). Besides this knowledge, it is important for the winemaker to have other practical, tangible, information, by asking some specific questions:

1. What are the weather conditions during the grape dehydration period?
2. What kind of facility is available to carry out dehydration?
3. What financial resources are available?
4. What is the situation in terms of available energy and relative cost?

First of all, we must keep in mind that grape berry is a living tissue which continues to live after harvest, meaning that it breathes, consuming oxygen (O_2) and eliminating carbon dioxide (CO_2) and heat. For the winemaking process, the faster the transfer of the harvested grapes from the vine to the winery, the better, but in the case of wine that requires grape dehydration, the postharvest life of the berry is very important and needs to be managed. But why do we speak about dehydration and not drying? We define *drying* as the rapid process of water removal from items containing water (processing technique), so raisins are dried and wood is dried. *Dehydration* indicates the physical principle by which, depending on the environmental conditions surrounding the item, water vapour is drawn off from that item. Finally, in the wine sector, we speak about *withering* (*appassimento* in Italian), which can be explained as a slow process of water removal from a berry where over-ripening, or even better, senescence, occurs. Thus, dehydration is the correct term to identify water loss from the berry to the surrounding air and thus our objective is to manage the dehydration because it means managing the water loss and, in turn, all the metabolisms affected by water loss, which causes *water stress* in the berry cells.

This is the point: in a living tissue like berry, with dehydration we create water stress and, as described in Chapter 4, dramatic changes occur in the berry cell's metabolism.

Sweet, Reinforced and Fortified Wines: Grape Biochemistry, Technology and Vinification, First Edition.
Edited by Fabio Mencarelli and Pietro Tonutti.

In this chapter we will provide information on the fundamentals of water loss based on environmental conditions and we will explain why the above questions are important.

3.2 BASIC CONCEPT AND PRACTICAL NEEDS

Food drying is based on three main concepts (Rozis, 1997):

- Supply of thermal energy (heat)
- Ability of the air surrounding the commodity to take the water vapour released by the commodity
- Air speed striking the commodity.

Moreover, according to Rozis, drying should not occur either excessively fast or at a high temperature. This suggestion is true because, as Wang and Brennan (1995) discovered, shrinkage occurs first at the surface and then gradually moves to the internal tissue as drying time increases; at a slow drying rate, the moisture content at the centre of a berry is not much higher than at the surface, the internal stresses are minimized, and the material shrinks fully onto a solid core. For commercial reasons, the faster the drying, the more profitable is the process. Raisins are dried quickly, exposed to the sun or kept in commercial ovens, because the process is more profitable (more berries per time unit) and also because there is no interest in having a final aromatic product but just one that is as sweet as possible, and maybe a bit chewy. However, when the final goal is to have not just sugars but a panorama of different characteristics useful for wine quality, then drying must be carefully controlled or, as we say, 'managed'.

The main processes involved in the drying process can be summarized in (1) heat transfer to the liquid, (2) mass transfer as liquid or vapour inside the solid, and (3) as vapour from the solid surface to the outside. So, depending on the drying plant, these processes can proceed faster or slower. Inside the solid, that is, a berry, the driving force which allows the matter (water) transfer is a concentration gradient (mainly sugars) and the movement of the water can occur through diffusion, due to VPD (vapour pressure deficit) or to continuous changes in evaporation and condensation.

If we take into account the water loss during the heating process, first of all water evaporation occurs from the surface due to VPD, and this evaporation cools down the cells; consequently, water vapour moves from the internal layers (cell and intercellular spaces) to the superficial ones, cools down and condenses. To evaporate this water, we must provide further heat. At the same time, the dry matter of superficial cells concentrates, drawing more water from internal layers. Thus, in the cell and in the intercellular spaces of the berry, we will have a continuous concentration gradient and a continuous evaporation–condensation process during dehydration, which in turn will affect the diffusion coefficient of water in the tissue.

The water diffusion inside the tissue is a slow movement whose speed decreases as the relative humidity (RH) of the tissue decreases. The moisture diffusion coefficient D ($m^2\ s^{-1}$) is based on Fick's second law (Barbanti *et al.*, 2008):

$$\ln\left(\frac{M - M_e}{M_o - M_e}\right) = \ln\left(\frac{6}{\pi^2}\right) - \left(\frac{\pi^2 Dt}{R^2}\right)$$

Table 3.1 Pick-up efficiency (g water vapour per m^3 of air) of air at different temperatures and relative humidity.

Relative humidity (%) \ Temperature (°C)	10	15	20	25	30	35	40	45
10						8.2	9.2	9.9
20					6.0	7.3	8.0	8.6
30				5.0	5.6	6.2	6.6	6.9
40	2.7	3.2	3.7	4.2	4.6	5.1	5.5	6.1
50	2.2	2.6	3.0	3.4	3.7	3.9	4.1	
60	1.8	2.1	2.3	2.6	2.9	3.1		
70	1.3	1.5	1.7	1.9	2.1			
80	0.9	1.0	1.1	1.2				
90	0.5	0.5	0.7					

where M = moisture content at any time during the berry dehydration process (kg water × kg dry matter^{-1}); Mo = initial moisture content (kg water × kg dry matter^{-1}); Me = equilibrium moisture content (kg water × kg dry matter^{-1}); R = berry radius at time 0 (m); t = withering time (days). Thus, the diffusion coefficient is directly proportional to the berry surface and inversely proportional to the withering (dehydration) time and water content.

Practically speaking, some basic questions on product dehydration are asked and relative answers provided in the drying manual (Rozis, 1997):

1. How much water should be removed from the product? The mass of water (M_w in kg of water per kg of fresh product to be dried) to be extracted during dehydration = (m_i% − m_f%) M_p/(100 − m_f%) where m_i % = percentage of moisture of the product before dehydration; m_f % = percentage of moisture of the product after dehydration; M_p = product mass.
2. How long should the drying last? It depends on the production rhythm given by the quantity of dried product to be processed during each cycle, divided by the length of each cycle; the second factor is how long the product can wait before being dry enough so that pathogen micro-organisms do not develop; the third factor is the pick-up rate which must not be too fast to avoid case hardening.
3. What is the moisture pick-up rate (V_{ap}) in the product? It is the ratio between M_w/T_d where T_d is the ideal drying time for a given product (hours).
4. What is the pick-up efficiency of the air? It is the maximum quantity of moisture that a cubic meter of air can pick up when in contact with the humid product. It depends on the water vapour density, the pressure conditions, and the air temperature.

The ability of air to catch the water vapour is shown in Table 3.1 and is directly proportional to the increase in temperature and to the decrease in RH, as expected. But the table is interesting for grape drying because this ability is similar under different temperature and relative humidity conditions; that is, at 10°C and 40% RH the value is 2.2, similar to 2.1 and 2.3, which we find matching 15 and 20°C and 60% RH or 30°C and 70% RH. If we observe these data mechanically, we conclude that keeping grapes under these conditions, the dehydration time is similar, so the quality of the final product is similar. This is true for non-living material such as a wooden bar, but with harvest grape we are dealing with a living being, so the effect of temperature first and RH second plays an important role in the metabolism.

All of the questions mentioned earlier, together with other basic questions such as the drying temperature to be used, the average moisture pick-up rate and the air flow to be used, are important for an industrial technique or in a natural (sun) drying plant where the final objective is to dry the product completely, in the shortest amount of time but avoiding case hardening when the temperature is adjustable. However, when we speak about grape dehydration for wine production, dehydration management is a complex system in which the operator must have in mind the type of grape (physical and physiological features), the final water percentage in the berry, the time to reach this value and, very important, what kind of dehydrated grape he or she wishes to obtain. In short, the operator must manage water stress and senescence of the berries and thus adapt the dehydration conditions to control these two events.

First of all, as mentioned earlier, we need to know the final quality of the dehydrated product that we wish to obtain. In this case, the amount of water loss, the rate of water loss and the temperature are fundamental. We usually calculate the following values, based on our scientific experience, for the amount of weight loss (%) and for the speed of dehydration (days).

Amount of weight loss (%) = 10–15%
= 20–30%
= >30%

We know through experience (Costantini *et al.*, 2006; Chkaiban *et al.*, 2007; Santonico *et al.*, 2010; Mencarelli *et al.*, 2010; Cirilli *et al.*, 2012) that significant change occurs in the grape metabolism with these weight loss percentage values. Regarding the dehydration speed, we list the following:

Very fast = 5–10 days
Fast = 2–3 weeks
Slow = 4–8 weeks
Very slow = >8 weeks

Some important dehydration parameters to check, which are useful for comparison, are given in Table 3.2.

The rate of weight loss (WLR) under carefully controlled atmospheric conditions is one of the more important parameters: Aleatico cultivar loses water rapidly and consequently the sugar content increase rate (SIR) is much higher than in the other cultivars. But if we look at the SIR-to-WLR ratio, Aleatico, Shiraz or Montepulciano cultivars have similar values, meaning that the increase in sugars during dehydration is not merely concentration but events such as sugar respiration (main) and sugar biosynthesis by gluconeogenesis ((sugars can be synthesized from malic acid under stress conditions, as is the case of high sugar concentration (Sweetman *et al.*, 2009)) occur. In contrast, cultivars such as Grechetto rosso (red cv.) and Roscetto (white cv.) have a lower ratio, suggesting that, most likely, the consumption of sugars is greater than their synthesis, partially reducing the concentration effect.

Table 3.2 Wine grape cultivars dehydrated under the same atmospheric conditions (15–17°C and 45–55% RH).

	Aleatico	Grechetto Rosso	Procanico	Roscetto	Montepulciano	Shiraz
Dehydration time (days)	12	12	16	16	33	33
°Brix at harvest	21	21	21	21	24	24
Final weight loss (%)	27	19	20.5	20	40	44
Final °Brix	27	24	25	24	32	34
WLR = rate of weight loss (%/day)	2.25	1.58	1.28	1.25	1.21	1.33
SIR = rate of sugar increase (°Brix/day)	0.50	0.25	0.25	0.19	0.24	0.30
SIR/WLR (°Brix/%)	0.22	0.16	0.19	0.15	0.20	0.23

3.3 ENDOGENOUS AND EXOGENOUS FACTORS AFFECTING WATER LOSS

How does water get transferred from inside to outside the berry? It depends on the environmental conditions and will be dealt with later in this chapter. In any case, with a RH (tissue water content) higher than 27%, first of all, the free water (the water which works like a solvent) evaporates, and subsequently, the osmotic water evaporates when the RH is between 13 and 27%; the evaporation of these two types of water cools down the cells. In the third step, when the RH is between 5 and 13%, the polymolecular water, highly adsorbed on dry matter (matrix), is removed with difficulty and cell heating occurs. Finally, with RH below 5%, water is very strongly bound through van der Waals bonds and is a single monomolecular layer (Rozis, 1997). This water cannot be removed.

The factors affecting the postharvest water loss of berry are *endogenous* (depending genetically on the cultivar and on the physical and biochemical characteristics induced by the climate conditions during berry growth), and *exogenous*, which are related to the environment where the bunch is kept after harvest for the dehydration process. Speaking of endogenous factors, the following are the main physical features which affect water loss:

- ratio of surface to mass or volume (berry size)
- berry size and shape
- berry skin surface
- micropores and cracks
- presence and physical condition of the rachid
- bunch density.

Technically speaking, the main factors for grape dehydration are: initial moisture content, sugar content, berry volume, berry surface area and surface-to-volume ratio (S/V) values (Barbanti *et al.*, 2008).

The ratio of surface to volume is very important for water loss; indeed it is the main factor affecting the vacuum cooling technique, which is based on rapid and controlled water loss to cool down vegetables (Thompson, 2004). In dehydration, we do not want to cool down the berry but we want to have easy water loss. In the same bunch, small berries lose water

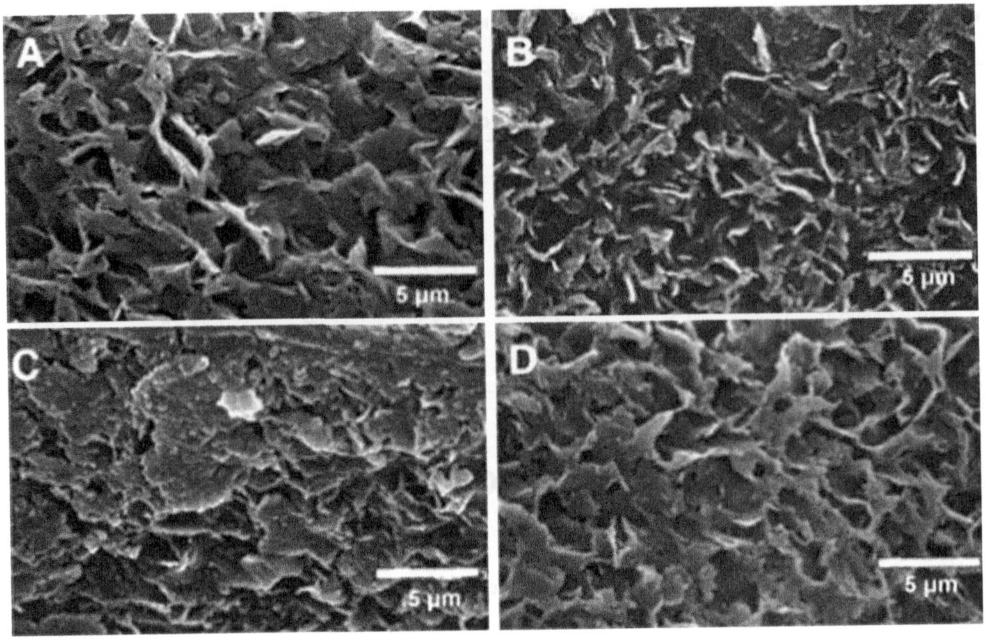

Figure 3.1 Scanning electron micrographs of epicuticular wax on berry surface at harvest. Cultivar Trebbiano Toscano: (A) extra-canopy; (B) intra-canopy; cv. Rossetto: (C) extra-canopy; (D) intra-canopy. Reproduced from Muganu *et al.* (2011), with permission from American Society for Enology and Viticulture.

faster than larger berries because the surface-to-volume ratio is higher; as a matter of fact, if the surface rises two-, three-, fourfold, the volume increases at a lower rate: 2x $\sqrt{2}$, 3x$\sqrt{3}$, 4x$\sqrt{4}$.

Moreover, the shape affects the water loss of berries with the same volume: the spherical berry loses water more slowly than the spheroidal one because the surface-to-volume ratio is higher for the latter than for the former.

Speaking about surface means speaking about the thickness of the wax (pruine), and how the wax is distributed on the surface because, depending on the cultivar, the pruine platelets are different. It is clear that this surface is strongly affected by the growing conditions of the berries in the vineyard, which mainly means sun exposure and irrigation, but also vine nutrition. It has been seen recently (Muganu *et al.*, 2011) that shaded berries (intra-canopy) dehydrate more slowly than berries exposed to the sun (extra-canopy) but the wax platelet content is greater in extra-canopy berries. It is likely that the berry anatomy and morphology characteristics and the morphology of the wax could be the reason (Figure 3.1).

The intactness of the berry surface is another important characteristic affecting the rate of water loss. It is very important not only to avoid micro-organism contamination but to manage water loss. The presence of microcracks alters the natural permeability of berry skin; abrasion due to vine stem or leaves favours water loss, removing pruine and causing microcracks. The grapes are harvested carefully (by hand) and plastic boxes are used for grape bunches intended for dehydration.

Grape berry is unusual compared to other fruits because the surface does not have stomatas but only a few lenticels and tiny micropores which are covered by a thick layer of pruine, a wax formed by oleanolic acid (2/3) and the rest by alcohols, esters, and long-chain fatty acids, which are produced by epidermal cells. Thus, apparently the flesh of the grape berry is almost completely isolated from the external environment, and the berry is similar to a controlled atmosphere room, gas-tight, where apples or kiwifruits are commercially stored for long-term use. This means that the gas concentration inside the berry is not the same as the outside air, and O_2 will be reduced and CO_2 increased. There are no data about this on grape berry due to the difficulty in measuring, but there are interesting findings on pears by Franck *et al.* (2007), who measured a concentration of 17.2% O_2 and 0.76% CO_2 in the pear core. Taking into account that pear peel is rich in stomata and lenticels, we can assume that the O_2 in the berry core (seed area) is much lower and CO_2 much higher than is found in pear, and this would explain the high ADH (alcohol dehydrogenase) activity found during berry ripening (Tesniere *et al.*, 1993). The concept of aerobic fermentation is today accepted as a way for the cell to find other sources of energy when the main pathway (Kreb's cycle) is blocked, as in the case of high sugar content of dehydrated berry, before the cell dies. In this case, even under normal atmospheric conditions, glycolysis goes fast while the Krebs cycle is blocked, thus pyruvic acid is converted to ethanol, completely modifying the aromatic panorama of the berry (Cirilli *et al.*, 2012).

Despite this apparently impermeable surface, grape berry loses water if it is kept under atmospheric conditions of low RH or high air flow. Becker and Knoche (2011) have shown that for transpiration, the berry surface contribution went from 58% between 26 and 59 DAFB (days after full blooming) to an average of 77% between 73 and 129 DAFB in cv. Riesling, and similar data was found in cv. Chardonnay and cv. Muller-Thurgau. Moreover, transpiration through the berry surface occurred at a higher rate at the stylar end (+44%) as compared to the cheek. No relationship was found between stylar scar area and transpiration rates and neither the rusty area nor the number of lenticels had a significant effect on transpiration rates. Thus, berry transpiration occurs mainly through the surface. The same authors showed that the permeability of the berry cuticle decreased throughout berry development. Thus, diffusion is the primary mechanism for water movement through the berry surface during transpiration. As explained by Schönherr (2006), diffusion occurs along a lipophilic and/or polar pathway; the former is formed by the amorphous wax fraction within the cuticle and the latter is formed by the orientation, clustering and hydration of polar functional groups within the cuticle, clusters forming an aqueous continuum such as micropores (not detectable under the microscope), which allow viscous water flow.

In addition, the berry is attached to the stem through a pedicel, so the bunch is a complex system of stems and pedicels. This green structure is rich in stomatas and lenticels, and does not have a thick cuticle like the berry. The periderm of the receptacle where the berry is joined to the pedicel, together with the pedicel, has a high permeability rate of 376.1 nm s^{-1} versus 6.7 nm s^{-1} (Becker and Knocke, 2011). Generally speaking, water loss from the green parts of the bunch is reduced by the presence of the boundary layer, but under postharvest environmental conditions when the RH is low and air flow is present, the high permeability of the rachid, therefore its high water loss, induces a pressure differential that draws water from the berry to the rachid, causing water loss from the berry, even though limited. This process continues until the pedicel is completely dry, an event that usually occurs at between 5 and 10% of water loss from the bunch of grapes. This is why, in Figure 3.2, berries detached from the rachid with 2 mm of pedicel lose less water than the whole bunch. To confirm this behaviour, the hypothesis for the mechanism of berry weight loss on vine proposed by Tilbrook and Tyerman (2008), that is that the membranes of the pericarp cells begin to lose semi-permeability at the same time as weight loss begins, so the large negative osmotic potential of the berry sap is no longer effective in opposing the xylem tensions developed by

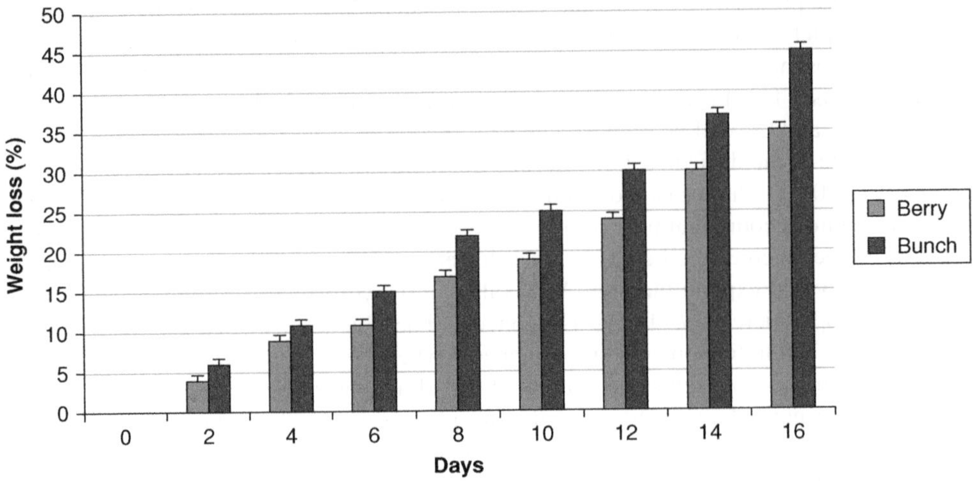

Figure 3.2 Weight loss of a sample of berries (cv. Aleatico) detached with 2 mm of pedicel (20 berries) or of the whole bunch (3 bunches), both kept in air flow of 1.5 m s^{-1}, 50% RH, and 20°C. Vertical bar indicates standard deviation. Reproduced with permission from Mencarelli *et al.* (2006).

leaves. Those cultivars that maintain higher xylem-to-berry hydraulic conductance, such as Shiraz, are therefore more vulnerable to backflow via the xylem.

If the environmental conditions change and the rate of weight loss (weight loss per day) is very low, this difference is not so significant.

Beyond the physical features, respiration affects water loss by heating the tissue, favouring water evaporation, and, in turn, dehydration affects the respiration rate. Furthermore, ethylene production, as a consequence of water stress, may affect the water loss and the respiration rate of the berry cells (Botondi *et al.*, 2011). Figure 3.3 shows the respiration rate of some wine grape cultivars during dehydration.

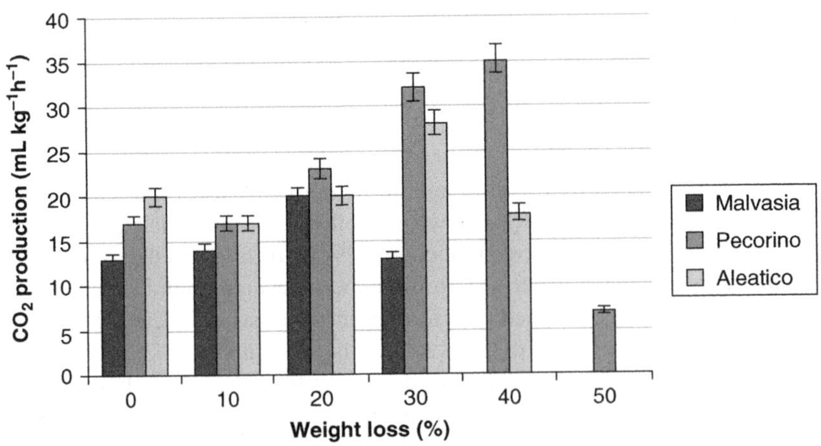

Figure 3.3 CO_2 production of three cultivars during dehydration in the same environmental conditions (20°C; 45% RH; 1.5 m s^{-1}).

The environmental factors affecting water loss are mainly temperature, RH, and air flow. Sunlight can be another factor for dehydration in open air but its role is a combination of infrared and UV radiation, which are not studied at all.

As mentioned beforehand, providing heat is the first factor affecting water loss. Thus temperature plays an important role. Until recently, no one paid attention to carefully controlling the temperature during grape dehydration. This lack of interest is based on the concept that, as in an industrial drying process, it is important to remove water quickly, adding to the use of high temperature, ventilation and/or removal of water vapour or, more simply, leaving the grape in the sun. This is a mechanical view of the dehydration process, which does not take into account that the berry cells are living, as we emphasized earlier.

What is the best temperature for the dehydration of grapes to produce wine? It depends on what we want to obtain from the grape metabolism. Bellincontro *et al.* (2009) have shown that the lower the temperature (with fixed RH and air flow), the better the grape structure during dehydration, and it is known that fruit pigment is more likely to remain intact with a lower dehydration temperature (Del Caro *et al.*, 2004). This is very important because the water loss must occur slowly, by transfer from cell to cell and to cell interspace. Speaking of low temperature, we mean temperatures between 5 and 10°C. These temperatures can be found naturally in the Verona area, where Amarone and Recioto wines are produced and where the dehydration facilities are built in such a way that when the outside air conditions are optimal for dehydration, the outside air is pulled inside the facility (see Chapter 13 'Amarone'). In a closed, insulated facility where all the ambient parameters can be carefully controlled, the temperature can be maintained at 10°C. If the cellular structure of the berry is maintained, we obtain the following advantages: first, water stress can be postponed and be less traumatic; second, the secondary metabolism for the production of polyphenols (Mencarelli *et al.*, 2010) and volatile compounds can proceed slowly (Santonico *et al.*, 2010), releasing important compounds into the cell sap; third, less oxidation occurs; and fourth, slower aerobic fermentation takes place with lesser formation of metabolites dangerous for berry quality such as ethanol, acetaldehyde and acetic acid (Cirilli *et al.*, 2012). These advantages do not mean that the grape is guaranteed to be better because there may even be a reason to use temperatures of 20 or 30°C, but we know that the higher the temperature, the more oxidation of compounds occurs. High temperature is a determining factor for sun-drying as in Pantelleria or the Aeolian Islands or in Andalusia (Malaga, Jerez, Montilla-Moriles). The must from such dried grapes mainly has the aroma of over-ripe or dried fruit with some toasty notes, due to the Maillard reaction (Franco *et al.*, 2004). An interesting paper by Piombino *et al.* (2010) on the processing of wine grape cv. Malvasia carried out in the Aeolian Islands to produce the Passito wine Malvasia delle Lipari, showed that the off-vine bunches dried under shaded conditions significantly affected the volatiles of Malvasia delle Lipari wine traditionally produced by sun-dried grapes. The wine coming from shaded dried berries was richest (67%) in free volatiles with 20% higher total bound terpenoids. It is interesting to note the high content of alcohols derived from amino acid catabolism in wine coming from shaded dried berries, probably due, as the authors suggest, to the lesser availability of free amino acids (less degradation due to the lower temperature) and even to the higher concentration of volatile phenols likely due to light degradation of the precursors in the sun-dried berries. Speaking of sun-drying, the effect of temperature and light is combined and there are no papers which discriminate between the two factors. When we speak of closed facilities for dehydration, as explained later, the light effect disappears and only the temperature effect becomes important. However, if we wish to study the effect of temperature on metabolism during dehydration, the closed system is the best. Sun radiation

Table 3.3 Changes in volatile terpene compounds (μg 100 g^{-1}) in Muscat of Alexandria (Zibibbo) on Pantelleria island.

	Before dehydration	After sun dehydration (40% weight loss)	After sun dehydration in plastic tunnel (40% weight loss)
Free terpenes	1285 ± 518	1246 ± 40	352 ± 21
Glycosylated terpenes	3814 ± 1607	2450 ± 1119	1840 ± 607
Free linalol	1192	392	158
Glycosylated OH-linanol	386	295	164

Source: Modified from Di Lorenzo (2007).

(UV and infrared) plays an important role in polyphenol biosynthesis (Rustioni *et al.*, 2007) and in the anthocyanin fraction (Tarara *et al.*, 2008) during berry growth. Most compounds important for flavour, such as methoxypyrazines, accumulate during the green stages of berry development and are gradually metabolized during maturation; the extent of this is dependent upon sun exposure and other microclimate factors (Lund and Bohlman, 2006). Since grape berry cells continue to live after harvest, the possibility of having the biosynthesis of important compounds for final wine quality will depend on the dehydration rate and on the amount of weight loss during sun exposure for drying. Sun exposure, however, is both complex and difficult (but not impossible) to control, and the need to attain a high sugar content as a unique quality parameter for dried grapes, results in no one taking care to manage the dehydration process during sun-drying, which cancels out the differences in the quality of different varieties of grapes, especially in terms of primary aroma. In particular, under high temperature conditions, strong oxidation occurs and the majority of the primary volatiles are oxidized with the formation of hydroxylated compounds such as the large increase in OH-geraniol, OH-linalool, OH-citronellol, and the consequent decrease in linalool, geraniol and citronellol (Di Lorenzo, 2007) in Muscat grape sun-dried on Pantelleria island. In Table 3.3 we can observe the great drop in free and glycosylated terpenes and linalool during sun dehydration under plastic tunnel and between sun dehydration in air or under plastic tunnel.

For this reason we believe that the use of high temperature has a negative impact on grapes and cancels their amazing aromatic diversity, giving sweet wines a similar aroma. This is why we have changed Rozis' first rule for food drying, reported earlier; instead of supplying thermal energy, we must remove thermal energy (heat), reducing the dehydration temperature. Figure 3.4 shows what happens in different aromatic and non-aromatic cultivars during dehydration at three different dehydration temperatures.

Aromatic cultivars (Muscat, Aleatico, Malvasia, etc.) dehydrated at 10°C keep the varietal aroma, but dehydrating them at high temperature causes them to lose the varietal aroma and shift to the aroma of Vinsanto and Passito wines from hot regions, because oxidative aroma becomes predominant. Vice versa, non-aromatic cultivars such as Trebbiano, Cesanese and Sagrantino can emphasize their own weak primary aroma if they are dehydrated at low temperature. This is only a general assessment to give an idea of how the management of dehydration can change the aroma profile, but the following vinification process plays an important role in guaranteeing the maintenance of the varietal volatile compounds.

In conclusion, as regards temperature, we can state that keeping the dehydration temperature at 10°C or lower, first of all, slows down the water stress response, then reduces the

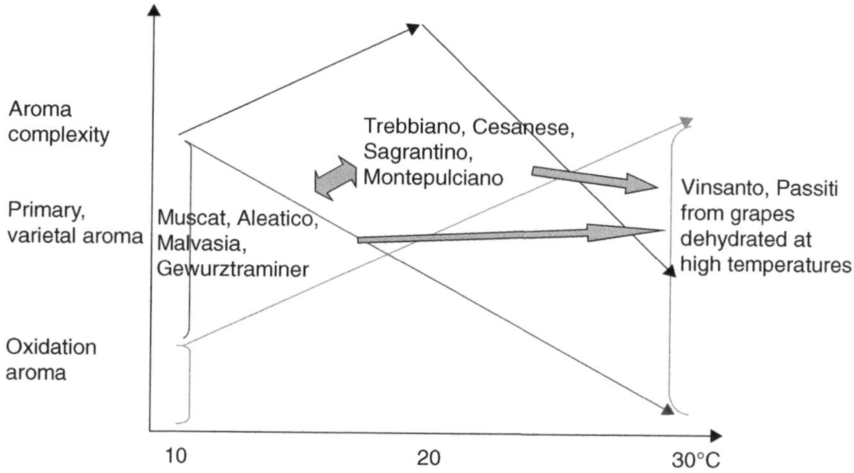

Figure 3.4 Changes of aromatic panorama of different grapes cultivars during dehydration (40% weight loss) at different temperatures based on experimental and commercial data (Mencarelli, unpublished data).

formation of oxidative compounds, maintains the varietal volatile compounds and delays the formation of volatile acidity; whereas, using a temperature of 20°C favours aroma complexity, increases volatile acidity and increases the phenolic fractions.

Besides temperature, the other very important factor for grape dehydration is the RH of the environment. The formula for RH is:

$$(\text{vapour pressure of air} \times 100)/\text{vapour pressure at saturation point.}$$

For all practical purposes, it gives an indication of the capacity of air to pick up water. The lower the RH of the environment, the faster the weight loss from the grapes, whatever temperature is used. We have already spoken about VPD as the driving force of weight loss: this factor increases when the RH decreases, keeping the temperature constant. Thus, maintaining a low RH is one of the goals of the dehydration process, but there are physiological and technical limitations.

Physiological limitation is related to the need to slow the dehydration process down. We saw earlier that the dehydration rate decreases when low temperature is used and this is very important for obtaining good grape characteristics. As we have already discussed, the water loss process must occur slowly to have continuous diffusion of water vapour from cell to intercellular space and to prevent the superficial cells of the berry from squeezing each other, making a continuous barrier against water loss. The interesting paper published by Barbanti *et al.* (2008), where different temperatures (10, 15, 20 and 25°C) and different relative humidities (53, 75, 95%) were used to follow the dehydration of Corvina, Corvinone and Rondinella grape cultivars, reported the withering kinetic rates. They increased with increasing temperature and decreasing RH (both above and below the critical moisture value). Moisture diffusion coefficients, calculated for each trial condition by solving a simplified equation based on Fick's second law, varied from 1.63×10^{-13} to 6.34×10^{-12} m^2 s^{-1}. Withering kinetic rates showed an opposite dependence on temperature and RH, as shown by multiple linear regression analysis results. The grape surface-to-volume ratio affected the withering kinetic rates for values above 3.8 and at low RH (53%). Chkaiban *et al.* (2007) showed that when the dehydration environment is not hygrothermally controlled, such as for the traditional dehydration of grapes for making Vinsanto wine in Tuscany, the continuous change in VPD causes a fluctuating change in relative weight loss. This behaviour causes an acceleration of water stress and a rapid rise in volatile acidity in the berry. As a comparison, they show that the maintenance of a hygrothermally controlled environment postpones water stress. Careful control of the environment plays an important role even in the evolution of volatile compounds in grape berry. In a study carried out on white wine cv. Pecorino, dehydrated under controlled or uncontrolled conditions, up to 25 and 45% weight loss, it was shown that the volatile compound panorama is quite different with similar percentages of groups of volatile compounds at 25% weight loss and with a significant prevalence of alcohols, mainly fuel alcohols, in berries from uncontrolled conditions at 45% weight loss (Figure 3.5).

On the other hand, if the low RH speeds the dehydration process up too quickly, the high RH during grape dehydration is extremely detrimental, especially if an efficient ventilation system is not used, as is the case in the traditional dehydration technique (*fruttaio*). High RH

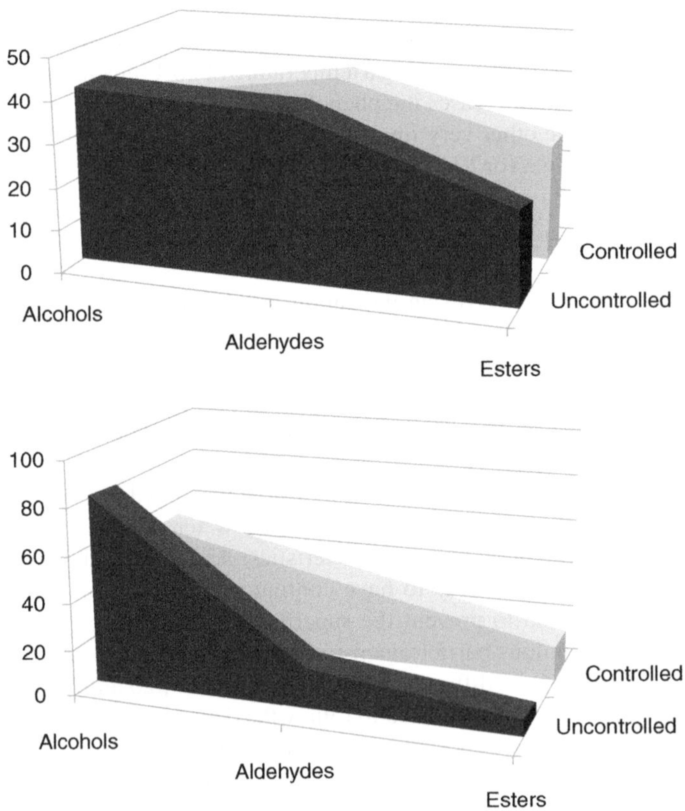

Figure 3.5 Total alcohols, aldehydes and esters in Pecorino grape berries dehydrated in uncontrolled or controlled conditions (10°C, 45% RH) up to 25% (up) and 45% (down) weight loss. Reproduced with permission from Prosperi (2007).

Figure 3.6 Excess of sulfur bentonite on grapes cv. Sagrantino placed over a metallic net for dehydration. (For a colour version of this figure, please see the colour plate section.)

not only prevents water loss but favours fungal attack, mainly by *Botrytis cinerea,* but also by other fungi like the *Aspergillus* family, which can produce ochratoxins.

As we mentioned earlier, besides the physiological limitation, there is a technological constraint. The reduction of RH in the environment where the grapes are placed for dehydration requires a great input of energy for the dehumidification apparatus, thus it represents one of the main costs of the dehydration process. For this reason the RH is often kept around 70–75% in the dehydration facility; this value is not always associated with temperatures below 10°C, and together with inadequate ventilation, makes the berry susceptible to fungal attack. To prevent this infestation and the development of mould, the operators spread sulfur-bentonite over the grape bunches, which results in a white, powdery blanket. This powder absorbs the water vapour from the surface of the berry, facilitating water loss but, at the same time, when the absorbent capacity is finished, it creates a barrier to prevent water vapour from escaping (Figure 3.6).

Rapid water loss and contact between sulfur and berry peel induces cell death, compromising the useful pool of secondary metabolites located in the berry peel, with loss of volatiles, browning of white grapes and bleaching of red grapes. In conclusion, keeping RH low (40–60%) during dehydration of the grapes is as important as keeping the temperature low, and the investment is repaid by the quality of the final grapes.

The last but not least factor affecting dehydration is ventilation, air flow. Fluid dynamics is a subdiscipline of fluid mechanics that deals with fluid flow, the natural science of fluids (liquids and gases) in motion. In traditional dehydration techniques, the air flow is natural

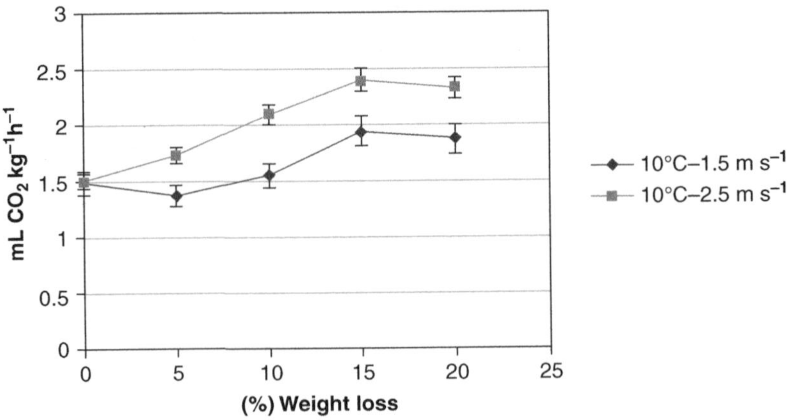

Figure 3.7 Respiration rate of grape cv. Pecorino during dehydration at different air flow. Reproduced with permission from Prosperi (2007).

and conditioned by the arrangement of windows and doors in the room where the grapes are placed to dehydrate. In a facility where the dehydration environment is hygrothermally controlled, the air flow is often insufficient, meaning that it does not reach and 'envelop' the berries, or it is too fast, causing rapid dehydration, or in other cases the air flow is not properly channelled and thus the grapes bunches do not receive the same amount of air. The effect on berry structure of different dehydration air speeds has been clearly shown through magnetic resonance imaging (MRI) by Bellincontro *et al.* (2009). They demonstrated that the faster the air flow at 10°C, the more decompartmentalized the texture of the berry, and the texture is similar to that of a berry kept at a higher temperature.

The effect of air flow on the main metabolism is emphasized by the influence on the respiration rate of the berries during dehydration (Figure 3.7).

Besides the effect on the main metabolism, the air flow affects the secondary metabolism. In Cesanese grape cultivar, dehydration with an air flow of 2.4 m sec^{-1} caused a loss of 30% of total polyphenols whereas, with an air flow of 1.2 m sec^{-1}, the loss is only 17%; no difference was observed for total anthocyanins (Tiberi *et al.*, 2008, unpublished data). It is clear that the effect is not related to the air flow but to the rapid water loss induced by higher air flow. We have observed that in detached berries placed in a dehydration environment, if the RH of the environment is 60%, an air speed of 3 m sec^{-1} accelerates the weight loss of the berries compared with the use of an air speed of 1.5 m sec^{-1}; in contrast, using an RH of 40%, the effect of the speed of the air flow is not noteworthy. By using whole bunches instead of detached berries, the difference observed at 60% of RH between the two air flows was not seen. Why this different response? The reason is found in fluid dynamics and, particularly, in the Reynolds number (RN), which points out when an air flow goes from laminar (the flow moves straight in one direction) to turbulent (the flow draws different paths but always in the same direction).

$$RN = (\text{humid air density} \times \text{air speed} \times \text{product diameter})/\text{viscosity coefficient}.$$

At 20°C the viscosity coefficient is 1×10^{-3} and 1.8×10^{-5} for water and air, respectively. For water, reducing the environmental temperature to 0°C, this coefficient increases

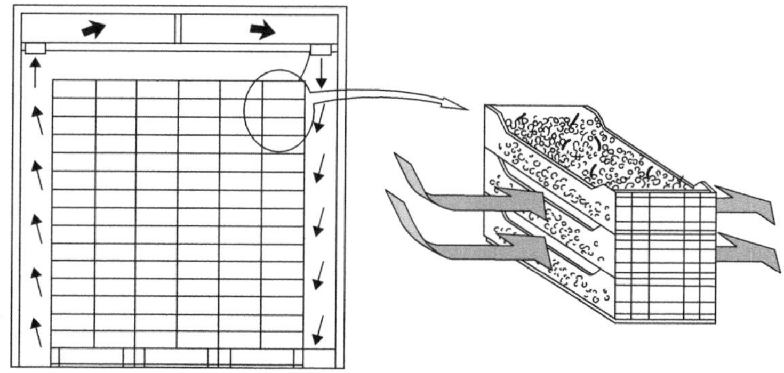

Figure 3.8 Drawing of a cross-section of a new tunnel for grape dehydration with air flow path. Image courtesy of Mercadini, Marvil Eng., Bozen, Italy (2004).

to 1.8, while for air it decreases to 1.7. Thus, the greatest change in the viscosity coefficient with modification of the environmental temperature is due to the water vapour content in the environment.

The higher the RN, the more turbulent the air flow. Being directly correlated with air speed, if the air speed is high, the air flow is more turbulent and strikes a compact bunch (greater mass than single berries), the turbulence increases further, flowing around the bunch and creating vortices in the face opposite to the one struck by the flow. This is not a good way to sweep water vapour off the berry surface. In contrast, if the air speed is lower, the RN becomes smaller and the flow becomes laminar, flowing into the bunch, through the berries, enveloping them.

Finally, besides the care that should be taken when choosing the air flow, which should be between 0.2 and 0.5 m sec^{-1} among the bunches in a box (meaning that the air flow striking the box should have a speed of 1.5–3 m sec^{-1}), it is important to take the following points into account:

- fill the box (or straw) with the bunches in a single layer;
- place the bunches close to each other without leaving empty spaces if using a new dehydration tunnel (Figure 3.8) where boxes with bunches create a double wall and the fans pull the air through the bunches into the double wall, such as a serpentine cooling system (Thompson, 2004); in the case of the traditional technique, the bunches must be placed far apart;
- if plastic boxes are used, use those made specifically for grape dehydration, perforated, with solid plastic, not flexible, with no sharp edges (Figure 3.9);
- stacking of boxes in closed ventilated facilities is very important; air must pass through the bunches; if it finds an open space, it flows into the open space and does not strike the bunches;
- if big fans (7–8000 m^3 h^{-1}) are used, their arrangement is very important to avoid air crossing or too strong a flow against one stack and no flow to another; for these kinds of fans it is better to leave a channel between two stacks of boxes that is as wide as the fan diameter, by creating a Venturi effect, which allows ventilation of the boxes by means of a negative pressure; furthermore, it is important to take into account the air arrow, meaning the distance reached by the air flow leaving the fan, which should be as straight as possible for the length of the stacking channel.

Figure 3.9 Perforated box unsuitable for grape dehydration: berries are pressed in the holes with sharp edges.

Some new grape dehydration facilities have an air flow that is carefully arranged through fan adjustment. The following is a list of the requirements to be considered when dimensioning a dehydration room (Accordini, 2009).

1. Technology employed: natural-traditional, integrated or with total control of the thermohygrometric conditions
2. Perforated plastic boxes of 4–5 kg or 9–10 kg capacity
3. Quantity of grapes to dehydrate
4. Duration of the dehydration
5. Stacking of the grapes in the dehydration room
6. Available electric power (kW)
7. Number of recirculations of room air volume
8. Air flow speed
9. Temperature to be used
10. RH to be used.

3.4 GRAPE DEHYDRATION/WITHERING TECHNIQUES

As mentioned earlier, the term withering is frequently and commercially used especially in the case of the process carried out under uncontrolled environmental conditions. Apparently, the withering process appears to be a more complex event of metabolic evolution (water stress and over-ripening or senescence) and modification of the grape berry as well as the cluster, which are significantly affected by the process. Zironi and Ferrarini (1987) suggested that wine-grape drying (withering) is a special application of the over-ripening processes that is prevalently characterized by water removal from the grape berry. This process, which can last from three weeks to up to four months, allows a sugar concentration of about 30–40% to be achieved, and a final product that is richer in aroma and glycerol and thus characterized by more structure and full of aromatic complexity. In this context, it is obvious how different methods for removing water from grapes greatly affect the dried grapes obtained at the end of the treatment.

As explained at the beginning of this chapter, the term which we use is dehydration and not drying. The grape dehydration techniques currently employed are varied; some of them are based on the oldest wine traditions and are closely related to the characteristic soil and climate environment in which the grapes are grown, others are the result of the latest technologies in this field. Fregoni (2005) suggested that the grape dehydration techniques used in Italy can be grouped into three different types: natural withering, on-vine withering, and forced withering. This breakdown can more or less be applied to all the grape dehydration techniques used throughout the world. As has been explained earlier in this chapter, withered grapes come from a natural process without controlled environmental conditions, very slow, such as for the production of Vinsanto or in the case of Sfursat in Valtellina. In contrast, when the process is fast, as in sun-dried, no over-ripening process occurs, the berries are intensely dried, and so we speak of a drying process (raisins). In the case of environmentally controlled conditions where we can go quickly or slowly through the process, the final product can be dry grapes or withered grapes, so the technique is a dehydration technique. Within each single class of this general grouping it is possible to include specific methods referring to the grape dehydration. Natural withering methods of dehydration include grapes exposed to the sun, placed on mats, on rush matting, on rocks, on roofs, and hanging in bunches. In contrast, late harvest, stem cutting, torsion of the stem, and application of dipping products are methods belonging to on-vine withering. Forced withering (dehydration) is usually obtained using ventilated (naturally or artificially) rooms called 'fruttai', or thermo-hygro conditioned rooms, together with forced ventilation. Another type of classification has been proposed by Valero *et al.* (2008) where a breakdown is reported in terms of grapes dehydrated by: sun-drying, warm chamber, fresh chamber, noble rot and ice.

It is particular and curious that the authors attributed the definition of 'warm chamber' to grape dehydration processes that take place in closed environments under uncontrolled conditions, and the definition of 'fresh chamber' to dehydration always conducted in controlled environments but under the control of the technological parameters affecting the drying process (e.g. temperature, RH and air ventilation).

3.4.1 On-vine withering

In regard to on-vine withering, all the practices adopted have the purpose, directly or indirectly, of delaying the harvest of wine grapes. In Italy, which is the country with the most varied typology of grape drying processes, on-vine withering is used for 17% of the total grapes treated (Fregoni, 2005). This technique is widespread and consists of leaving the grapes on the vine beyond the regular stage of technological ripening.

During the process it is usually possible to observe an increase in sugar content as a consequence of water loss and a decrease in acidity, due to the degradation processes mainly on the part of malic acid. However, the procedure of leaving the grapes on the vine after the optimal ripening stage by inducing an over-ripening effect is just one of the possible on-vine withering techniques. In fact, the producer can choose the option for grape drying directly on the vine employing technical tools capable of interrupting the natural evolution of the physiological ripening. The practices used for on-vine grape withering, such as engraved ring, torsion of the pedicle, double reasoned maturation (Cargnello *et al.*, 2006), are all techniques that partially or completely interrupt sieve vessel continuity and inhibit water flow to the bunches; these effects can induce an acceleration of regular grape dehydration. Another example of on-vine grape withering is represented by the method used for Eiswein or Ice wine production. The drying processes caused by the presence of *Botrytis cinerea* in the form of noble rot can also be counted among the processes belonging to the on-wine withering methods; in this case, the drying effect is strictly related to the fungal metabolism.

3.4.2 Natural withering

The general term of natural withering describes all the procedures of wine grape dehydration that are based on the favourable actions of the environmental conditions like direct grape exposure to the sun or to wind, together with the use of house garrets and house floors arranged for grape drying. In Italy, natural withering covers 78% of the total drying procedures for grapes destined for wine production (Fregoni, 2006), considering that the specific techniques employed differ from country to country and from vineyard to vineyard. In Italy, when grape withering is still carried out in closed and ventilated rooms, these environments are typically called fruttai, and this is probably the most common method for natural grape drying (Amati *et al.*, 1983). Grape bunches collected and suitably selected are placed in a single layer on specific supports, which in most cases are made of wood, at least as regards the frame, and are usually called 'graticci' (Figure 3.10). These structures can be stacked or not and left in ventilated rooms or in external environments for the withering processes. The specific base on which the grapes are placed for dehydration can differ in terms of structure and construction materials. For instance, it is possible to find dehydration procedures obtained by leaving grape bunches on straw mats (like for Vinsanto or other types of Vino Passito produced in Italy) or on iron wires. In both of these methods the grape bunches are turned over many times during the withering process.

Another commonly used technique is to hang the grape bunches; the bunches can be suspended in different ways from supports made of different materials. In the Italian Picolit wine production area (the Friuli–Venezia Giulia region), producers hang the grape bunches, still with their branch, from vertical mesh nets. In other cases, the bunches are twisted around strings (Torcolato wine), or around wires, as in the case of some grape varieties destined for the production of Vinsanto. It is also possible to find horizontal wires used, to which the bunches are suspended from 'S'-shaped hooks, as in grape withering for the production of Erbaluce wine or for Vinsanto in Tuscany (Figure 3.11).

In all these cases the reason for suspending the bunches is both to achieve a healthy air flow passing through them and to facilitate the selection operations with the elimination of damaged and/or mouldy bunches. However, despite the care taken in grape bunch management, it is not unusual to find the development of moulds during withering under natural conditions, when environmental parameters (temperature, RH and ventilation) are not properly controlled.

Figure 3.10 Fruttaia of Avignonesi winery in Tuscany for the production of Vinsanto. Photograph by Mencarelli with permission of the owner (2005). (For a colour version of this figure, please see the colour plate section.)

Figure 3.11 Erbaluce (left) and Trebbiano and Malvasia (right) bunches dried for Passito in Piedmont (photograph by Mencarelli, with permission of the owner, Ciek winery, 2010) and Tuscany (photograph by Mencarelli with permission of the owner, Frescobaldi Winery, Pomino, 2006), respectively, hung on wires or on a wooden frame. (For a colour version of this figure, please see the colour plate section.)

Figure 3.12 Muscat of Alexandria (Zibibbo) bunches dried in the sun in a yard on Pantelleria island. (For a colour version of this figure, please see the colour plate section.)

If climatic conditions are favourable, grape withering is carried out with grapes exposed to the sun, as is the case of Passito di Pantelleria in Sicily (Italy) and Pedro Ximénez in Andalusia (Spain). In the drying procedure used for Muscat of Alexandria (Zibibbo), which is the grape variety used for the production of Passito di Pantelleria, or Malvasia delle Lipari in the Aeolian Islands, the grapes are usually harvested slightly over-ripe. The grapes bunches are then quality selected and placed over plastic nets laid on the ground between the vines or over the house roof or in the yard (Figure 3.12); sometimes wooden frames with metal net or plastic boxes are used, as is the case of Salina island (Figure 3.13).

Figure 3.13 Malvasia bunches drying in the sun on the Aeolian Islands. (For a colour version of this figure, please see the colour plate section.)

Figure 3.14 Zibibbo grapes drying in plastic tunnel on Pantelleria island. (For a colour version of this figure, please see the colour plate section.)

In this way the grapes are subjected to sun exposure for the dehydration process, which lasts approximately 10–15 days, obviously depending on the environmental conditions. It depends on the specific area of production and on the weather conditions, especially overnight; the grapes may be covered in order to prevent negative effects caused by high humidity or possible rainfall. To avoid the problem of rainfall, tunnels covered with polyethylene film have recently been used (Figure 3.14). Unfortunately, this method leads to a huge increase in berry temperature, significantly compromising the primary aroma of the grapes.

In Spain, as a result of the dry weather, the grape bunches are placed in a single layer on sheets which are laid directly on the ground. The sheets arranged in this way, one next to the other, can cover entire fields, which are typically called *pasera* in Spanish. Because of the high temperature, the dry climate and the strong sun exposure, the drying procedures generally last 7–10 days. When natural withering processes are carried out under external conditions and, in particular, by exposing grapes to the sun, a possible negative effect, which is also favoured by the high RH and good temperature conditions, is the spreading of fungi and the possible consequent contamination of the grapes by ochratoxins (Valero *et al.*, 2008).

3.4.3 Forced withering

The definition of forced withering is usually used to describe the grape drying procedures that take place in closed environments which frequently have technology installed for the partial or total control of air ventilation, RH and temperature (Fregoni, 2006).

Indeed, the term 'forced' does not seem to be totally appropriate to define the process, as total control of the environmental parameters might not necessarily favour acceleration of the drying effect on grapes but could be aimed solely at control of the process. As explained in the first part of this chapter, the possible modulation of the parameters affecting dehydration could also be used for the control or slowing-down of the drying event (e.g. the use of low temperature, or high humidity or low ventilation, alone or combined). On the basis of this consideration, the definition of 'controlled withering' might be more appropriate to describe this kind of wine grape dehydration. The use of technologies for controlling environmental conditions is not only aimed at modulation of the grape dehydration process, with important consequences on the formation of useful metabolites for wine (Chkaiban *et al.*, 2007; Mencarelli *et al.*, 2010; DeSanctis *et al.*, 2012), but is also aimed at preventing pathogen attacks, rot diffusion and consequentially, possible mycotoxin contamination. Technologies for the partial control of environmental conditions were proposed based mainly on high ventilation and relatively high humidity (Amati *et al.*, 1983; Corte *et al.*, 2001). The production of Amarone (dry) and Recioto (sweet) wines, which are famous Italian wines obtained from the same dehydrated grape varieties, is also based on grape dehydration that takes place in closed environments (fruttai) where, however, there is a combination of forced ventilation and more or less dry air (Paronetto and Dellaglio, 2011). When the climate is quite hot and dry, the windows of the fruttaio are opened and the dry air from outside is ventilated into the room, to favour grape dehydration and the partial control of mould spreading. Vice versa, when the external air is humid, the windows are closed. It is obvious that such techniques do not guarantee accurate control of grape dehydration and grape sanitation. Temperature, as well as RH, can be more or less controlled by artificial systems, but in some case the large size of the dehydration facilities, such as for Amarone wine, and the large amount of grapes to be dehydrated, in addition to the need to save energy for the environmental control of the facility, makes these systems not perfectly suited to obtaining an excellent final dehydrated product. As regards ventilation, large fans (1.5 m in diameter) with a delivery capacity of around 3000 m^3 and 0.75 kW h^{-1} of electricity consumption, are used in these facilities. The spatial distribution on the floor of these fans among the stacks of pallets with grapes inside the large facilities is done with the objective of removing as much air as possible. Unfortunately, when the ceiling of the facility is very high (5–10 m), the air flow of the fans placed on the floor is unable to remove the air over the fans and under the ceiling, with consequent non-uniformity in temperature and RH. A new technology for guaranteeing uniform air distribution has been developed by an Italian company, Antonelli, with a vertical stack of fans which moves forward and backward through the pallet stacks (Figure 3.15). This guarantees a more uniform temperature and RH in the dehydration environment.

Recently, even traditional producers of Amarone and Recioto wines have started to use dehydration systems in which environmental parameters are totally under control. Studies conducted over the last 10 years proposed and demonstrated the effectiveness of controlled dehydration in terms of possible modulation of the dehydration process (Bellincontro *et al.*, 2006). For instance, the possible use of different temperature conditions (low or relatively high) together with a lengthening or shortening of the duration of the process, can have a significant effect on the quality attributes of the grapes destined for wine production (Bellincontro *et al.*, 2004; Pietromarchi *et al.*, 2007; Botondi *et al.*, 2008; Bellincontro *et al.*, 2009). New facilities by Italian Companies, built as cold storage rooms but with a combination of coil and condenser inside the dehydration room, have the aim of reducing

Figure 3.15 New technology for air ventilation in a grapes dehydration facility. (For a colour version of this figure, please see the colour plate section.)

temperature and, at the same time, RH, and are becoming widespread in Italy (Figure 3.16). These rooms provide very accurate control of the environment climate by allowing the use of a wide range of dehydration temperatures (5–30°C) and relative humidities (15–95%), as well as the possibility of managing air speed. This room can even be used for controlled botrytization treatment on grapes (Guenzi *et al.*, 2011).

Figure 3.16 Dehydration rooms for grapes with accurate control of the environmental parameters. (For a colour version of this figure, please see the colour plate section.)

ACKNOWLEDGEMENTS

Part of the research was supported by the Ministry of Education, University and Research, PRIN08, coordinator F. Mencarelli. We would like to thank Dr Katherine Prucha for her valuable contribution in the English revision.

REFERENCES

Accordini, D. 2009. L'appassimento delle uve per l'Amarone in Valpolicella. Lecture at the 3rd Course of Appassimento delle Uve e Qualità dei Vini, Pisa, 21–22 May 2009. CD available from DIBAF, University of Tuscia, Viterbo, Italy.

Amati, A., R. Ferrarini, C. Riponi, *et al.* 1983. Una nuova tecnologia per l'appassimento delle uve. *Vigne e Vini*, 10(9):27–35.

Barbanti, D., B. Mora, R. Ferrarini, *et al.* 2008. Effect of various thermo-hygrometric conditions on the withering kinetics of grapes used for the production of 'Amarone' and 'Recioto' wines. *Journal of Food Engineering*, 85(3):350–358.

Becker, T. and M. Knoche. 2011. Water movement through the surfaces of the grape berry and its stem. *American Journal of Enology and Viticulture*, 62(3):340–50.

Bellincontro, A., R. Botondi, D. De Santis, *et al.* 2006. Il controllo del metabolismo dell'uva in appassimento per la determinazione della qualità del passito. *Phytomagazine*, Eds. Phytoline, 15:35–46.

Bellincontro, A., D. De Santis, R. Botondi, *et al.* 2004. Different postharvest dehydration rates affect quality characteristics and volatile compounds of Malvasia, Trebbiano and Sangiovese grape for wine production. *Journal of Science and Food Agriculture*, 8(13):1791–1800.

Bellincontro, A., I. Nicoletti, M. Valentini, *et al.* 2009. Integration of non-destructive techniques with destructive analyses to study postharvest water stress of wine grapes. *American Journal of Enology and Viticulture*, 60(1):57–65.

Botondi, R., L. Lodola, and F. Mencarelli. 2011. Postharvest ethylene treatment affects berry dehydration, polyphenol and anthocyanin content by increasing the activity of cell wall enzymes in Aleatico wine grape. *European Food Research Technology*, 232:679–685.

Botondi, R., G. Antelmi, M.T. Frangipane, *et al.* 2008. Influenza della temperatura di appassimento sulla qualità delle uve varietà Montepulciano. *L'Enologo*, 9:87–92.

Cargnello, G., L. Pezza, G. Gallo, *et al.* 2006. D.M.R. ('Double Reasoned Maturing'): innovative technique of agronomic ecologic control of grey mould on grapevine. Trials and various considerations. *Communications in Agricultural and Applied Biological Sciences*, 71(3B):1055–1061.

Chkaiban, L., R. Botondi, A. Bellincontro, *et al.* 2007. Influence of postharvest water stress on lipoxygenase and alcohol dehydrogenase activities, and on the composition of some volatile compounds of Gewürztraminer grapes dehydrated under controlled and uncontrolled thermohygrometric conditions. *Australian Journal of Grape and Wine Research*, 13:142–149.

Cirilli, M., A. Bellincontro, D. De Santis, *et al.* 2012. Temperature and water loss affect ADH activity and gene expression in grape berry during postharvest dehydration. *Food Chemistry*, 132:447–454.

Corte V., D. Oliva, M. Ragusa, *et al.* 2001. Aspetti tecnici, microbiologici e chimici connessi con i sistemi di appassimento delle uve. *L'Enologo*, 12: 87–97.

Costantini, V., A. Bellincontro, D. De Santis, *et al.* 2006. Metabolic changes of Malvasia grapes for wine production during postharvest drying. *Journal of Agriculture and Food Chemistry*, 54:3334–3340.

Del Caro, A., A. Piga, I. Pinna, *et al.* 2004. Effect of drying conditions and storage period on polyphenolic content, antioxidant capacity, and ascorbic acid of prunes. *Journal of Agriculture and Food Chemistry*, 52:4780–4784.

Di Lorenzo, R. 2007. Problematiche dell'appassimento in campo di uve autoctone siciliane per la produzione di vini dolci. Lecture at the 2nd course of Appassimento delle Uve e Qualità dei Vini, Marsala, 19–20 April 2007. [CD available from DIBAF, University of Tuscia, Viterbo, Italy.]

DeSanctis, F., M.G. Silvestrini, R. Luneia, *et al.* 2012. Postharvest dehydration of wine white grapes to increase genistein, daidzein and the main carotenoids. *Food Chemistry*, 135:1619–1625.

Franck, C., J. Lammertyn, Q.T. Ho, *et al.* 2007. Browning disorders in pear: a review. *Postharvest Biology and Technology*, 43:1–13.

Franco, M., R.A. Peinado, M. Medina, *et al.* 2004. Off-vine grape drying effect on volatile compounds and aromatic series in must from Pedro Ximénez grape variety. *Journal of Agriculture and Food Chemistry*, 52:3905–3910.

Fregoni, M. 2005. *Viticoltura di qualità. VI Fisiologia della vite*, Ed. L'Informatore Agrario, pp. 161–240.

Fregoni, M. 2006. Caratteri genetici e viticolo-ambientali caratterizzanti l'idoneità all'appassimento delle uve. *Phytomagazine*, Ed. Phytoline, 15:5–6.

Guenzi S., F. Mencarelli, A. Bellincontro, *et al.* 2011. Botritizzazione controllata. *VQ*, 1:46–49.

Lund, S.T. and J. Bohlman. 2006. The molecular basis for wine grape quality – a volatile subject. *Science*, 311:804–805.

Mencarelli, F., A. Bellincontro, I. Nicoletti, *et al.* 2010. Chemical and biochemical changes of healthy phenolic fractions in winegrape by means of postharvest dehydration. *Journal of Agriculture and Food Chemistry*, 58:7557–7564.

Mencarelli, F., D. De Santis, A. Bellincontro, *et al.* 2006. Controllo accurato dell'ambiente di appassimento per la valorizzazione di uve di varieta' autoctone dell'Italia centrale destinate alla produzione di vini dolci. Convegno Nazionale 'Vitigni Autoctoni Minori', Turin, 30 November–1 December (ed. V. Gerbi), Fondazione Fotobiotech. ISBN: 88-6136-001-7.

Muganu, M., A. Bellincontro, F.E. Barnaba, *et al.* 2011. Influence of bunch position in the canopy on berry epicuticular wax during ripening and on weight loss postharvest dehydration. *American Journal of Enology and Viticulture*, 62:91–98.

Paronetto, L. and F. Dellaglio. 2011. Amarone: a modern wine coming from an ancient production technology. *Advances in Food and Nutrition Research*, 63:285–306.

Pietromarchi, P., D. Tiberi, A. Bellincontro, *et al.* 2007. Cesanese passito? In tunnel e a bassa temperatura. *VQ*, 8:58–62.

Piombino, P., A. Genovese, A. Gambuti, *et al.* 2010. Effects of off-vine bunches shading and cryomaceration on free and glycosilated flavours of Malvasia delle Lipari wine. *International Journal of Food Science and Technology*, 45:234–244.

Prosperi, P. 2007. Studio del metabolismo di uve var. Pecorino durante l'appassimento con sistema tradizionale e con una tecnologia innovativa. Master's Thesis, University of Tuscia, Viterbo, Italy.

Rozis, J.F. 1997. *Drying Foodstuffs: Techniques, Processes, Equipment – Technical Guidebook*, Backhuys Publ., Leiden, the Netherlands, p. 311.

Rustioni, L., M. Rossoni, G. Cola, *et al.* 2007. Accumulo e profilo antocianico in relazione al microclima termico e luminoso in Croatina, Sangiovese, Syrah e Nebbiolo. *Italus Hortus*, 14(3):176–180.

Santonico, M., A. Bellincontro, D. De Santis, *et al.* 2010. Electronic nose to study postharvest dehydration of wine grapes. *Food Chemistry*, 121: 789–796.

Schönherr, J. 1982. Resistance of plant surfaces to water loss: Transport properties of cutin, suberin and associated lipids. In *Encyclopedia of Plant Physiology*, New Series. Vol. 12B (eds O.L. Lange *et al.*), Springer, Berlin, pp. 153–179.

Sweetman, C., L.G. Deluc, G.R. Cramer, *et al.* 2009. Regulation of malate metabolism in grape berry and other developing fruits. *Phytochemistry*, 70:1329–1344.

Tarara, J.M., J. Lee, S.E. Spayd, *et al.* 2008. Berry temperature and solar radiation alter acylation, proportion, and concentration of anthocyanin in Merlot grapes. *American Journal of Enology and Viticulture*, 59(3):235–247.

Tesniere, C.M., C. Romieu, and M.E. Vayda. 1993. Changes in gene-expression of grapes in response to hypoxia. *American Journal of Enology and Viticulture*, 44:445–451.

Thompson, J.F. 2004. Precooling and storage facilities. In *The Commercial Storage of Fruits, Vegetables, and Florists and Nursery Stock* (eds K.J. Gross, C.Y. Wang, M.E. Saltveit), Agriculture Handbook No. 66, USDA. http://www.ba.ars.usda.gov/hb66/ (last accessed 1 December 2012).

Tilbrook, J. and S.D. Tyerman. 2008. Cell death in grape berries: varietal differences linked to xylem pressure and berry weight loss. *Functional Plant Biology*, 35:173–184.

Valero, A., S. Marín, A.J. Ramos, *et al.* 2008. Survey: Ochratoxin A in European special wines. *Food Chemistry*, 108:593–599.

Wang, N. and J.G. Brennan. 1995. Changes in structure, density and porosity of potato during dehydration. *Journal of Food Engineering*, 24:61–67.

Zironi, R. and R. Ferrarini. 1987. La surmaturazione delle uve destinate alla vinificazione. *Vigne e Vini*, 14(4):31–45.

4 Biochemistry and Physiology of Dehydrating Berries

Pietro Tonutti[1] and Claudio Bonghi[2]

[1] *Institute of Life Science, Scuola Superiore Sant'Anna, Pisa, Italy*
[2] *Department of Agronomy, Food, Natural Resources, Animals and Environment, University of Padova, Legnaro, Italy*

4.1 INTRODUCTION

The composition of grape berries used in dessert, sweet or reinforced wines is the result of the interaction between physiological processes genetically programmed during the developmental cycle and responses to imposed environmental parameters that can cause pronounced stress conditions. As in other fleshy fruit, grape berry development is characterized by marked changes in biochemical-physical properties, which are particularly evident during ripening (Ollat *et al.*, 2002; Conde *et al.*, 2007).

Fruit ripening is a highly coordinated, genetically programmed syndrome, occurring at an advanced stage of development and involving a series of physiological, biochemical, structural and compositional changes, all leading to the development of an edible fruit with certain desirable qualities (Brady, 1997; Giovannoni, 2001, 2004). Ripening is traditionally perceived as a distinct stage of fruit development, although it can also be considered as an arbitrary point of maximum palatability or optimal composition, in a continuous process that eventually leads to senescence and the death of the fruit. Due to the dynamic nature of the ripening process, it is essential to know the rate of the changes and the length of time required to ripen once the process begins, which in grape berries is generally defined as the veraison stage.

With specific regard to grape berries undergoing extended ripening (late harvest) and/or postharvest dehydration, it is also crucial to explore and understand: (i) the effects of environmental parameters on the ripening metabolism and the evolution of senescence in attached berries; and (ii) the behaviour of detached berries under stress conditions and their responses to different levels of dehydration. Recent approaches based on stress, ripening and postharvest physiology combined with molecular analyses are providing vital new information and major advances in our understanding of the aforementioned processes.

Sweet, Reinforced and Fortified Wines: Grape Biochemistry, Technology and Vinification, First Edition.
Edited by Fabio Mencarelli and Pietro Tonutti.

4.2 PHYSIOLOGY OF RIPENING, OVER-RIPENING AND SENESCING GRAPE BERRIES

Since the early works by Hale (1968) and Coombe (1976), several reviews, whole books and individual chapters have been published on the physiology of grape berry ripening. Thanks to the release of the genome sequence (Jaillon *et al.*, 2007; Velasco *et al.*, 2007) and recent developments in the large-scale analysis of gene transcript levels, the *Vitis vinifera* fruit is one of the best-known in terms of molecular mechanisms that modulate and regulate ripening. Various techniques (cDNA differential screening, microarrays, and, more recently, RNAseq) have demonstrated that the expression of a number of genes changes in relation to veraison, and throughout ripening (Davies and Robinson, 2000; Terrier *et al.*, 2005; Waters *et al.*, 2006; Deluc *et al.*, 2007; Grimplet *et al.*, 2007; Pilati *et al.*, 2007; Zamboni *et al.*, 2010, Zenoni *et al.*, 2010). At a transcriptional level, the majority of these genes are regulated by a complex network of interactions in which hormones and other mobile signals (e.g. small RNAs) play a major role.

The grape berry is classified as a non-climacteric fruit due to the lack of an increase in respiration rate and ethylene production at ripening, events which in contrast occur in climacteric fruit (Seymour *et al.*, 1993). However, there is much evidence that the endogenous levels of ethylene increase at veraison and that the exogenous application of ethylene accelerates the inception of ripening (Chervin *et al.*, 2008). The application of ethylene has been shown to positively affect the accumulation of anthocyanins, by inducing the transcription of the main biosynthetic genes and their positive transcription regulators (El-Kereamy *et al.*, 2003; Jeong *et al.*, 2004; Chervin *et al.*, 2008) as well as grape berry expansion. This thus enables cell elongation in the pulp and skin, inducing gene-encoding aquaporins and cell wall hydrolases (Chervin *et al.*, 2008; Sun *et al.*, 2010). Similarly to ethylene, abscisic acid (ABA) and brassinosteroids (BRs) have been claimed to be ripening inducers (Symons *et al.*, 2006; Sun *et al.*, 2010), and the synergic interaction between ethylene and ABA, in particular, appears to be a key factor in triggering grape ripening (Sun *et al.*, 2010).

The expression of genes involved in ethylene biosynthesis, perception and signal transduction (*VvACO1*, *VvETR2*, *VvCTR1*) have been detected just before veraison, and the trace of ethylene may induce the expression of *VvNCED1*, which catalyses the first committed step in ABA biosynthesis (Finkelstein and Rock, 2002). On the other hand, as in other fruit, auxins are considered as strong inhibitors of grape berry ripening (Davies and Böttcher, 2009). Applications of indole-3-acetic acid (IAA) or synthetic auxin (such as naphthalene-1-acetic acid (NAA)) before veraison, delay ripening, as indicated by the dynamics of ripening-related physiological processes (Davies *et al.*, 1997; Böttcher *et al.*, 2010; Ziliotto *et al.*, 2012), thus greatly modifying the transcription of key genes involved in phenylpropanoids biosynthesis, sugar metabolism and cell wall turnover (Jeong *et al.*, 2004). In addition, auxin treatments have been shown to negatively affect ABA metabolism, while no information is available on the possible interactions between auxins and other hormones (Davies and Böttcher, 2009). In addition to hormones, other important players in the regulatory transcriptional network are small RNAs (sRNAs). A RNA-seq approach, via 454 technology, allowed the identification of sRNAs actively transcribed during berry development (Carra *et al.*, 2009; Wang *et al.*, 2011). Most of the sRNA-predicated targets are plant-specific transcription factors such as AP2, NAC, SBP, NF-YA, GRAS, Zinc finger and ARF family. Among these, transcription factors known for regulating genes involved in flower development, disease resistance, signal transduction, energy, and anthocyanin metabolism (such as flavonol reductase and cinnamoyl-CoA reductase) were identified.

A certain typology of wines (e.g. late harvest) is produced using berries that undergo over-ripening and reach senescence on the vine. Less information is available on the biochemical and molecular aspects of these stages, as physiological studies can be complicated by the alterations which, in some varieties, can occur in the hydraulic conductance for flow in and out of individual berries (Tyerman *et al.*, 2004). In cv. Shiraz, xylem backflow with the concomitant reduced functionality of phloem results in weight (water) loss during the later stages of ripening (~7% of berry volume per day) (Tilbrook and Tyerman, 2009). The altered water potential can affect the general metabolism of the berry, inducing specific responses and accelerating transition to the senescent stage. The over-ripening stage in on-vine berries leads to a magnification of the changes occurring during ripening, in particular concerning sugar concentration, organic acid evolution, aroma and polyphenol compound profiling.

A further element affecting the metabolism of over-ripe berries is possible attack by noble rot, which results in a solute concentration and in a unique fruit composition. A delay in harvesting results in the expression of specific genes associated with the advanced stage of ripening and/or senescence. Using a cDNA-AFLP approach, Rizzini *et al.* (2006) preliminarily identified 58 and 24 transcripts that accumulate more and less abundantly, respectively, in the epicarp of the red-skinned cv. Raboso Piave, when samples collected at technological maturity (TM) were compared with those undergoing on-vine over-ripening (TM+7 days). Some of these transcripts were specifically induced in the late-harvested berries and not in a parallel trial of postharvest dehydration, suggesting a specific role of the corresponding genes in the over-ripening/senescence processes.

Using the same technical approach but coupled with qRT-PCR, Zamboni *et al.* (2008) analysed the transcriptional profiling of red-skinned cv. Corvina berries in four different stages of on-plant over-ripening (up to 74 days after TM). A negligible water loss was detected in these grapes at the end of the whole period of observation, confirming that different genotypes (e.g. Shiraz and Corvina) behave differently in terms of hydraulic conductance in and out of the berries at ripening. A total of 86 transcripts were specifically modulated and, among the annotated tags, a bZIP and a putative WD-40 repeat protein showed a transient up-regulation. In addition two chlorophyll a–b binding protein transcripts had accumulated in specific stages of the on-vine over-ripening process. Genes involved in protein metabolic processes, such as polyubiquitin and cysteine protease showed a significant down-regulation starting from the early over-ripening stages.

In cv. Chardonnay, keeping the grapes for 10 days longer (TH+10) than the theoretical harvest (TH) affects wine composition by inducing changes in terms of proline, sugar, alcohol and total acid content in the decanted must and base wine (Guillaumie *et al.*, 2011). Compared to TH, the over-ripe TH+10 grapes display differential transcript profiling of 53 out of the 14 562 probes present in the AROS V1 (Operon, USA) microarray. These genes belong to several categories including aroma-, desiccation-, and pathogenesis-related genes, as well as hormonal control/metabolism, stress-, growth and development- related genes. Other important categories are phenylpropanoid/lignin, cell-wall modifying, and transporter and trafficking genes. By comparing the expression patterns of some of these 53 genes during the late stages of Chardonnay and Cabernet Sauvignon berries, seven genes can be identified as ripening indicators, and one, in particular (*VvValCS*, which is involved in sesquiterpene biosynthesis), as a late-ripening marker. This molecular information confirms that, as observed in other fruit or plant organs, grape berry over-ripening and senescence are, at least initially, strictly controlled at the gene expression level.

4.3 THE METABOLISM OF DETACHED BERRIES AND THE MOLECULAR RESPONSES TO POSTHARVEST DEHYDRATION

Besides the wines produced using grapes that complete the ripening process and undergo over-ripening on the vine (late harvest), some other wines are made from berries which, after detachment from the mother plant, are allowed to dehydrate up to weight losses (WL) of 40–50% or even more.

From a physiological and metabolic point of view, two interacting aspects must be considered: the first is that grape berries, like all perishable plant products, are alive and, therefore, continue to function metabolically even after detachment and carry on doing so until cellular death takes place. The second element is that the metabolism occurring in the postharvest phase is not identical to that occurring in fruit that is still attached to the plant. This is due to: (i) the lack of water, mineral and energy supply provided through the vascular connections, and (ii) the environmental conditions that are present throughout the postharvest period and the effects of varying the level and source of stress. Grape berries are non-climacteric, and this has a great impact on postharvest behaviour. In fact, unlike climacteric fruit, where ripening can proceed or even start after detachment (if the fruits have reached physiological maturity and the ability to ripen), non-climacteric fruit enter the senescence stage.

A crucial role in triggering the onset of senescence in detached berries seems to be played by ABA, as a result of an increased expression of the *VvNCED1* gene (Sun *et al.*, 2010). Thus postharvest water loss represents a further element that greatly affects the general metabolism and the physiology of the berries. This makes it difficult to distinguish between senescence-related and water stress-induced processes. There is quite a strict relationship between ABA levels and water loss in harvested berries, considering that an increase in ABA content is observed in cv. Malvasia grapes at 10% WL (Costantini *et al.*, 2006), but not in detached cv. Muscat Hamburg berries when the pedicel is supplied with water (Sun *et al.*, 2010). Although ABA has been traditionally associated with plant/organ senescence, this evidence suggests that in addition to the crucial role played by ABA at the onset and early stage of grape ripening (Wheeler *et al.*, 2009; Sun *et al.*, 2010), it is actually more involved in the responses to water stress rather than in the senescence of detached grape berries. In orange fruit undergoing postharvest dehydration, the altered expression of genes involved in ABA synthesis and perception, ABA-dependent transcription factors, and gene encoding ABA-responsive proteins clearly indicates that the phytohormone is involved in the responses to the water loss of fruit tissues (Romero *et al.*, 2012).

Marked physiological changes occur during postharvest withering. This directly and/or indirectly affects the composition of the berry and chemical-physical properties, which are only partly the result of concentration and loss of cellular turgor. The changes observed in partially dehydrated berries (e.g. the altered glucose-to-fructose and malate-to-tartrate ratios, the modified profiles of aroma and polyphenol compounds, the different composition and properties of membrane and cell wall structures) reflect the modulation of the metabolism that acts in response to the decrease in cell water potential. Note that if a concentration of cell solutes is a common response of postharvest dehydration in all grape varieties, the effects of these treatments on metabolic processes may vary in relation to the genotypes, the fruit tissue (flesh or skin), the environmental conditions, and the intensity and rate of the stress.

That postharvest dehydration is effective in modulating berry metabolism through changes in gene expression was initially demonstrated by Versari *et al.* (2001) and Tonutti *et al.*

(2004), who reported increases in the expression of stilbene synthase (*STS*) and phenylalanine ammonia lyase (*PAL*) genes, respectively. Using an AFLP-TP approach, Zamboni *et al.* (2008) identified 699 withering/over-ripening-related genes, 167 of which are unique to off-plant dehydration. These genes were classified as putatively involved in stress protection (dehydrin) and osmolite accumulation, hexose metabolism and transport, cell wall composition, and secondary metabolism (particularly the phenolic and terpene compound pathways). The effects of postharvest dehydration on transcriptome changes in a red-skinned variety (Raboso Piave) was first analysed via microarray by Rizzini *et al.* (2009), who used the AROS V1 platform. This approach enabled up to 309 genes to be identified, which were differentially expressed in skins of samples dehydrated up to a 30% WL. Compared to the harvested berries, the expression of 26 genes was affected similarly by both levels of stress (10 and 30% WL) and both rates (slow, S and rapid, R) of water loss. These genes are involved in the response to biotic and abiotic stresses, primary and secondary metabolism, cell wall modification, transport, transcription factors (*WRKY, MybA*), and hormone (ethylene) metabolism, and are highly representative of the postharvest dehydration of grapes.

Worthy of note is the up-regulation of ACC oxidase (*ACO*), a key gene in the biosynthesis of ethylene. The induction of this gene is accompanied by a down-regulation of AUX/IAA, acting as negative regulator of auxin action, and an up-regulation of auxin-induced genes. Cross-talk between ethylene and auxin has been demonstrated in several fruits (both non-climacteric and climacteric) at ripening (Barry *et al.*, 2000; Trainotti *et al.*, 2007; Choudhury *et al.*, 2008; El-Sharkawy *et al.*, 2008; Ziliotto *et al.*, 2011). Besides the interaction between ABA and ethylene observed in grape berries after detachment (Sun *et al.*, 2010), there may be an interplay between auxin and other hormones in detached ripe fruit as a reaction to stress conditions. The possible involvement of ethylene in response to the postharvest dehydration of grape berries is also suggested by the altered expression of recognized ethylene-dependent genes. In addition to those with a role in ethylene perception (*ERS* and *ETR-like*) and action (*EIN3* and *ERFs*), genes involved in the cell wall metabolism (pectimethylesterases (*PMEs*); extensins (*EXTs*)) and stress response (*stress-related genes)* are also present (Guillaime *et al.*, 2011).

Enhancing dehydration levels from 10 to 30% WL is associated with an increase in the total number of differentially expressed probes, thus indicating that the cells of the skin are still reactive at a transcriptional level in the advanced stages of postharvest dehydration. Considering that, from a proteomics approach, only a few changes in protein expression have been observed throughout the three stages of the withering process (cv. Corvina) up to about 20% WL (Di Carli *et al.*, 2011), it is likely that post-transcriptional regulatory mechanisms also play an important role.

As well as quantitative aspects, different dehydration levels also affect the distribution of differentially expressed genes within functional categories. In fact, compared to the Raboso Piave samples which dehydrated up to 10% WL, a significantly increased incidence was detected for probes belonging to the protein binding function for both slow and rapid 30% WL. In addition to the intensity of WL, the rate of dehydration also has a significant effect on transcriptomic profiling. In fact, in skin samples undergoing rapid dehydration, an increase in the percentage incidence of differentially expressed genes involved in secondary metabolism was observed in the 10% WL dehydrated samples. On the other hand, cell rescue, defence and virulence, and interaction with the environment increased in the samples that dehydrated rapidly up to 30% WL (Rizzini *et al.*, 2009). These increases occurring in the most rapidly dehydrated samples may be the consequence of pronounced damage to the cell and membrane systems, thus resulting in the modulation of stress-related genes, with important consequences in terms of berry composition and wine properties. Quality characteristics and volatile compounds of Malvasia, Trebbiano and Sangiovese (Bellincontro *et al.*, 2004) and Aleatico (Cirilli *et al.*, 2012) grapes changed markedly following different postharvest dehydration rates. These rates are also effective in modulating the activity of alcohol dehydrogenase (ADH) and lipoxygenase (LOX) enzymes and the expression of ADH and carotenoid cleavage dioxygenase 1 (CCD1) genes (Chkaiban *et al.*, 2007; Cirilli *et al.*, 2012).

A genome-wide transcriptional analysis was carried out using a microarray (Combimatrix) containing 24 571 non-redundant probes, on dehydrated (up to about 20% WL) deseeded (pulp + skin) Corvina berries, used for the production of Amarone wine. This analysis highlighted that, besides genes with a role in the aforementioned processes, withering triggers specific osmotic and oxidative stress response genes and changes in the polyphenol metabolism (Zamboni *et al.*, 2010). Transcriptional data were coupled with proteomic data, obtained by two-dimensional difference gel electrophoresis (2D-DIGE), and metabolomic data, obtained by HPLC coupled to mass spectrometry (MS). The O2PLS-discriminant analysis (DA) technique was used to analyse each data set in terms of the different withering intervals and then to derive putative transcript, protein, and metabolite biomarkers. This integrated approach indicated that transcriptomic, proteomic and metabolomic changes represent a stress response to dehydration and possible pathogen attack. It also enabled putative biomarkers defined as molecules to be identified, indicating or correlating with the physiological changes that occur during withering. Fifteen putative increasing (six transcripts, four proteins and five metabolites) and four decreasing (two transcripts and two proteins) biomarkers were identified, and four of the increasing transcripts may have a role in gene regulation during withering. Two of the putative increasing protein biomarkers were identified, one representing an osmotin-like protein and the other a thaumatin-like protein, both of which are known to be involved in stress responses. A specific proteomics approach clearly demonstrated that the induction of stress response-associated proteins is one of the most pronounced changes occurring during postharvest withering (Di Carli *et al.*, 2011). The three putative increasing metabolic biomarkers include two taxifolins (dihydroquercetin) and a tetrahydroxyflavanone-O-deoxyhexoside, which may be involved in the withering stress response (Zamboni *et al.*, 2010).

The integration of 'omics' data with data from physiological and technological approaches sheds light on the regulation of specific metabolic processes induced and/or affected by postharvest dehydration in wine grape berries. One of the observed responses is the increase in respiration which shifts to the fermentation pathway (Bellincontro *et al.*, 2004, 2009; Costantini *et al.*, 2006) leading to the accumulation of compounds such as ethanol, acetaldehyde and ethyl acetate. Reactions related to hexose aerobic/anaerobic respiration, hexose conversion to malate, gluconeogenesis and malate respiration have been observed in postharvest dehydrating berries (Zironi and Ferrarini, 1987; Bellincontro *et al.*, 2006; Chkaiban *et al.*, 2007). Changes in the expression of glycolitic genes encoding aldolase, phosphoglycerate kinase (both up-regulated), and pyruvate kinase (down-regulated) have been observed in cv. Corvina (Zamboni *et al.*, 2008), whereas transcripts corresponding to a pyruvate decarboxylase (PDC) and a lactate dehydrogenase were shown to increase in cv. Raboso Piave (Rizzini *et al.*, 2006, 2009). In their proteomics approach, Di Carli *et al.* (2011) found that glycolitic enzymes showed an increase in expression during the late withering stages.

Considering that a reduction in phosphoenolpyruvate carboxykinase transcripts has also been observed in Corvina berries (Zamboni *et al.*, 2008), the respiratory metabolism in particular, which concerns the fate and metabolism of pyruvate, seems to be markedly affected by postharvest dehydration. Activation of the fermentative pathway has been observed in several trials with a general increase in ethanol and acetaldehyde contents following dehydration in different conditions, accompanied by higher levels of ADH activity (Bellincontro *et al.*, 2004, 2009; Chkaiban *et al.*, 2007; Cirilli *et al.*, 2012) and protein amounts (Di Carli *et al.*, 2011). In *Vitis vinifera*, ADH is encoded by a multigene family made up of at least six members (Tesniere and Abbal, 2009).

ADH genes show a different expression trend during postharvest withering. In Raboso Piave berries dehydrated up to 30%WL, a significant decrease in the transcript of a putative ADH was observed by Rizzini *et al.* (2009), whereas Cirilli *et al.* (2012) found no expression of *VvADH1* and an up-regulation of *VvADH2* in particular in Aleatico grapes, in relation to 10 and 20% WL. These data clearly demonstrate that intensity and the rate of dehydration are crucial factors in modulating both ADH gene expression and enzyme activity, which seem to be highly reactive to temperature, RU and air flow (Chkaiban *et al.*, 2007; Bellincontro *et al.*, 2009; Cirilli *et al.*, 2012) during the withering process.

Since *VvADH2* is the isogene that is up-regulated under anaerobiosis (Tesniere and Abbal, 2009), an increase in expression during grape berry withering has been correlated with the presence of hypoxic conditions and the shift from aerobic metabolism to anaerobic fermentation as a result of changes in gas concentration in the fruit tissue (Cirilli *et al.*, 2012). However, a general increase in ADH transcript accumulation, in particular of *VvADH2* mRNA, has been observed during the normal ripening of grape berries (DeLuc *et al.*, 2007; Sweetman *et al.*, 2007; Tesniere and Abbal, 2009). The production of ethanol by PDC and ADH occurs in ripening fruit under a normal atmosphere if conditions inside the cells become too acidic, known as aerobic fermentation (Longhurst *et al.*, 1990). The induction of aerobic fermentation during ripening is likely to be due to an excess of sugars, which may also be the case of berries undergoing postharvest dehydration, where a high concentration of sugar is present due to concentration effects. It can thus be hypothesized that the initial increase in ADH gene expression detected in withering berries is not related to hypoxic conditions, but is a necessary step in removing acetaldehyde (produced by base levels of PDC activity and the increasing concentration of sugars). The fact that an accumulation of PDC transcripts has only been detected in advanced stages of postharvest dehydration (Rizzini *et al.*, 2006), supports the hypothesis that increases in ethanol and acetaldehyde content detected in early withering are the result of aerobic fermentation. ADH acts as a dimer and relies on a NAD(P) co-factor to interconvert ethanol and acetaldehyde. In addition to the fermentative pathway, it is also involved, under 'normoxic' conditions, in the so-called PDH (pyruvate dehydrogenase) complex (PDH bypass) to form acetate and then acetyl CoA (Strommer, 2011) with the activation of aldehyde dehydrogenase (ALDH). This pathway, originally detected in pollen grains (Bucher *et al.*, 1995) but recently also observed in sporophytic tissues (Wei *et al.*, 2009), is characterized by the interconversion of ethanol and acetaldehyde. The increase in ADH activity (in the ethanol to acetaldehyde oxidation) detected under several dehydration conditions and in different genotypes (Chkaiban *et al.*, 2007; Bellincontro *et al.*, 2009; Cirilli *et al.*, 2012) could support the hypothesis that the PDH bypass is active in dehydrated berries, at least in the early stages of dehydration. A decrease in ADH (ethanol to acetaldehyde) activity has been observed in Aleatico berries after a WL of about 30% (Cirilli *et al.*, 2012); in Raboso Piave berries, at this dehydration level, probes corresponding to two ALDH (that oxydize acetaldehyde to acetate) were down-regulated, and one lactate dehydrogenase gene was induced (Rizzini *et al.*, 2009) together with an increase in PDC transcripts (Rizzini *et al.*, 2006). Thus, it can be hypothesized that while in the early stages of dehydration, ripening-related aerobic fermentation and PDH bypass pathways are both activated, in the advanced stages of dehydration when the gas exchange between the fruit tissues and the surrounding atmosphere is impaired (due, for example, to a thickening of the epicarp cell wall, discussed later), anaerobic fermentation is induced with a high level of ethanol accumulation (Santonico *et al.*, 2010).

ADH is also involved in the production of C6 and C9 aldehydes and alcohol volatiles produced through the LOX pathway. Dehydration stress is a dramatic metabolic event that affects membrane composition, as observed in other fruit species (Lester and Bruton, 1986; Maalekuu *et al.*, 2005, 2006) and the production of a number of aroma compounds, some of which are caused by the activation of LOX. The increase in specific aldehydes produced from membrane lipids has been associated with the induction of LOX and ADH activity in withering Gewürztraminer berries (Chkaiban *et al.*, 2007). During Malvasia grape berry drying, an increase in LOX activity and the concomitant production of C6 compounds such as hexen-1-ol, hexanal and (E)-hex-2-enal was reported by Costantini *et al.* (2006). Little information is available at a molecular level: one *LOX* encoding transcript was identified as being induced in Corvina berries but only at an advanced stage of postharvest withering (Zamboni *et al.*, 2008) and, in fact in Gewürztraminer grapes, LOX specific activity only showed an increase after a WL of 20% (Chkaiban *et al.*, 2007). These data suggest that the production of C6 and C9 aldehydes and alcohol volatiles through the LOX pathway is not an early event in postharvest dehydration. More limited and fragmentary information is available concerning the expression of genes involved in norisoprenoids (Cirilli *et al.*, 2012) and terpenoids (Zamboni *et al.*, 2008).

Of the various secondary metabolic processes affected by postharvest dehydration, those concerning phenol compounds are of paramount importance in terms of the quality traits of the resulting wines. Although there are conflicting views in the literature – probably due to the use of different genotypes, fruit tissue, dehydration conditions (in particular temperature) and stress intensity – total phenol content appears to decrease, with respect to dry weight, after prolonged postharvest dehydration. This occurs despite transcripts of *PAL*, the first step of the phenylpropanoid pathway, in general increase during postharvest dehydration (Tonutti *et al.*, 2004; Rizzini *et al.*, 2009; Zamboni *et al.*, 2010; Bonghi *et al.*, 2012). However, two members of the PAL gene family were shown to be down-regulated in Raboso Piave withering berries (Bonghi *et al.*, 2012). This suggests that the expression regulation of the numerous members of the *Vitis vinifera PAL* multigene family is complex and may represent a key step in the multiple physiological responses to postharvest stress. A general activation of the phenylpropanoid pathway in dehydrated samples has also been confirmed by the up-regulation of cinnamate 4-hydroxylase (C4H) and 4-coumarate-CoA ligase (4CL) genes (Zamboni *et al.*, 2008, 2010; Bonghi *et al.*, 2012).

The effects of withering vary according to the specific phenol category, and this is evident when both compositional and molecular data are considered. One of the earliest pieces of evidence concerning the effects of postharvest dehydration on the modulation of the gene expression of specific phenol compounds was provided by Versari *et al.* (2001). They demonstrated an up-regulation of the *STS* gene accompanied by an accumulation of resveratrol in cv. Corvina berries. The marked effects of postharvest withering at the level of *STS* gene expression, was then confirmed by transcriptomic profiling approaches in the same cultivar (Zamboni *et al.*, 2008, 2010) and in Raboso Piave berries (Bonghi *et al.*, 2012).

Integrating metabolic and molecular data highlights that a strict relationship exists between the intensity of the stress and the induction of the stilbene pathway (leading to the accumulation of trans-resveratrol, trans- and cis-piceid, and ε-viniferins). Considering the biological role played by stilbene compounds, the induction of *STS* gene expression can be seen as the berry's response to stress. The synthesis of stilbenes, such as resveratrol and ε-viniferin, correlates with the increased expression of *STS*, but also that of peroxidase (*POD*), laccase (*LAC*), and polyphenol oxidase (*PPO*) (Zamboni *et al.*, 2010; Bonghi *et al.*, 2012).

Genes involved in oxidative stress responses generally appear to be induced during postharvest dehydration. This is the case of transcripts encoding catalase, glutathione-S-transferase, glutaredoxin, nudix hydrolase and a thaumatin-like protein in cv. Corvina (Zamboni *et al.*, 2010). The role of PPO, LAC and POD in polyphenol metabolism is also related to their possible involvement in the oxidative dimerization of resveratrol to viniferins, the polymerization of monolignols, and the oxidation of flavan-3-ols. As far as the monolignols and lignans pathways are concerned, a general increase in the expression of cinnamoyl CoA reductase (*CCR*) and cinnamyl alcohol dehydrogenase (*CAD*) genes has been particularly observed at advanced stages of postharvest dehydration (Bonghi *et al.*, 2012) when POD and LAC may play an important role in monolignol polymerization (Whetten and Sederoff, 1995). Activation of the lignin pathway is likely to be a reaction of skin cells to an advanced stage of dehydration, which could represent a barrier for the gas exchange between the cells and the surrounding environment. In still metabolically active fruit tissues (skin, in particular), this stress condition activates specific reactions aimed at stress adaptation and the minimization of microbial attacks.

Turning now to the flavonoid pathway this is probably negatively affected by postharvest dehydration. This is because two genes encoding chalcone synthase (*CHS*), the first committed step in flavonoid biosynthesis, were shown to be down-regulated in Raboso Piave berries (Bonghi *et al.*, 2012). The marked decrease in flavan-3-ols concentration could be related to oxidation (up-regulation of *PPO* and *LAC* genes) in the absence of new synthesis (no changes in leucoanthocyanidin reductase (*LAR*) expression) (Bonghi *et al.*, 2012). Changes in proanthocyanidins could affect the degree of polymerization, which, in Pinot Noir berries, shows a decreasing trend (Moreno *et al.*, 2008).

One category of flavonoids, however, appears to be specifically induced by postharvest dehydration, namely flavonols. Flavonol synthase (*FLS1*) is up-regulated in Raboso Piave samples dehydrated up to 30% WL, and this was accompanied by a general increase in quercetin concentration (Bonghi *et al.*, 2012). A similar effect was observed by Ziliotto *et al.* (2011) in grape berries of Raboso Piave sampled from clusters maintained for 45 days on the plants but in the absence of vascular connections as result of cane cutting (a technique called double reasoned maturation (DMR)). In this case too, an increase of quercetin was observed, while catechin concentration showed a significant drop. The decrease in catechin could result in changes in the synthesis of tannins and/or their degree of polymerization.

This event correlates with the reduction in astringency detected in Raboso wines obtained from DMR berries, as demonstrated by sensorial analysis. The increase in quercetin has an important impact on quality parameters considering that this flavonol contributes to the bitter taste and colour of red wine by stabilizing anthocyanin pigments (Cheynier *et al.*, 2006).

Also taking into consideration the results reported by Bellincontro *et al.* (2009) on Cesanese, and Mencarelli *et al.* (2010) on Aleatico grapes, the induction of a flavonol biosynthetic pathway could be seen as one of the main metabolic events characterizing grape berries undergoing postharvest dehydration. As reported above, Zamboni *et al.* (2010) suggest that flavonol taxifolin (dihydroquercetin) can be considered as a reliable biomarker of postharvest dehydration in Corvina grapes. The increase in flavonol concentration in response to a wide array of stress agents, in particular UV-B irradiance, is well documented (Pollastri and Tattini, 2011). Their anti-oxidant role seems to be related to the capacity to reduce reactive oxygen species (ROS) formation, a result of different oxidative stresses including desiccation (Rothschild and Mancinelli, 2001). A possible involvement of flavonoids as scavengers of H_2O_2 generated during severe stress conditions has also been hypothesized (Mubarakshina *et al.*, 2010; Pollastri and Tattini, 2011). Based on molecular and biochemical data, in grape berries undergoing postharvest water stress this role may be played by flavonols.

In dessert, sweet, reinforced wines obtained from red-skinned berries undergoing dehydration, a crucial parameter defining quality is represented by the red colour. Data on the concentration of anthocyanins in dehydrated skins are quite controversial. No significant changes have been detected in cv. Raboso Piave (Bonghi *et al.*, 2012) and Pinot Noir (Moreno *et al.*, 2008). Bellincontro *et al.* (2009) found a general decrease in different anthocyanins in Cesanese berries dehydrated up to 20% WL, whereas Toffali *et al.* (2011) highlighted that in cv. Corvina berries, postharvest dehydration led to a marked increase in anthocyanins in their acylated forms, whereas non-acylated forms accumulated to a much lesser extent. Considering gene expression data, it seems unlikely that a new synthesis of anthocyanins takes place, at least in the advanced stage of dehydration, since no changes in the expression of the *UFGT* gene have been detected in transcriptomic analyses (Zamboni *et al.*, 2010; Bonghi *et al.*, 2012), thus confirming previous specific expression analyses carried out by Tonutti *et al.* (2004). This seems to be in contrast with the expression of *MybA*, the main transcription factor known to control *UFGT* expression (Walker *et al.*, 2007), which appeared to be down-regulated in dehydration trials on Raboso Piave berries. The steady level of the *UFGT* transcript could be the result of a different expression of the MBW ternary complex members (Hichri *et al.*, 2011) induced by dehydration.

4.4 CONCLUDING REMARKS

The unique traits of wines obtained from over-ripe and/or dehydrated berries are the results of processes and events, some of which are strictly regulated in terms of gene expression. Several pathways appear to be affected by delaying the harvest and, in particular, when berries undergo postharvest withering. Based on the transcriptional analyses carried out on different genotypes, more information is now available concerning physiological and hormonal mechanisms affected by dehydration (Figure 4.1). The interplay between ABA and ethylene and, most likely, auxin, appears to be a key element in modulating the levels and the involvement of TFs such as ERF, DREB, MYB, NAC and WRKY. This results in a cascade of metabolic responses through the different and selective modulation of the expression of genes involved in pathways/processes (glycolisis/fermentation, phenylpropanoid, oxidative stress) affecting the berry composition. The interaction between water stress intensity and rate seems to play a major role in modulating molecular responses involving not only transcriptional but also post-transcriptional and post-translational regulatory mechanisms. The development and implementation of genomic tools will greatly help to discover and/or improve our knowledge of the metabolic pathways and the function of key genes involved in stress responses. This approach, coupled with the mass analysis of metabolic changes, should be able to provide solid support for the optimization of the techniques and protocols used for the production of these special wines.

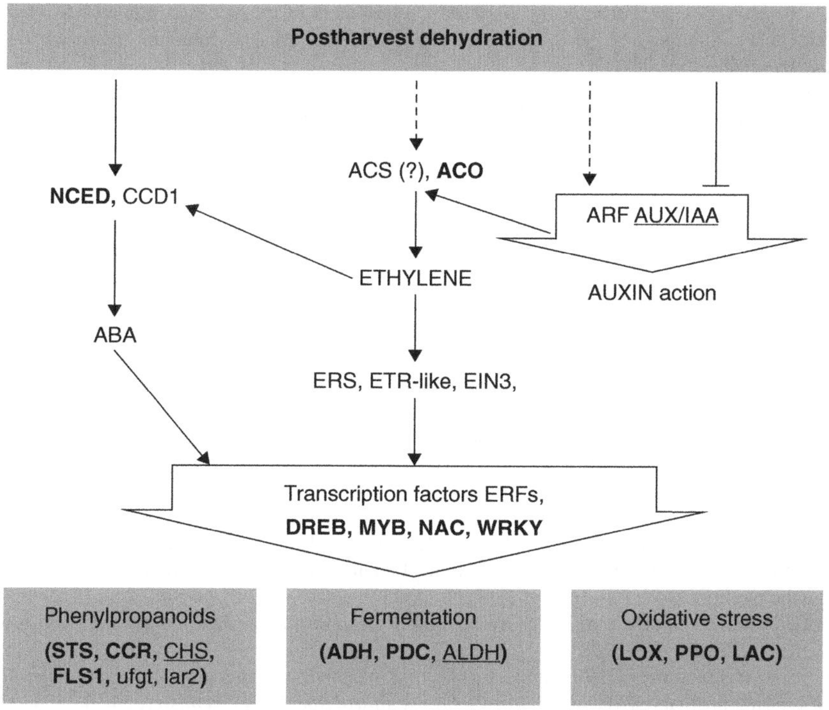

Figure 4.1 Hypothetical model explaining molecular events associated with postharvest water losses. This model was assembled based upon the available transcriptomic data described in the text. The first event is a positive effect on the biosynthesis of ABA and ethylene and a re-programming of auxin action by modulating regulatory elements (such as ARF and AUX/IAA). The modified hormonal network induces transcriptional changes of genes involved in different physiological processes (such as fermentation and oxidative stress) and metabolic pathways (phenylpropanoids), throughout the modulation of a set of transcription factors. Conventional symbols are used to describe hypothetical positive and negative interactions. The solid lines indicate a direct effect of postharvest water loss; dotted lines indicate an indirect effect; and the inverted T line, a repression. Induced and repressed genes are bolded and underlined, respectively, while small letters are used to indicate genes not significantly affected by dehydration.

REFERENCES

Barry, C.S., M.I. Llop-Tous, and D. Grierson. 2000. The regulation of 1-aminocyclopropane-1-carboxylic acid synthase gene expression during the transition from system-1 to system-2 ethylene synthesis in tomato. *Plant Physiology*, 123:979–986.

Bellincontro, A., D. De Santis, R. Botondi, *et al.* 2004. Different postharvest dehydration rates affect quality characteristics and volatile compounds of Malvasia, Trebbiano and Sangiovese grapes for wine production. *Journal of the Science of Food and Agriculture*, 84:1791–1800.

Bellincontro, A., A. Fardelli, D. De Santis, *et al.* 2006. Postharvest ethylene and 1-MCP treatments both affect phenols, anthocyanins, and aromatic quality of Aleatico grapes and wine. *Australian Journal of Grape and Wine Research*, 12:141–149.

Bellincontro, A., I. Nicoletti, M. Valentini, *et al.* 2009. Integration of nondestructive techniques with destructive analyses to study postharvest water stress of winegrapes. *American Journal of Enology and Viticulture*, 60:57–65.

Bonghi, C., F.M. Rizzini, A. Gambuti, *et al.* 2012. Phenol compound metabolism and gene expression in the skin of wine grape (*Vitis vinifera* L.) berries subjected to partial postharvest dehydration. *Postharvest Biology and Technology*, 67:102–109.

Böttcher, C., K. Harvey, C.G. Forde, *et al.* 2010. Auxin treatment of pre-veraison grape (*Vitis vinifera* L.) berries both delays ripening and increases the synchronicity of sugar accumulation. *Australian Journal of Grape and Wine Research*, 17:1–8.

Brady, C.J. 1987. Fruit ripening. *Annual Review of Plant Physiology*, 38:155–178.

Bucher, M., K.A. Brander, S. Sbicego, *et al.* 1995. Aerobic fermentation in tobacco pollen. *Plant Molecular Biology*, 28:739–750.

Carra, A., E. Mica, G. Gambino, *et al.* 2009. Cloning and characterization of small non-coding RNAs from grape. *The Plant Journal*, 59(5):750–763.

Chervin, C., A. Tira-umphon, N. Terrier, *et al.* 2008. Stimulation of the grape berry expansion by ethylene and effects on related gene transcripts, over the ripening phase. *Physiologia Plantarum*, 134:534–546.

Cheynier, V., M. Duenas-Paton, E. Salas, *et al.* 2006. Structure and properties of wine pigments and tannins. *American Journal of Enology and Viticulture*, 57:298–305.

Chkaiban, L., R. Botondi, A. Bellincontro, *et al.* 2007. Influence of post-harvest water stress on lipoxygenase and alcohol dehydrogenase activities, and the composition of some volatile compounds of Gewürztraminer grapes dehydrated under controlled and uncontrolled termohygrometric conditions. *Australian Journal of Grape and Wine Research*, 13:142–149.

Choudhury, S.R., S. Roy, and D.N. Sengupta. 2008. Characterization of transcriptional profiles of *MA-ACS1* and *MA-ACO1* genes in response to ethylene, auxin, wounding, cold and different photoperiods during ripening in banana fruit. *Journal of Plant Physiology*, 132:447–454.

Cirilli, M., A. Bellincontro, D. De Santis, *et al.* 2012. Temperature and water loss affect ADH activity and gene expression in grape berry during postharvest dehydration. *Food Chemistry*, 132(1):447–454.

Conde, C., P. Silva, N. Fontes, *et al.* 2007. Biochemical changes throughout grape berry development and fruit and wine quality. *Food*, 1:1–22.

Coombe, B.G. 1976. Development of fleshy fruits. *Annual Review of Plant Physiology and Plant Molecular Biology*, 27:207–228.

Costantini, V., A. Bellincontro, D. De Santis, *et al.* 2006. Metabolic changes of Malvasia grapes for wine production during postharvest drying. *Journal of Agricultural and Food Chemistry*, 54(9):3334–3340.

Davies, C. and C. Böttcher. 2009. Hormonal control of grape berry ripening. In *Grapevine Molecular Physiology & Biotechnology*, 2nd edn (ed. K.A. Roubelakis-Angelakis), Springer Science+Business Media, pp. 229–261.

Davies, C. and S.P. Robinson. 2000. Differential screening indicates a dramatic change in mRNA profiles during grape berry ripening. Cloning and characterization of cDNAs encoding putative cell wall and stress response proteins. *Plant Physiology*, 122:803–812.

Davies, C., P.K. Boss, and S.P. Robinson. 1997. Treatment of grape berries, a nonclimateric fruit with a synthetic auxin, retards ripening and alters the expression of developmentally regulated genes. *Plant Physiology*, 115:1155–1161.

Deluc, L.G., J. Grimplet, M.D. Wheatley, *et al.* 2007. Transcriptomic and metabolite analyses of Cabernet Sauvignon grape berry development. *BMC Genomics*, 8, 429.

Di Carli, M., A. Zamboni, M.E. Pè, *et al.* 2011. Two-dimensional differential in gel electrophoresis (2D-DIGE) analysis of grape berry proteome during postharvest withering. *Journal of Proteome Research*, 10(2):429–446.

El-Kereamy, A., C. Chervin, and J.P. Roustan. 2003. Exogenous ethylene stimulates the long-term expression of genes related to anthocyanin biosynthesis in grape berries. *Physiologia Plantarum*, 119:175–182.

El-Sharkawy, I., W.S. Kim, S. Jayasankar, *et al.* 2008. Differential regulation of four members of the ACC synthase gene family in plum. *Journal of Experimental Botany*, 59:2009–2027.

Finkelstein, R.R. and C.D. Rock. 2002. Abscisic acid biosynthesis and response. In *The Arabidopsis Book* (eds C.R. Sommerville and E.M. Meyerowitz), American Society of Plant Biologists, Rockville, MD.

Giovannoni, J.J. 2001. Molecular biology of fruit maturation and ripening. *Annual Review of Plant Physiology and Plant Molecular Biology*, 52:725–749.

Giovannoni, J.J. 2004. Genetic regulation of fruit development and ripening. *Current Opinion in Plant Biology*, 10:238–289.

Grimplet, J., L.G. Deluc, R.L. Tillett, *et al.* 2007. Tissue-specific mRNA expression profiling in grape berry tissues. *BMC Genomics*, 8:187.

Guillaumie, S., R. Fouquet, C. Kappel, *et al.* 2011. Transcriptional analysis of late ripening stages of grapevine berry. *BMC Plant Biology*, 11:165.

Hale, C.R. 1968. Growth and senescence of grape berry. *Australian Journal of Agricultural Research*, 19: 939–945.

Hichri, I., F. Barrieu, J. Bogs, *et al.* 2011. Recent advances in the transcriptional regulation of the flavonoid biosynthetic pathway. *Journal of Experimental Botany*, 62:2465–2483.

Jaillon, O., J.M. Aury, B. Noel, *et al.* 2007. The grapevine genome sequence suggests ancestral hexaploidization in major angiosperm phyla. *Nature*, 44:463–465.

Jeong, S.T., N. Goto-Yamamoto, S. Kobayashi, *et al.* 2004. Effects of plant hormones and shading on the accumulation of anthocyanins and the expression of anthocyanin biosynthetic genes in grape berry skins. *Plant Science*, 167:247–252.

Lester, G.E. and B.D. Bruton. 1986. Relationship of netted muskmelon fruit water loss to postharvest storage life. *Journal of American Society Horticultural Science*, 111:727–731.

Longhurst, T.J., H.F. Tung, and C.J. Brady. 1990. Developmental regulation of the expression of alcohol dehydrogenase in ripening tomato fruits. *Journal of Food Biochemistry*, 14:421–433.

Maalekuu, K., Y. Elkind, A. Leikin-Frenkel, *et al.* 2006. The relationship between water loss and lipid content, membrane integrity and LOX activity in ripe pepper fruit after storage. *Postharvest Biology Technology*, 42:248–255.

Maalekuu, K., Y. Elkind, S. Tuvia-Alkalai, *et al.* 2005. Characterization of physiological and biochemical factors associated with postharvest water loss in ripe pepper fruit during storage. *Journal of American Society Horticultural Science*, 130:735–741.

Mencarelli, F., A. Bellincontro, I. Nicoletti, *et al.* 2010. Chemical and biochemical change of healthy phenolic fractions in winegrape by means of postharvest dehydration. *Journal of Agriculture and Food Chemistry*, 58:7557–7564.

Moreno, J., F. Cerpa-Calderòn, S. Cohen, *et al.* 2008. Effect of postharvest dehydration on the composition of Pinot Noir grapes (*Vitis vinifera* L.) and wine. *Food Chemistry*, 109:755–762.

Mubarakshina, M.M., B.N. Ivanov, I.A. Naydov, *et al.* 2010. Production and diffusion of chloroplastic H_2O_2 and its implication to signalling. *Journal of Experimental Botany*, 61:3577–3587.

Ollat, N., P. Diakou-Verdin, J.P. Carde, *et al.* 2002. Grape berry development: A review. *Journal International des Sciences de la Vigne et du Vin*, 36:109–131.

Pilati, S, M. Perazzolli, A. Malossini, *et al.* 2007. Genome-wide transcriptional analysis of grapevine berry ripening reveals a set of genes similarly modulated during three seasons and occurrence of an oxidative burst at veraison. *BMC Genomics*, 8:428.

Pollastri, S. and M. Tattini. 2011. Flavonols: old compounds for old roles. *Annals of Botany*, 108:1225–1233.

Rizzini, F.M., C. Bonghi, and P. Tonutti. 2006. Approcci molecolari per lo studio della fisiologia delle uve in appassimento. *www.phytomagazine.com*, 15:21–33.

Rizzini, F.M., C. Bonghi, and P. Tonutti. 2009. Postharvest water loss induces marked changes in transcript profiling in skins of wine grape berries. *Postharvest Biology and Technology*, 52(3):247–253.

Romero, P., M.J. Rodrigo, F. Alferez, *et al.* 2012. Unravelling molecular responses to moderate dehydration in harvested fruit of sweet orange (*Citrus sinensis* L. Osbeck) using a fruit-specific ABA-deficient mutant. *Journal Experimental Botany*, 63:2753–2767.

Rothschild, L.J. and R.L. Mancinelli. 2001. Life in extreme environments. *Nature*, 409:1092–1101.

Santonico, M., A. Bellincontro, D. De Santis, *et al.* 2010. Electronic nose to study postharvest dehydration of wine grapes. *Food Chemistry*, 121(3):789–796.

Seymour, G.B., J.E. Taylor, and G.A. Tucker. 1993. *Biochemistry of Fruit Ripening*, Chapman and Hall, London.

Strommer, J. 2011. The plant ADH gene family. *The Plant Journal*, 66(1):128–142.

Sun L., M. Zhang, J. Rien, *et al.* 2010. Reciprocity between abscisic acid and ethylene at the onset of berry ripening and after harvest. *BMC Plant Biology*, 10:257.

Sweetman, C., L.G. Deluc, G.R. Cramer, *et al.* 2009. Regulation of malate metabolism in grape berry and other developing fruits. *Phytochemistry*, 70:1329–1344.

Symons, G.M., C. Davies, Y. Shavrukov, *et al.* 2006. Grapes on steroids. Brassinosteroids are involved in grape berry ripening. *Plant Physiology*, 140:150–158.

Terrier, N., D. Glissant, J. Grimplet, *et al.* 2005. Isogene specific oligo arrays reveal multifaceted changes in gene expression during grape berry (*Vitis vinifera* L.) development. *Planta*, 222:832–847.

Tesniere, C. and P. Abbal. 2009. Alcohol dehydrogenase genes and proteins in grapevine. In *Grapevine Molecular Physiology & Biotechnology*, 2nd edn (ed. K.A. Roubelakis-Angelakis), Springer Science+Business Media, pp. 141–160.

Tilbrook, J. and S.D. Tyerman. 2009. Hydraulic connection of grape berries to the vine: varietal differences in water conductance into and out of berries, and potential for backflow. *Functional Plant Biology*, 36:541–550.

Toffali, K., A. Zamboni, A. Anesi, *et al.* 2011. Novel aspects of grape berry ripening and post-harvest withering revealed by untargeted LC-ESI-MS metabolomics analysis. *Metabolomics*, 7:424–436.

Tonutti, P., G.B. Tornielli, and G. Cargnello. 2004. Characterization of 'territories' throughout the production of wine obtained with withered grapes: the cases of 'Terra della Valpolicella' (Verona) and 'Terra del Piave' (Treviso) in northern Italy. *Proceedings of the International Conference on Viticultural Zoning*, Cape Town, South Africa, November 2004.

Trainotti, L, A. Tadiello and G. Casadoro. 2007. The involvement of auxin in the ripening of climacteric fruits comes of age: the hormone plays a role of its own and has an intense interplay with ethylene in ripening peaches. *Journal of Experimental Botany*, 58:3299–3308.

Tyerman, S.D., J. Tilbrook, C. Pardo, *et al.* 2004. Direct measurement of hydraulic properties in developing berries of *Vitis vinifera* L. cv. Shiraz and Chardonnay. *Australian Journal of Grape and Wine Research*, 10:170–171.

Velasco, R., A. Zharkikh, M. Troggio, *et al.* 2007. A high quality draft consensus sequence of the genome of a heterozygous grapevine variety. *PLoS One*, 2:e1326.

Versari, A., G.P. Parpinello, G.B. Tornielli, *et al.* 2001. Stilbene compounds and stilbene synthase expression during ripening, wilting, and UV treatment in grape cv. Corvina. *Journal of Agricultural and Food Chemistry*, 49:5531–5536.

Walker, A.R., E. Lee, J. Bogs, *et al.* 2007. White grapes arose through the mutation of two similar and adjacent regulatory genes. *The Plant Journal*, 49:772–785.

Wang, C., X. Wang, N.K. Kibet, *et al.* 2011. Deep sequencing of grapevine flower and berry short RNA library for discovery of novel microRNAs and validation of precise sequences of grapevine microRNAs deposited in miRBase. *Physiologia Plantarum*, 143(1):64–81.

Waters, D.L.E., T.A. Holton, E.M. Ablett, *et al.* 2006. The ripening wine grape berry skin transcriptome. *Plant Science*, 171:132–138.

Wei, Y., M. Lin, D.J. Oliver, *et al.* 2009. The roles of aldehyde dehydrogenases (ALDHs) in the PDH bypass of Arabidopsis. *BMC Biochemistry*, 10:7.

Wheeler, S., B. Loveys, C. Ford, *et al.* 2009. The relationship between the expression of abscisic acid biosynthesis genes, accumulation of abscisic acid and the promotion of *Vitis vinifera* L. berry ripening by abscisic acid. *Australian Journal of Grape and Wine Research*, 15(3):195–204.

Whetten, R. and R. Sederoff, 1995. Lignin biosynthesis. *Plant Cell*, 7:1001–1013.

Zamboni, A., M. Di Carli, F. Guzzo, *et al.* 2010. Identification of putative stage-specific grapevine berry biomarkers and omics data integration into networks. *Plant Physiology*, 154(3):1439–1459.

Zamboni, A., L. Minoia, A. Ferrarini, *et al.* 2008. Molecular analysis of post-harvest withering in grape by AFLP transcriptional profiling. *Journal of Experimental Botany*, 59:4145–4159.

Zenoni, S., A. Ferrarini, E. Giacomelli, *et al.* 2010. Characterization of transcriptional complexity during berry development in *Vitis vinifera* using RNA-seq. *Plant Physiology*, 152:1787–1795.

Ziliotto, F., F.M. Rizzini, G. Teo, *et al.* 2011. Sensorial, biochemical and molecular changes on Raboso Piave grape berries harvested by applying double reasoned maturation and late harvest techniques. *Proceedings of the PAV-GiESCO Meeting*, pp. 193–195.

Ziliotto, F., M. Corso, F.M. Rizzini, *et al.* 2012. Grape berry ripening delay induced by a pre-veraison NAA treatment is paralleled by a shift in the expression pattern of auxin- and ethylene-related genes. *BMC Plant Biology*, 12:185.

Zironi R. and R. Ferrarini. 1987. La surmaturazione delle uve destinate alla vinificazione. *Vignevini*, 4:31–45.

5 Changes in Volatile Compounds

Claudio D'Onofrio

Department of Agriculture, Food and Environment, University of Pisa, Pisa, Italy

5.1 INTRODUCTION

The grape aroma, with its vast range of components, better characterizes the wine quality and the technological potential of a grapevine variety and it allows the varieties and areas of production to be identified by means of wine tasting. Therefore, some components of the aromatic fraction may be regarded as technological markers in both the conservation and storage of wines, as well as playing a role in determining the authenticity and protection of regional produce, and contributing to the recovery and valorization of minor grapevine varieties and the preservation of biodiversity.

The term 'aroma' refers to the detection of 'volatiles' by nasal olfactory receptors (smell), while the term 'flavour' indicates the combination of both odour and taste. Hundreds of secondary volatile metabolites have been identified in wines which potentially contribute to the aroma, and according to the origin they are classified as:

- primary aroma, already present in the grape
- secondary aroma, generated during the pre-fermentation and fermentation processes
- tertiary aroma, generated during the maturation and evolution of the wine.

The primary aroma is also known as the 'varietal aroma' because its VOCs (volatile organic compounds) give to the wine the characteristic and distinctive aroma of grapevine varieties. In grape berries, the majority of aroma compounds are present in the odourless glycoconjugate form, while their odorous aglycones in free form are liberated into the wine by acidic and enzymatic hydrolysis. The so-called 'aromatic' cultivars (such as Muscats and aromatic Malvasias) are characterized by grapes with a concentration of free aroma higher than the threshold perception level, which could consequently be detected by berry tasting; on the other hand, in the grape of the so-called 'non-aromatic', 'non-floral' or 'neutral' cultivars, the free form of aromas are below the threshold perception level.

5.2 VARIETAL AROMA

The main groups of the primary aroma metabolites include terpenoids (monoterpenes, sesquiterpenes and C_{13}-norisoprenoids), shikimate pathway derivatives (or volatile phenols

Sweet, Reinforced and Fortified Wines: Grape Biochemistry, Technology and Vinification, First Edition.
Edited by Fabio Mencarelli and Pietro Tonutti.

or benzene derivatives), aliphatic C6 volatile compounds (aldehydes and alcohols), volatile thiols (or mercaptans) and methoxypyrazines. In this section the key information on the above classes of aroma reported in the Italian review by D'Onofrio (2011) are summarized.

Terpenes are the largest class of plant secondary metabolites and include many volatile compounds: hemiterpenes (at 5 carbons; C5), monoterpenes (C10) and sesquiterpenes (C15), and even some diterpenes (C20) are volatile under normal atmospheric conditions.

Monoterpenes are typically varietal aromas and generally predominate in the 'aromatic' varieties. About 70 different monoterpenes have been identified in grapes as hydrocarbons, aldehydes, alcohols (monoterpenols), acids and esters, with odour thresholds varying from a fraction to hundreds of $\mu g\ L^{-1}$ of must or wine. The most common of these are linalool and its related compounds (RCs) (such as pyranic and furanic oxides, linalool hydroxides, diols), geraniol and RCs (such as isogeraniol, nerol, nerol oxide, rose oxide), terpineol and citronellol. Monoterpenes are responsible for the aroma resembling rose, camphor, coriander and citric characters, but at a high concentration level undesirable herbaceous notes may also be present. Usually, the glycosylated monoterpenes are 3–10 times more abundant than free aglycones, even in aromatic grapes.

Sesquiterpenes are present in grapes as free forms. Farnesol is one of the most abundant sesquiterpenes in flowers and young berries, and α-caryophyllene, α-copaene, α-cubebene, muurolene, calamenene, and other sesquiterpenes have been detected in the berries of some German wine varieties, of the cultivars Baga, Shiraz, Cabernet Sauvignon and Riesling, as well as in berry cell suspensions. Recently, a relationship has been identified between the sesquiterpene 'rotundone' and the typical pepper aromas of Syrah wines, which has also been detected in Vespolina grapes.

The norisoprenoids originate from the oxidative degradation of carotenoids (tetraterpenes; C40) and in the grape the most abundant are at 13 carbons (C_{13}-norisoprenoids), which are mainly present as glycoside compounds and are responsible for some floral, fruity and spicy characters, particularly important in some 'neutral' cultivars, like Sangiovese. C_{13}-norisoprenoids are classified in megastigmanes and non-megastigmanes. The megastigmanes are oxygenated on the carbon 7 (damascene group) or 9 (ionone group) and the most important of this group are β-damascenone (fruity-floral, honey-like and stewed apple characters) and β-ionone (sweet and violet notes) that have a very low threshold perception (2 $ng\ L^{-1}$ and 7 $ng\ L^{-1}$, respectively). Other megastigmanes are the 3-hydroxy-β-damascone (tobacco characters), β-damascone (fruity and tobacco notes) and 3-oxo-α-ionolo (tobacco characters). Among the non-megastigmanes, the most interesting are 1,6-trimethyl-1,2-dihydronaphthalene (TDN), (*E*)-1-(2,3,6-trimethylphenyl)buta-1,3-diene (TPB) and actinidols. TDN has a kerosene-like character and is generally considered to have a detrimental effect on wines. The TPB has floral and geranium notes or undesirable insecticide and plastic characters at a high concentration level.

The benzene derivatives are responsible for the spicy, tobacco, citrus, honey, vanilla and floral characters, but at high concentration levels give undesirable chemical and phenolic characters. Benzenoids, whose detection threshold varies from $\mu g\ L^{-1}$ up to over 10 $mg\ L^{-1}$, are present in both free and glycosylated forms, the most remarkable being benzyl alcohol, vanillin, methyl vanillate, acetovanillone, homovanillic alcohol, dihydroconiferyl alcohol and zingerone.

The thiols are present in grapes exclusively as cysteine and glutathione S-conjugates and they are responsible not only for important wine aromas but also for undesirable notes at high concentrations. The most important thiols identified in grapes are: 3-mercaptohexanol (3MH), with passion fruit and grapefruit characters; 3-mercaptohexyl-O-acetate (3MHA), with boxwood, grapefruit peel and passion fruit characters; 4-mercapto-4-methylpentan-2-one (4MMP), with boxwood and broom characters; and 4-mercapto-4-methylpentan-2-ol (4-MMPOH), with citrus peel characters.

Methoxypyrazines have been mainly detected in the Sauvignon blanc, Cabernet-Sauvignon, Cabernet franc, Merlot and Semillon grapevine varieties and they contribute to vegetative, herbaceous, bell pepper and earthy wine aromas. The methoxypyrazines detection threshold is very low (2–15 ng L^{-1}), and at a concentration over 50 ng L^{-1}, these compounds emit undesirable characters. The main methoxypyrazines identified in grapes are: 2-methoxy-3-isobutylpyrazine (IBMP), with its typical green capsicum characters; 2-methoxy-3-isopropylpyrazine (IPMP), with its asparagus or sweet pea characters; and 2-methoxy-3-sec-butylpyrazine (SBMP) responsible for galbanum oil characters.

The main aldehydes and alcohols aliphatic compounds at 6 carbons detected in grapes have herbaceous, unripe fruits and crumpled leaf characters. The most remarkable C6 aliphatic alcohols in grapes are hexanol and its isomers (*Z*-3-hexenol and *E*-2-hexenol), while the aldehydes are hexanal, *Z*-3-hexenal and *E*-2-hexenal, heptanal and *E*-2-heptenal.

5.3 AROMA EVOLUTION DURING BERRY DEVELOPMENT AND OVER-RIPENING

The berry growth curve is a double-sigmoid with three specific phases: phase I, the 'green phase' or 'berry formation phase'; phase II, the 'lag and veraison phase'; phase III, the 'ripening phase'. While berry formation is characterized by cell division, during berry ripening growth is exclusively due to cell enlargement. Additionally, phase IV, the 'over-ripening phase', characterized by berry evaporative water loss ('withering') under hot and dry conditions, can occur and can be particularly useful for the production of Passito wines (Figure 5.1). 'Flavour ripening' is the development of attractive berry flavours during the ripening phase, as a result of changes in the sugar-to-acid balance, phenols and aroma concentrations, while the term 'engustment' mainly indicates the part of the berry flavour detected as aroma by berry tasting during ripening (Coombe and McCarthy, 1997).

Grape aroma compounds and aroma precursors are produced during the whole berry development and the final mixture depends on variables, which include the grapevine variety, the environmental conditions during the growing season, the management of the vineyard and the degree of ripening. The grapevine variety is the most important factor in determining the grape aroma composition. In aromatic varieties, such as Moscato bianco (White Muscat, Muscat blanc) and Aleatico, two varieties largely used for the production of Passito wines, there is a higher prevalence of monoterpenes, particularly in ripened grapes, over the other classes of aroma compounds, and a concentration of free monoterpenes above the threshold perception level that covers the characters of the other classes of aroma compounds. In non-floral varieties, such as Sangiovese, the concentration of monoterpenes, C_{13}-norisoprenoids and benzene derivatives is quite similar, consequently there is a high complexity of wine aromas (D'Onofrio *et al.*, 2008; D'Onofrio *et al.*, 2010b). Environmental conditions and vineyard management strongly influence the grape aroma composition, which, together with the winemaking techniques, are responsible for the aromatic differentiation of wines obtained from the same grapevine variety. Temperature and light exposure levels are the most important microclimatic parameters influencing the grape aroma biosynthesis, and consequently the removal of shoot basal leaf, especially early shoot leaf removal (at pre-flowering, during

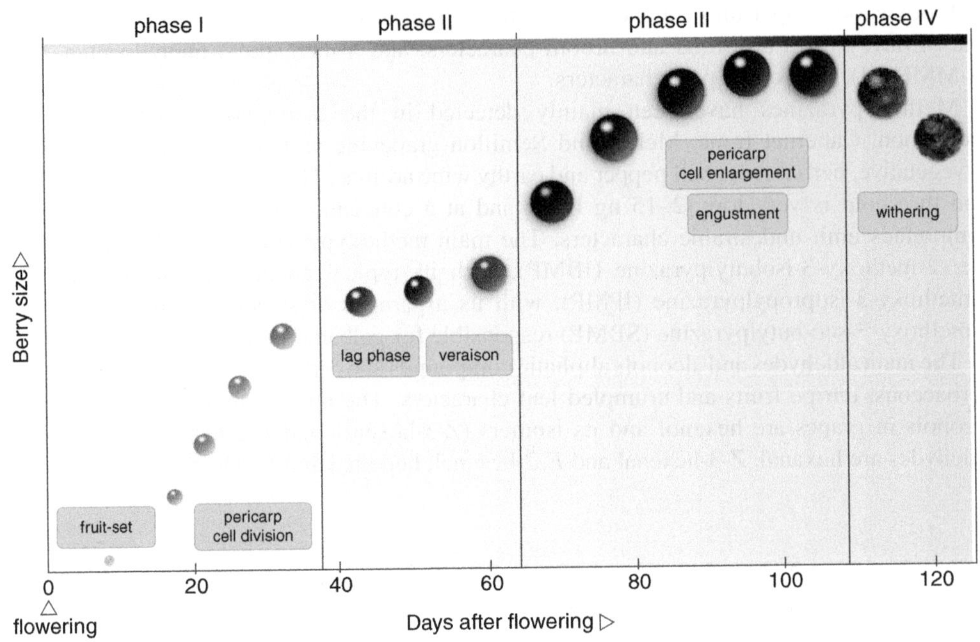

Figure 5.1 Diagram showing the four phases of berry development (D'Onofrio, unpublished). (For a colour version of this figure, please see the colour plate section.)

flowering or just after fruit-set), influences the grape aroma composition, increasing the concentration of monoterpenes and particularly of C_{13}-norisoprenoids (D'Onofrio *et al.*, 2012).

In general, the concentration of terpenes and benzene derivatives aroma compounds is high in flowers, then decreases during fruit-set and increases again during the ripening phase (D'Onofrio *et al.*, 2010a, b; D'Onofrio, 2011). Typically, the concentration of carotenoids, precursors of C_{13}-norisoprenoids, increases from fruit-set to the end of phase I and then decreases during ripening, when the concentration of C_{13}-norisoprenoids increases (Baumes *et al.*, 2002; D'Onofrio *et al.*, 2012). Some other classes of aromas, such as methoxypyrazines, accumulate during the green phase and then decrease during ripening, while some other specific compounds, like aliphatic esters and thiols, are synthesized exclusively during ripening (Ryona *et al.*, 2008; Kalua and Boss, 2009; Kobayashi *et al.*, 2010).

One important consideration when we speak of aroma is: how to express the concentration of aroma in grapes? From an enological point of view, it is opportune to express the aroma concentration per berry weight (usually $\mu g\ g^{-1}$ or $\mu g\ kg^{-1}$), because this parameter is directly correlated to the potential metabolites concentration in wines. While, from a viticultural point of view and if the objective is to study the biosynthesis of aroma during berry development, it could be more appropriate to express the aroma concentration per single berry (μg berry^{-1}) to avoid the dilution effect due to the rapid accumulation of soluble solids and water during ripening. In other words, in Moscato bianco where over 90% of VOCs are represented by monoterpenes during all berry development phases, the displaying of aroma concentrations as $\mu g\ g^{-1}$ of tissue in flowers and berries at 10–11 day intervals from fruit-set to over-ripening (Figure 5.2), indicates a high monoterpenes concentration in flower tissues, a great decrease during fruit-set, and subsequently an accumulation that starts during phase II; an apparent lag-phase associated with the fast berry growth at the beginning of phase III is shown, and

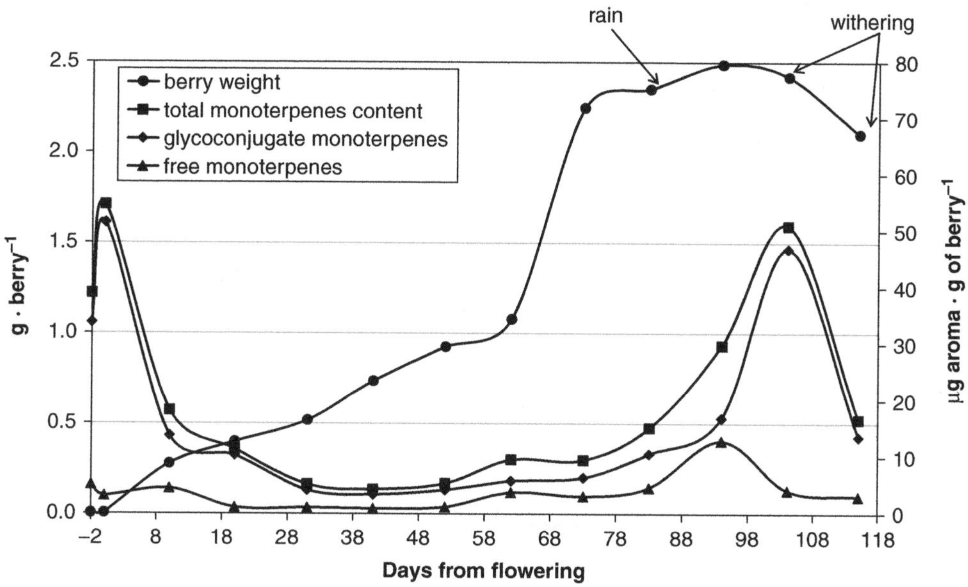

Figure 5.2 Monoterpenes concentration in flowers (flower buds and open flowers) and during berry development of Moscato bianco, expressed as µg g^{-1} of berry, in a vineyard in the Brunello di Montalcino area, Siena province, central Tuscany (D'Onofrio *et al.*, 2010b).

a subsequent increase until a peak of concentration during the end of ripening when 95% of aromas in the berry are monoterpenes (D'Onofrio *et al.*, 2010a, b; D'Onofrio, 2011). If the same aroma content is displayed as µg per single flower or berry (Figure 5.3), the concentration of aromas in flowers and young berries is very low and the accumulation begins to be significant at the end of phase I with a subsequent continuous increase until the end of ripening without any lag-phase (D'Onofrio *et al.*, 2010a, b; D'Onofrio, 2011). In addition, considering that berry and wine aroma characters depend on the relative proportion of the several aroma compounds, it could be useful to evaluate the concentration factor (proportion of variation of the concentration of a single metabolite) to study the effect of treatments on grape and wine aroma characteristics.

In grapes, the peak of aroma accumulation often precedes the technological ripening, as observed in Aleatico, Ciliegiolo, Moscato bianco and Sangiovese (D'Onofrio *et al.*, 2008; D'Onofrio *et al.*, 2010a, b), Muscat Hamburg (Fenoll *et al.*, 2009), Airén, Chardonnay and Macabeo (Garcia *et al.*, 2003), table grape (Yang *et al.*, 2011), while the concentration decreases during over-ripening.

In the Moscato bianco grapes described earlier, an over-ripening of 10 days, with a loss of berry weight of about 15%, reduced the amount of monoterpenes per berry weight by about 65% and by about 75% per single berry (D'Onofrio *et al.*, unpublished). In detail, there was a decrease in the concentration per berry of the free and glycosylated fractions of each single monoterpene, except for an increase in the free fraction of cis-furan linalool oxide and particularly of geraniol, which increased by about 30% and 120%, respectively, if the variation is calculated on the concentration per berry. Of particular importance was the high increase in the concentration of free geraniol, one of the most abundant monoterpenes in Moscato bianco, which is responsible for the rose flower character (D'Onofrio *et al.*, unpublished).

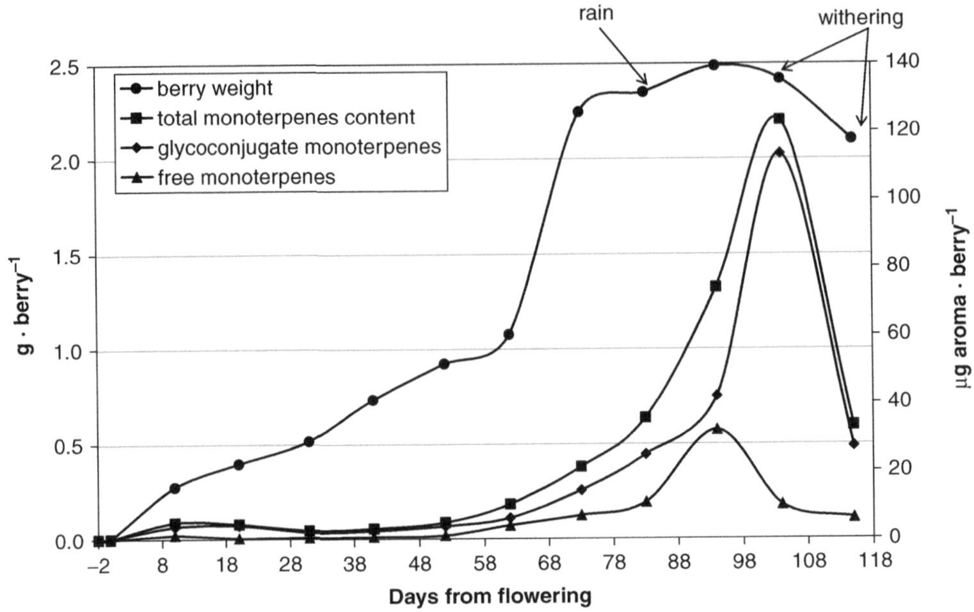

Figure 5.3 Monoterpenes concentration in flowers (flower buds and open flowers) and during berry development of Moscato bianco expressed as µg per single berry in a vineyard in the Brunello di Montalcino area, Siena province, central Tuscany (D'Onofrio *et al.*, 2010b).

Similar effects were also observed in the over-ripening of the Aleatico grape, another aromatic grapevine variety similar to Moscato bianco, which is used in the production of 'Passito' wines. In Aleatico, as in Moscato bianco, there is a prevalence of monoterpenes (usually about 85% of the total aroma content in ripened grapes, of which about 10% are in free form and 90% in glycosylated form). However, unlike Moscato bianco in ripened berries, the geraniol and RCs (hydroxylate derivatives and geranic acid) plus nerol and RCs (hydroxylate derivatives) highly prevail over linalool and related (hydroxylate) compounds (the sum of geraniol, geranic acid and nerol usually represent around the 90% of monoterpenes) (Figure 5.4) (D'Onofrio *et al.*, unpublished).

A slight over-ripening by two weeks of Aleatico grapes in two vineyards (Val di Cornia, Tuscany central coast, and Alberese, Tuscany south coast), with a loss of berry weight of 8% and 12%, respectively, induced a decrease of total aroma per berry weight in both vineyards (about 8% and 10%, respectively), which resulted in a decrease of aroma concentration per single berry of 2% and 10%, respectively (D'Onofrio *et al.*, unpublished). The total concentration of monoterpenes per berry decreased by almost 1% in the Val di Cornia vineyard, and by 8% in the Alberese vineyard. In particular, as concerns the concentration of geraniol and RCs, the concentration of free fractions increased or decreased (about +45% and −12% in the Val di Cornia and Alberese vineyards, respectively) while the glycosylated fraction decreased in both vineyards (about 10% and 5%), which resulted in a decrease of total geraniol and RCs of about 2% and 6%, respectively.

Boselli *et al.* (2009) reported that a slight over-ripening of Aleatico grapes on the island of Elba produced an increase in both free and glycosylated fractions of geraniol and nerol, a significant reduction in free linalool and citronellol concentrations, and a slight increase in their glycosylated fractions. The same authors clearly reported that the aroma concentration of

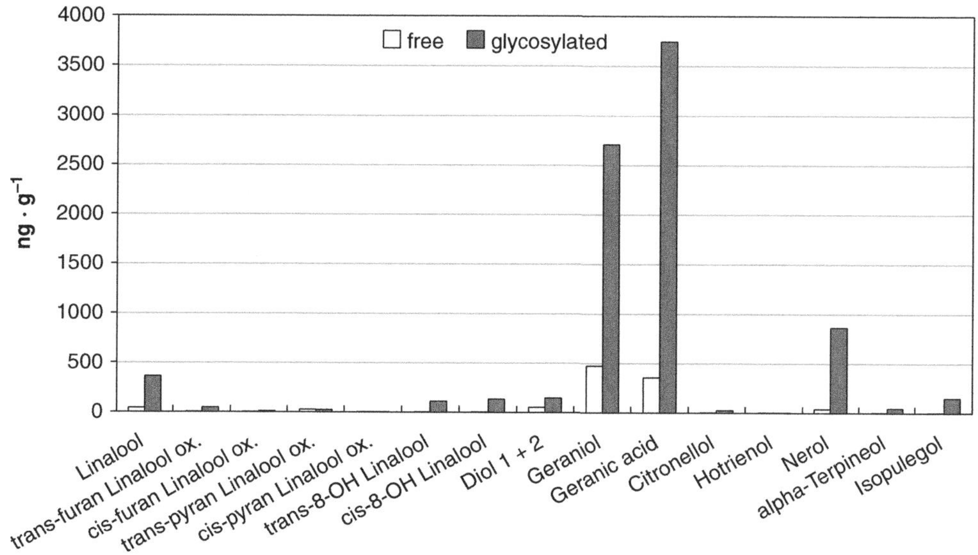

Figure 5.4 Monoterpenes profile of Aleatico ripe berries from a vineyard in the Val di Cornia area, Livorno province, Tuscany central coast (D'Onofrio *et al.*, unpublished).

Aleatico berries strongly depends on the environment, evidence that supports the differences in response to the over-ripening of the Aleatico grapes in the different areas and years.

So, in general it appeared that the over-ripening of aromatic grapes induced a decrease of the concentration of monoterpenes mainly due to the high reduction of glycosylated fractions, while the concentration of total free monoterpenes could increase or decrease depending on the environment. The decrease of the effective concentration (μg berry^{-1}) of each aroma compound clearly indicates that there is a prevalence of catabolic activity, and the increase of free fractions and the decrease of glycosylated fractions, mainly for geraniol and RCs, and nerol and RCs, is probably due to a hydrolysis rate of glycosylated forms higher than the catabolism rate of free forms.

The over-ripening of the Aleatico grape with the application of cane incision ('cane-cut') in the Alberese vineyard described earlier, gave interesting results on the composition of grape aroma (D'Onofrio *et al.*, unpublished). In detail, at the beginning of over-ripening the shoots bearing bunches were totally cut at the base as reported in Scalabrelli *et al.* (2008), a treatment that induced fast berry dehydration: 30% of berries weight reduction within 2 weeks, instead of 12% in the ordinary over-ripened grapes, and a further 28% reduction after 2 additional weeks. Unlike ordinary berry over-ripening, the 2 weeks of berry dehydration subsequent to cane-cut induced (compared to berries harvested at technological maturity) an increase in total aroma content per berry weight of about 17% (the total concentration of almost all classes of aroma increased, except the C6 aliphatic alcohols) (Figure 5.5), but a decrease in aroma per single berry of about 27% (around 30% of free fraction and 25% of glycosylated fraction). The 27% decrease results from a reduction in concentration in all the aroma categories. In detail, as concerns the monoterpenes, the concentration of free and glycosylated fractions of geraniol and RCs expressed per berry, decreased by about 30% and 25%, respectively. All free and glycosylated compounds decreased in concentration except for the free fraction of citronellol, which increased by about 80%.

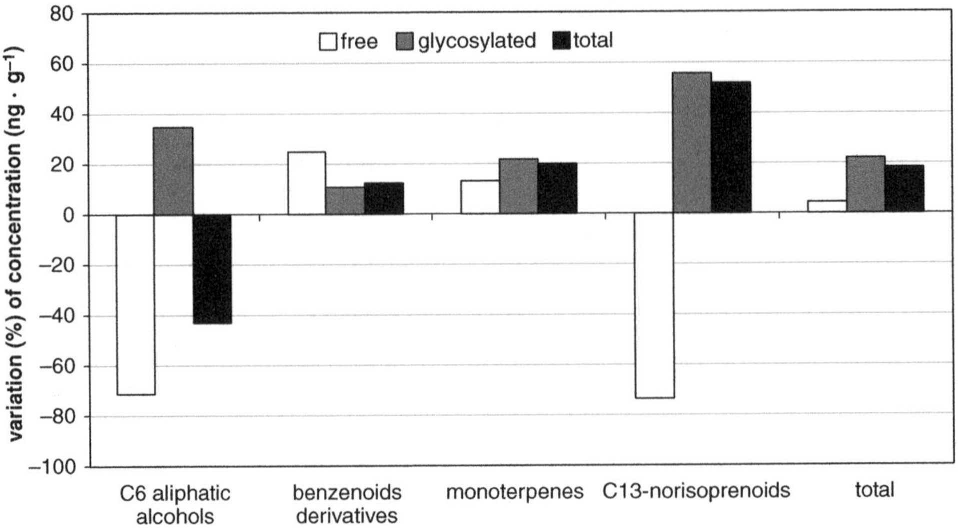

Figure 5.5 Variation (compared to berries harvested at technological maturity) in the concentration (referred to berry weight) of classes of aroma in Aleatico grapes induced by an over-ripening of 2 weeks with 'cane-cut' in a vineyard in the Alberese area, Grosseto province, Tuscany south coast (D'Onofrio *et al.*, unpublished).

As concerns C6 aliphatic alcohols, unlike the Aleatico of the Val di Cornia where the free fraction greatly increased (about +60%) and the glycosylated fractions decreased (about −10%), in the over-ripened Aleatico grape from the Alberese vineyard, with or without cane-cut, there was a reduction in total content of up to 70% of both free and glycosylated fractions. In Cabernet-Sauvignon, C6 volatile compounds changed from acetate esters to aldehydes and finally to alcohols during ripening, with a reduction of berry herbaceous characters; in addition, during fermentation, the alcohols could be converted into esters with desirable fruit characters (Kalua and Boss, 2009). Consequently, the reduction of C6 aliphatic alcohols concentration in the over-ripened Aleatico grape cultivated at Alberese, as a result of their conversion into esters, has a positive impact on wine aroma. A reduction of C6 aliphatic alcohols concentration was also observed during open-air and tunnel dehydration of Malvasia, Trebbiano and Sangiovese grapes (Bellincontro *et al.*, 2004), and in dehydrated Pedro Ximénez grapes (Ruiz *et al.*, 2010). Moreover, both late vintage and shoot thinning reduce the concentration of C6 compounds in grape (Sun *et al.*, 2011).

The concentration of benzenoids and C_{13}-norisoprenoids, responsible for agreeable aromas, in over-ripened Aleatico grapes decreased but a lesser reduction occurred in over-ripened Aleatico grapes with cane-cut.

5.4 CHANGES OF PRIMARY AND SECONDARY AROMAS DURING POSTHARVEST DEHYDRATION

The dehydration in tunnel with control of temperature (12–16°C; mean of 12°C) and RH (50–70%; mean of 60%), of Aleatico grapes in the Val di Cornia area (in the same vineyard described earlier in the over-ripening Aleatico experiments), produced a 31% reduction in

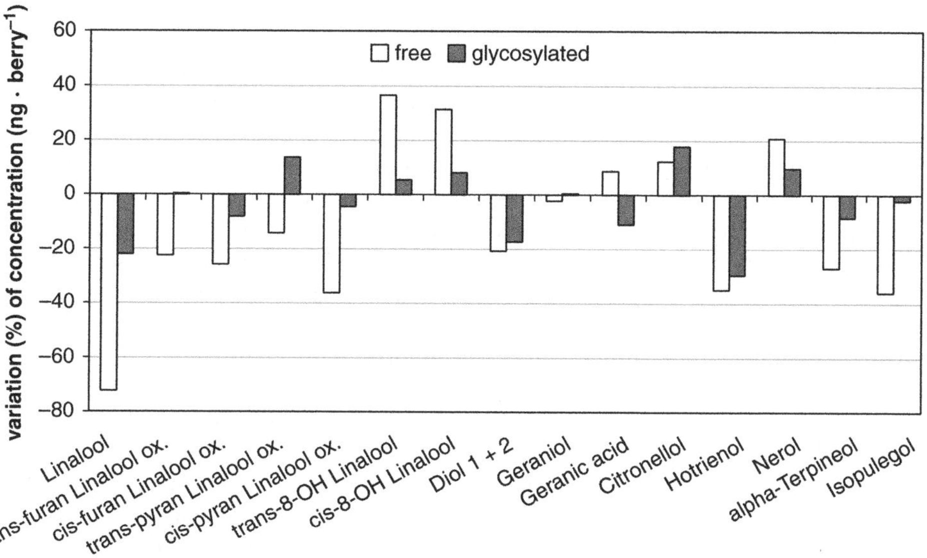

Figure 5.6 Variation (compared to berries harvested at technological maturity) in the concentration of monoterpenes induced by tunnel controlled dehydration in Aleatico grapes from a vineyard in the Val di Cornia area, Livorno province, Tuscany central coast (D'Onofrio *et al.*, unpublished).

berry weight within 2 weeks and induced an increase of aroma per berry weight of about 35% (45% of free fraction and 33% of glycosylated fraction, respectively), while showing a slight decrease of about 5% in aroma concentration per single berry (data not shown). With regard to monoterpenes content expressed per single berry, there was a decrease of about 4%. In particular, as concerns geraniol and RCs, which represented 90% of monoterpenes, the free fractions slightly increased (about 3%) whilst the glycosylated ones decreased (about 6%) (Figure 5.6). Both the free and glycosylated fractions of nerol and RCs increased (about 8% in total) while the free fraction of linalool and RCs strongly decreased by about 15%.

The dehydration in open air and in shade (temperature 16–37°C, mean of 22°C; RH 30–65%, mean of 52%) of the same Aleatico grape in the Val di Cornia, produced a reduction of berry weight of 26% in 1 week, and induced a reduction of aroma for both berry weight (15%) and single berry (35%) of around 28% for C6 aliphatic alcohols, 8% for benzenoids and RCs, 40% for monoterpenes, and 15% for C_{13}-norisoprenoids per single berry. So, this reduction was mainly due to the reduction of the concentration of monoterpenes. The diminution in the concentration of monoterpenes during dehydration in open air could be caused by the high uncontrolled temperature as previously observed during the dehydration of the Aleatico grape in tunnel at a different temperature. Chkaiban *et al.* (2007) reported the negative effect of dehydration on the aroma metabolism when the dehydration condition is uncontrolled. At a lower temperature, the concentration of terpenols increased slowly and the amount was maintained until 40% of berry weight loss, while with the increase of temperature, the terpenols accumulated quickly but then decreased to a lower level at the same intensity (40% of weight loss) of dehydration (Cirilli *et al.*, 2012). Terpenols were also more abundant in freeze-dried berry skin of Carménère and Cabernet-Sauvignon than in oven-dried skin (de Torres *et al.*, 2010), in Montepulciano grapes dehydrated at 10°C than in the same grapes dehydrated at 20°C (Santonico *et al.*, 2010), and also in wines of Malvasia delle Lipari obtained with grapes dehydrated in the shade rather than in the sun (Piombino *et al.*, 2010).

In tunnel at a constant temperature of 21°C with RH of 42% and an air speed of 1–1.5 m s^{-1}, the postharvest dehydration of Malvasia and Trebbiano until a berry weight loss of 50% and 34%, respectively, revealed an increase in the concentration of ethanol, esters and high alcohols, all secondary volatile compounds produced by fermentative processes (Bellincontro *et al.*, 2004). In detail, both ethyl acetate and acetic acid, almost absent at the beginning of dehydration, increased rapidly during the first 2 weeks of dehydration, particularly in open-air dehydrated grapes and in the Trebbiano variety; these compounds are indicators of an advanced fermentative process due to the oxidation of ethanol and are usually very high in Passito and Vinsanto wines. Also in Sangiovese grapes these volatiles increased during both open-air and tunnel dehydration (Bellincontro *et al.*, 2004) due to an induction of metabolic processes.

In more recent experiments of controlled dehydration of Malvasia grapes, the concentration of ethanol and acetaldehyde increased with the increase of alcohol dehydrogenase activity, and then decreased with the increase of ethyl acetate, showing a clear biphasic pattern: a first metabolic stress response of up to 11.7% of bunch weight loss and a second stress response above 19.5% of weight loss (Costantini *et al.*, 2006). During the dehydration of the Gewürztraminer grape the ethyl acetate slowly increased with a fluctuating pattern of change (double peak) due to the alternate activity of alcohol dehydrogenase in relation to the process of oxidation of ethanol to acetaldehyde and acetic acid (Chkaiban *et al.*, 2007).

A different evolution pattern of acetaldehyde, ethanol, acetic acid and ethyl acetate as a function of temperature and dehydration ratios have been observed in the Aleatico grape during tunnel dehydration; the temperature influenced the expression and activity of alcohol dehydrogenase and, consequently, the concentration of ethanol and oxidized RCs (Cirilli *et al.*, 2012). In detail, the low temperature (10°C) reduces the oxidation of volatile compounds compared with dehydration at higher temperatures (20°C and 30°C). At a low temperature, the alcohol dehydrogenase activity (ethanol to acetaldehyde direction) started to increase from 20% of berry weight loss and continued to rise even with a 40% weight loss, meaning that ethanol oxidation was delayed, and acetaldehyde increased rapidly. While at a high temperature (30°C), the alcohol dehydrogenase activity increased rapidly but acetaldehyde did not rise because of its oxidation to acetic acid, which increased together with ethyl acetate. At 20°C, the behaviour was intermediate in relation to the other temperatures. Anaerobic metabolites such as ethanol, acetaldehyde and ethyl acetate were also much higher at 20°C than at 10°C during Montepulciano grapes dehydration, where alcohols and esters also prevailed at 20°C while at 10°C larger abundances of aldehydes and terpene alcohols were observed (Santonico *et al.*, 2010). Secondary metabolites of alcoholic fermentation were also affected by bunch light exposure during postharvest dehydration, increasing particularly when grapes were dried in the shade (+67%) rather than in the sun (Piombino *et al.*, 2010).

An increase in volatile acidity was also observed during the dehydration of Pedro Ximénez grapes, used for the production of Pedro Ximénez sweet sherry wines, due to the switch from aerobic to anaerobic metabolism and to increased alcohol dehydrogenase activity (Ruiz *et al.*, 2010). An increase in the production factor was particularly observed (i.e. the ratio between the average concentration of each compound at the end of treatment and its starting concentration) in a group of volatile fermentative compounds in dehydrated Pedro Ximénez grapes, and among these, acetoin was the odorant that showed the highest increase. In addition, the must from Pedro Ximénez grapes dehydrated in tunnel exhibited a higher odour activity value than must from Pedro Ximénez grapes dried in the open air (Ruiz *et al.*, 2010).

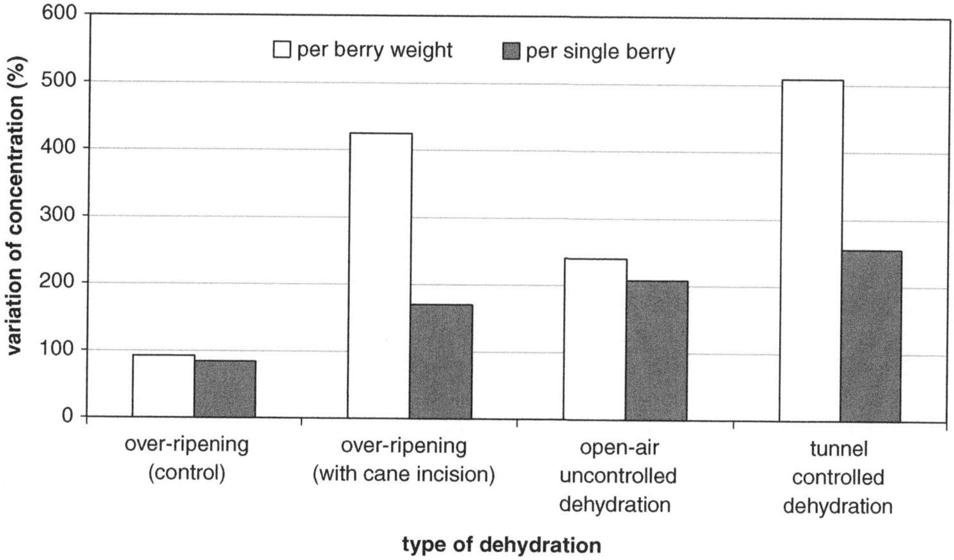

Figure 5.7 Variation (compared to berries harvested at technological maturity) in isoamyl alcohol concentration induced by over-ripening with or without cane incision and in controlled and uncontrolled postharvest grape dehydration conditions in Aleatico grapes from the Tuscan coastal areas (D'Onofrio *et al.*, unpublished).

Grape dehydration also affected the amino acids catabolism and, consequently, it induced an increase in the concentration of methylated higher alcohols in both Malvasia and Sangiovese grapes, especially in those dehydrated in tunnel (Bellincontro *et al.*, 2004). Also, dehydration of Aleatico grapes induced (compared to samples harvested at technological maturity) an increase in the concentration of some higher alcohols, such as isoamyl alcohol, which increased particularly during the over-ripening with cane incision and the controlled dehydration in tunnel (Figure 5.7) (D'Onofrio *et al.*, unpublished).

5.5 CONCLUSIONS

Berry aroma accumulates mainly during the ripening phase with a peak in concentration often preceding the technological ripening (with the exception of some particular classes of aroma, such as methoxypyrazines).

The aroma concentration per single berry, which indicates the effective aroma quantity and is independent from the degree of dehydration, typically decreases during grape over-ripening, but the concentration of aroma per berry weight may decrease, as during postharvest uncontrolled dehydration in open air, or may increase as during grape dehydration with cane-cut or, particularly, during postharvest dehydration with the control of temperature and humidity (tunnel dehydration) where the concentration per single berry of all classes of aroma also increases. In general, during postharvest dehydration the concentration of monoterpenes decreases significantly during the dehydration as the temperature increases.

Another important effect of berry dehydration is the change in the proportions among the aroma compounds and between the free and glycosylated fractions of each one, indicating a possible differential biosynthetic and catabolic activity for each aroma compound. In the Aleatico grape in particular, both on-vine and postharvest dehydration usually induce a strong reduction in the proportion of linalool and RCs and an increase in the proportions of free fractions of geraniol and nerol with their RCs, which alter the concentration factors of each aroma compound and, consequently, change the berry aroma character, improving the typical aroma characteristics of Passito wines.

Others positive effects of berry dehydration could be the reduction of C6 aliphatic alcohols responsible for undesirable green odour and herbaceous-flavoured wines, and the simultaneous increase of concentrations of benzenoids and C_{13}-norisoprenoids responsible for desired aroma, which were obtained by over-ripening with cane-cut.

The grape dehydration also induces an increase in the effective concentration (concentration per single berry) of fermentative and oxidative secondary volatile compounds, such as ethanol, acetic acid, esters and high alcohols, usually at high concentrations in Passito and Vinsanto wines when grapes are dehydrated in uncontrolled conditions.

ACKNOWLEDGEMENTS

Review paper published within the framework of the project PRIN 2008, coordinator F. Mencarelli, supported by the Italian Ministry of Education, University and Research.

The author would like to thank all the staff at the Laboratory of Viticulture and Enology, University of Pisa for their support on Aleatico grape dehydration experiments.

REFERENCES

Baumes, R., J. Wirth, S. Bureau, *et al.* 2002. Biogeneration of C13-norisoprenoid compounds: experiments supportive for an apo-carotenoid patway in grapevines. *Analytica Chimica Acta*, 45:3–14.

Bellincontro, A., D. De Santis, R. Botondi, *et al.* 2004. Different postharvest dehydration rates affect quality characteristics and volatile compounds of Malvasia, Trebbiano and Sangiovese grapes for wine production. *Journal of the Science of Food and Agriculture*, 84:1791–1800.

Boselli, M., R. Bandinell, G. Scalabrelli, *et al.* 2009. Caratterizzazione biogronomica e aromatica del vitigno Aleatico. Accademia Italiana della Vite e del Vino, Tornata all'Isola D'Elba, Portoferraio, 24 October 2003, pp. 31–49.

Chkaiban, L., R. Botondi, A. Bellincontro, *et al.* 2007. Influence of postharvest water stress on lipoxygenase and alcohol dehydrogenase activities, and on the composition of some volatile compounds of Gewürztraminer grapes dehydrated under controlled and uncontrolled thermohygrometric conditions. *Australian Journal of Grape and Wine Research*, 13:142–149.

Cirilli, M., A. Bellincontro, D. De Santi, *et al.* 2012. Temperature and water loss affect ADH activity and gene expression in grape berry during postharvest dehydration. *Food Chemistry*, 132:447–454.

Coombe, B.G. and M.G. McCarthy. 1997. Identification and naming of the inception of aroma development in ripening grape berries. *Australian Journal of Grape and Wine Research*, 3:18–20.

Costantini, V., A. Bellincontro, D. De Santis, *et al.* 2006. Metabolic changes of Malvasia grapes for wine production during postharvest drying. *Journal of Agricultural and Food Chemistry*, 54:3334–3340.

D'Onofrio, C. 2011. Caratterizzazione funzionale della biosintesi degli aromi delle uve durante lo sviluppo dell'acino e controllo della qualità aromatica delle uve. *Italus Hortus*, 18:39–61.

D'Onofrio, C., G. De Lorenzis, and P.K. Boss. 2010a. Molecular characterization of aroma genes in *Vitis vinifera* var. 'Moscato bianco'. Selected papers and abstracts from the 8th International Symposium on Grapevine Physiology and Biotechnology. *Australian Journal of Grape and Wine Research*, 16(S1):A21.

D'Onofrio, C., G. De Lorenzis, and G. Scalabrelli. 2008. Studies of aroma biosynthesis in clusters of 'Sangiovese' grapevine. *Third International Symposium on 'Sangiovese'*, Florence, 3–5 December 2008.

D'Onofrio, C., G. Bezzo, G. De Lorenzis, *et al.* 2012. Effetto della defogliazione sulla componente aromatica delle uve (Effects of leaf removal on grape aroma). *Italus Hortus*, 3:108–114.

D'Onofrio C., F. Matarese, G. Scalabrelli, *et al.* 2010b. Functional characterization of terpene synthases of 'aromatic' and 'non-aromatic' grapevine varieties. *10th International Conference on Grapevine Breeding and Genetics*, Geneva, NY, 1–5 August 2010.

de Torres, C., M.C. Díaz-Marotoa, I. Hermosín-Gutiérrez, *et al.* 2010. Effect of freeze-drying and oven-drying on volatiles and phenolics composition of grape skin. *Analytica Chimica Acta*, 660:177–182.

Fenoll, J., A. Manso, P. Hellín, *et al.* 2009. Changes in the aromatic composition of the *Vitis vinifera* grape Muscat Hamburg during ripening. *Food Chemistry*, 114:420–428.

García, E., J.L. Chacón, J. Martínez, and P.M. Izquierdo. 2003. Changes in volatile compounds during ripening in grapes of Airén, Macabeo and Chardonnay white varieties grown in La Mancha region (Spain). *Food Science and Technology International*, 9:33–39.

Kalua, C.M. and P.K. Boss. 2009. Evolution of volatile compounds during the development of Cabernet Sauvignon grapes (*Vitis vinifera* L.). *Journal of Agricultural and Food Chemistry*, 57:3818–3830.

Kobayashi. H, H. Takase, K. Kaneko, *et al.* 2010. Analysis of S-3-(Hexan-1-ol)-Glutathione and S-3-(Hexan-1-ol)-L-Cysteine in *Vitis vinifera* L. cv. Koshu for aromatic wines. *American Journal of Enology and Viticulture*, 61:176–185.

Piombino, P., A. Genovese, A. Gambuti, *et al.* 2010. Effects of off-vine bunches shading and cryomaceration on free and glycosylated flavours of Malvasia delle Lipari wine. *International Journal of Food Science and Technology*, 45:234–244.

Ruiz, M.J., L. Zea, L. Moyano, and M. Medina. 2010. Aroma active compounds during the drying of grapes cv. Pedro Ximénez destined to the production of sweet Sherry wine. *European Food Research and Technology*, 230:429–435.

Ryona, I., B.S. Pan, D.S. Intrigliolo, *et al.* 2008. Effects of cluster light exposure on 3-isobutyl-2-methxypyrazine accumulation and degradation pattern in red wine grapes (*Vitis vinifera* L. cv. Cabernet Franc). *Journal of Agricultural and Food Chemistry*, 56:10838–10846.

Santonico, M., A. Bellincontro, D. De Santis, *et al.* 2010. Electronic nose to study postharvest dehydration of wine grapes. *Food Chemistry*, 121:789–796.

Scalabrelli G., G. Ferroni, A. Zinnai, *et al.* 2008. Caratteristiche sensoriali di vini ottenuti da vinificazioni sperimentali di uve di Malvasia Istriana. *Rivista di Viticoltura e di Enologia*, 2-3-4::261–280.

Sun, Q., S. Sacks, and J.E. Vanden-Heuvel. 2011. Impact of shoot thinning and harvest date on yield components, fruit composition, and wine quality of Marechal Foch. *American Journal of Enology and Viticulture*, 62:32-41.

Yang, G., Y. Wanga, B. Wua, *et al.* 2011. Volatile compounds evolution of three table grapes with different flavour during and after maturation. *Food Chemistry*, 128:823–830.

6 Changes in Phenolic Compounds

Danilo Corradini and Isabella Nicoletti

National Research Council, Institute of Chemical Methodologies, Montelibretti, Rome, Italy

6.1 INTRODUCTION

For a long time, the dehydration of grapes for the production of raisins or for making either dessert or fortified wines, has been mainly considered as a process for increasing the sugar content of grape berries with little or no attention paid to the variation in other compounds induced by the modification of the cell metabolism accompanying the dehydration process. This aspect, which can be considered of minor importance for the production of raisins, is gaining increasing interest for the postharvest withering of grape berries used in the production of either dessert or fortified wines, which is conducted to improve the must quality, besides increasing the sugar content. Both postharvest dehydration and dry-on-vine processes induce cell wall enzyme activity, variations in respiration and ethylene production, in addition to causing changes in the cell metabolism with consequent variations in the levels of secondary metabolites, including volatile (e.g. terpenes and aldehydes) and phenolic compounds.

Phenolic compounds are essentially ubiquitous in the plant kingdom and serve a number of purposes, including protection from UV radiation, pigmentation (colouration), defence against invading pathogens (anti-fungal properties), nodule protection, and attraction of pollinators and seed dispersers. Although the phenolic compounds are not classed as essential nutrients, they have a wide range of health benefits in the human body. Several studies have shown that a large number of phenolic compounds can fight harmful free radicals, which are dangerous compounds released into cells during oxygen-related reactions leading to ageing processes (Shrikande, 2000; Nassiri-Asl and Hosseinzadeh, 2009; Vauzour *et al.*, 2010; Guilford and Pezzuto, 2011). Other studies have reported evidence that phenolic compounds may exhibit a beneficial effect against cancer, vasorelaxation (Yang *et al.*, 2001; Fresco *et al.*, 2006), and allergy (Singh *et al.*, 2011).

Phenolic compounds are composed of one or more aromatic benzene rings with one or more hydroxyl groups and can be found as free molecules (aglycones), in the form of conjugates (usually as glycosides) and esters, or as oligomers or polymers. Examples of simple aromatic benzene ring phenolic compounds are the phenolic acids, which comprise two distinguishing constitutive carbon frameworks, namely the hydroxycinnamic and hydroxybenzoic structures, having the basic skeleton C6-C3 and C6-C1, respectively (Figure 6.1).

The most numerous phenolic compounds found throughout the plant kingdom are the flavonoids, which are largely planar molecules that possess 15 carbons with two aromatic rings connected by a three-carbon bridge (C6-C3-C6 structure). Multiple combinations of

Sweet, Reinforced and Fortified Wines: Grape Biochemistry, Technology and Vinification, First Edition.
Edited by Fabio Mencarelli and Pietro Tonutti.

Main phenolic compounds found in grapes and wines

Class	Structure	R1	R2	R3	Compound
Phenolic acids (C_6-C_3) (C_6-C_1)	***Hydroxybenzoic acids*** 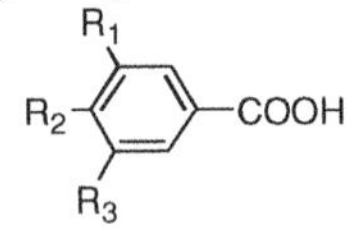	OH	OH	OH	Gallic acid
		H	OH	OH	Protocatechuic acid
		H	OH	OCH_3	Vanillic acid
		H	OH	H	p-Hydroxybenzoic acid
		OCH_3	OH	OCH_3	Syringic acid
	Hydroxycinnamic acid derivatives 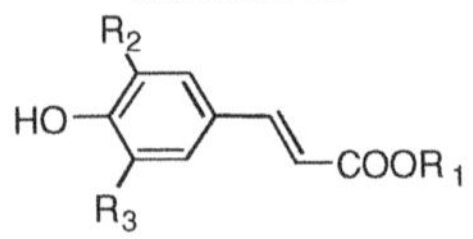	H	H	H	p-Coumaric acid
		H	H	OH	Caffeic Acid
		H	H	OCH_3	Ferulic acid
		H	OCH_3	OCH_3	Sinapic acid
		Tartaric acid	H	OH	Caftaric acid
		Tartaric acid	H	H	Coutaric acid
		Tartaric acid	H	OCH_3	Feftaric acid
Flavonoids ($C_6-C_3-C_6$)	***Flavan-3-ols*** 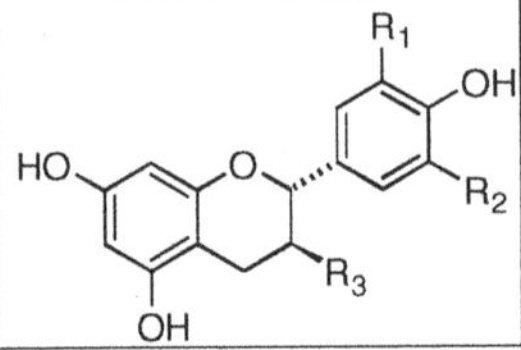	H	OH	OH	(+)-Catechin
		H	OH	OH	(-)-Epicatechin
		OH	OH	OH	(-)-Epigallocatechin
		H	OH	galloyl	(-)-Epicatechin gallate
		OH	OH	galloyl	(-)-Epigallocatechin gallate
	Flavonols 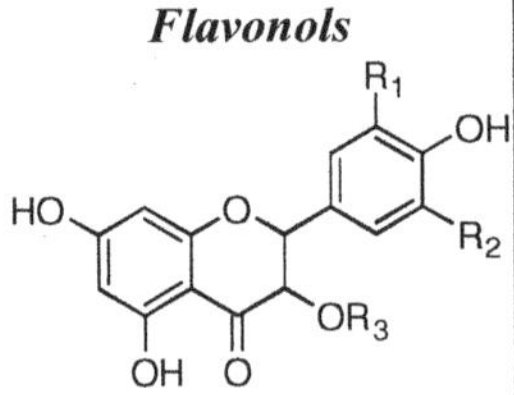	OH	H	H	Quercetin
		H	H	H	Kaempferol
		OH	OH	H	Myricetin
		OCH_3	H	H	Isorhamnetin
		OH	H	glucose	Quercetin-3-*O*-glucoside
		OH	H	glucuronic ac.	Quercetin-3-*O*-glucuronide
		OH	H	rutinose	Quercetin-3-*O*-rutinoside
		H	H	glucose	Kaempferol-3-*O*-glucoside
		H	H	rutinose	Kaempferol-3-*O*-rutinoside
		OH	OH	glucose	Myricetin-3-*O*-glucoside
		OCH_3	H	glucose	Isorhamnetin-3-*O*-glucoside
	Anthocyanins 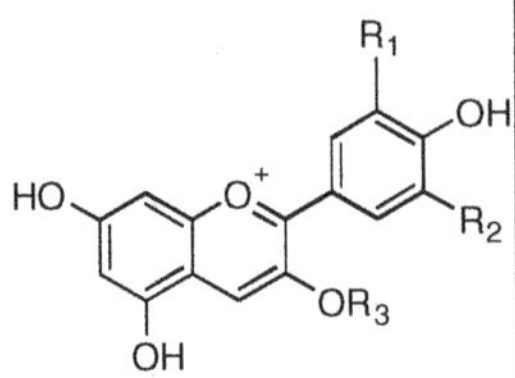	OH	OH	glucose	Delphinidin-3-*O*-glucoside
		OH	H	glucose	Cyanidin-3-*O*-glucoside
		OCH_3	OH	glucose	Petunidin-3-*O*-glucoside
		OCH_3	H	glucose	Peonidin-3-*O*-glucoside
		OCH_3	OCH_3	glucose	Malvidin-3-*O*-glucoside
		OH	OH	acetyl-glucose	Delphinidin-3-*O*-acetilglucoside
		OH	H	acetyl-glucose	Cyanidin-3-*O*-acetilglucoside
		OCH_3	OH	acetyl-glucose	Petunidin-3-*O*-acetilglucoside
		OCH_3	H	acetyl-glucose	Peonidin-3-*O*-acetilglucoside
		OCH_3	OCH_3	acetyl-glucose	Malvidin-3-*O*-acetilglucoside
		OH	OH	coumaroylgluc	Delphinidin-3-*O*-coumaroylglucoside
		OH	H	coumaroylgluc	Cyanidin-3-*O*-coumaroylglucoside
		OCH_3	OH	coumaroylgluc	Petunidin-3-*O*-coumaroylglucoside
		OCH_3	H	coumaroylgluc	Peonidin-3-*O*-coumaroylglucoside
		OCH_3	OCH_3	coumaroylgluc	Malvidin-3-*O*-coumaroylglucoside
		OCH_3	OCH_3	caffeoylgluc	Malvidin-3-*O*-caffeoylglucoside
Stilbenes ($C_6-C_2-C_6$)	***Resveratrol*** 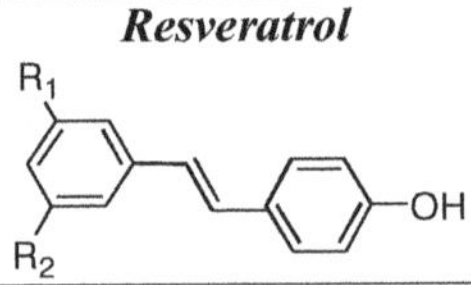	OH	OH		*trans*-Resveratrol
		OH	glucose		*trans*-Piceid

Figure 6.1 Nomenclature and structure of phenolic compounds.

hydroxyl groups, sugars, oxygens, and methyl groups attached to the C6-C3-C6 structure create the various subclasses of flavonoids, which comprise flavonols, flavanones, flavones, flavan-3-ols (catechins), anthocyanins and isoflavones. Polymeric forms and acetylated derivatives of flavonoids are also commonly found. Other non-flavonoid phenolic compounds equally found in grape and wine are stilbenes, which consist of two benzene rings connected by an ethane bridge.

Most of the phenolic compounds found in *Vitis vinifera* L., which include phenolic acids, flavan-3-ols, flavonols, flavanones, proanthocyanidines, anthocyanins and stilbenes, are expected to play important roles in the quality of grape and wines (Downey *et al.*, 2006). Anthocyanins are directly responsible for colour in grapes and young wines, whereas astringency and structure of wines seem to be mainly influenced by catechins and proanthocyanidins, which, as well as flavonols, are also believed to be responsible for their bitterness (Hufnagel and Hofmann, 2008). Occurrence and concentration of the various phenolic compounds in grape berries depend on the variety of grapevine and are influenced by viticultural and environmental factors (Cantos *et al.*, 2002; Ojeda *et al.*, 2002).

Several studies have concluded that flavonols and anthocyanins may be ascribed as molecular markers for the classification and differentiation of grape cultivars and single cultivar wines (Hermosín-Gutiérrez and García-Romero, 2004; Castello-Muñoz *et al.*, 2007). According to these studies, either the flavonols or the anthocyanins profile seem to be closely related to the genetic characteristic of grape. However, other studies have evidenced variations in the occurrence of either anthocyanins or flavonols with seasonal conditions (Ryan and Revilla, 2003), viticultural practice (Esterban *et al.*, 2001), and the postharvest dehydration process of the winegrape (Bellincontro *et al.*, 2009; Mencarelli *et al.*, 2010).

6.2 NOTES ON COMMON METHODS FOR THE ANALYSIS OF PHENOLIC COMPOUNDS IN GRAPE BERRIES, MUST AND WINE

Traditionally, total phenolic compounds are estimated using spectrophotometric methods based on the Folin-Ciocalteu reaction (Singleton and Rossi, 1965), which generate a coloured product as the result of the oxidative titration of phenolate anions by phosphotungstate and phosphomolybdate. However, these methods usually overestimate the content of phenolic compounds due to interference by other non-phenolic compounds, such as ascorbic acid which may also reduce the Folin reagent.

Other spectrophotometric methods are employed to determine the content of a specific class of phenolic compounds, such as the quantification of the total content of flavonoids based on the complexation of these phenolic compounds with Al (III) (Huang *et al.*, 2009) or the total anthocyanins that can be determined by spectroscopic methods based on their characteristic behaviour under acidic conditions. These methods are based on the transformation of all anthocyanins to the red-coloured flavilium cation when the pH of the sample solution is lowered to values ranging between 0.5 and 0.8 (Lee *et al.*, 2005).

Nuclear magnetic resonance spectroscopy is widely employed as an analytical tool to provide fingerprinting or profiling of phenolic compounds and other secondary metabolites. Such approach is usually applied in combination with multivariate statistical methods, such as principal components analysis, and can be used for the classification of wines according to the grape variety and its geographical origin as well as to evaluate the effects of environmental factors and grape vintage on wine quality (Lee *et al.*, 2009). Near-infrared (NIR) and mid-infrared (MIR) spectroscopy also provide specific molecular fingerprints and structural information which make these non-destructive techniques powerful tools for the analysis of biomolecules (Huck *et al.*, 1999). In addition, infrared spectroscopy offers the possibility to measure chemical and physical properties and quality parameters in food crops, such as optimal ripening stage (Bellincontro *et al.*, 2011) and to monitor grape dehydration processes (Bellincontro *et al.*, 2012).

6.2.1 High performance liquid chromatography

The individual identification and quantification of the phenolic compounds occurring in plant extracts is generally carried out by either gas chromatography, liquid chromatography, or capillary electromigration techniques coupled to a suitable detector. Among these approaches, high performance liquid chromatography (HPLC) is the analytical separation method of choice because of its high separation power, versatility, precision and relatively low cost. Most of HPLC methods employ reversed-phase (RP) analytical size columns (4.0–4.6 mm ID), packed with either C-18 or C-8 microparticulate bonded silica. Alternative column formats comprise narrow-bore columns, with an internal diameter of 2.0 mm, which allow efficient separations, sensitive spectrophotometric detection and the use of reduced sample size (Nicoletti *et al.*, 2008). Additional advantages of using a narrow-bore reversed-phase column lie in the values of mobile phase flow rate fully compatible with mass spectrometer detection and in the low consumption of eluent and, consequently, of organic solvents, with beneficial effects on the environment and a drastic reduction in the cost of their purchase and disposal. Also employed are monolithic columns, consisting of a continuous rod of chromatographic support with bimodal porosity, which operate at higher mobile phase flow rates with lower back pressures than conventional columns (Castellari *et al.*, 2002; Vian *et al.*, 2005). Generally, all the columns employed in RP-HPLC are operated under gradient elution mode with the starting eluent and the gradient former consisting of a water-rich and an organic solvent-rich solution, respectively, both containing a buffer or a suitable acid to control the protonic equilibrium at acidic pH value, in order to keep carboxyl and hydroxyl groups of the analysed phenolic compounds in their protonated form.

The primarily used detection method in RP-HPLC of phenolic compounds is based on the absorption of UV or visible light, as in the case of anthocyanins, whereas less popular is the use of fluorescence detection, which is employed for detecting the limited number of phenolic compounds that naturally fluoresce or that are chemically modified to produce molecules containing a fluorescent tag, usually using on-line postcolumn derivatization methods. Also limited is the use of chemiluminescence detection, usually employed in combination with the on-line postcolumn derivatization reaction of the phenolic compounds with terbium (III), which has been recently reported for the detection of hydroxybenzoic acids and catechins (Rodríguez-Díaz *et al.*, 2006).

The hyphenation of HPLC with spectroscopic techniques which, besides the identification, allow the elucidation of the chemical structure of the separated phenolic compounds, such as mass spectrometry (MS) and nuclear magnetic resonance spectroscopy (NMR) has been widely pursued in recent years. Such approaches, combining the separation efficiency of HPLC and the information-rich detection capability of either MS or NMR, are emerging as essential diagnostic tools for the analysis of phenolic compounds, as well as of other low-molecular weight and macromolecular compounds of interest in phytochemistry and plant biology. Several types of ionization sources can be employed for the on-line hyphenation of

HPLC with MS, which include electrospray ionization (ESI), atmospheric pressure chemical ionization (APCI), atmospheric pressure photochemical ionization (APPI), matrix-assisted laser desorption ionization (MALDI), inductively coupled plasma (ICP), and fast atom bombardment (FAB) (Allwood and Goodacre, 2010). At present, the most common approaches are based upon the atmospheric pressure ionization (API) interfaces APCI and ESI, whereas the earliest arrangements based on FAB interfaces, which are affected by several drawbacks, such as unstable electrical current, have been progressively abandoned.

Also promising is the hyphenation of HPLC with nuclear magnetic resonance spectroscopy (NMR), which is widely employed for identification and structural elucidation of organic compounds unattainable by other analytical methods (Andrade *et al.*, 2002; Valls *et al.*, 2009). Direct on-line hyphenation can be realized using flow NMR probes, either of the double saddle Helmholtz coil design or of solenoidal microcoil design (Exarchou *et al.*, 2005; Kühnle *et al.*, 2009), also in combination with methods designed to improve the otherwise low detection sensitivity of NMR (Brkljača and Urban, 2011), which is negatively affected by the background absorption of the liquid phase employed in the separation step.

The hyphenation of HPLC with infrared spectroscopy (IR) is also gaining importance. However, sample detection by infrared IR spectroscopy in HPLC can be negatively affected by the background absorption of the liquid phase employed in the separation step. This disadvantage can be overcome by employing IR spectroscopy as an off-line detection methods (Kuligowski *et al.*, 2010) or employing IR-transparent solvents and more intense IR light sources, such as quantum cascade lasers (Brandstetter *et al.*, 2010).

Several factors may affect the accuracy of the RP-HPLC methods employed for the identification and quantification of the individual phenolic compounds found in grapes of *Vitis vinifera* L., which are a rich source of these compounds either in the monomeric or polymeric form. This plethora of compounds comprises phenolic acids, stilbenes and flavonoids, which cover a large number of subclasses, such as flavonols, flavanols and anthocyanins. Several of these compounds are in the form of esters, while others are glycosylated. Most of the RP-HPLC methods reported in the literature target only a part of the major phenolic compounds that can be found in the different parts of the grape berries (seeds, pulp and skin). Nevertheless, the simultaneous separation, identification and quantification by RP-HPLC of the most abundant phenolic compounds occurring in whole berries of *Vitis vinifera* L. has been reported (Nicoletti *et al.*, 2008). An example of simultaneous RP-HPLC separation and identification of phenolic compounds occurring in whole berries of *Vitis vinifera* variety Aleatico, comprising phenolic acids, flavonols, catechins, stilbenes and anthocyanins, is displayed in Figure 6.2.

Number and quantity of the phenolic compounds identified in grapes are strongly affected by the sample preparation method utilized prior to RP-HPLC analysis and by the proper selection of the HPLC column and experimental conditions employed for the separation, identification and quantification of the target phenolic compounds. Acetonitrile is generally preferred to methanol as the organic modifier because of its lower UV cut-off and viscosity, which makes it more suitable than methanol for low wavelength UV detection and for limiting the back pressure associated with the low permeability of the narrow-bore HPLC columns and of the capillary tube used in the electrospray interface for the hyphenation of HPLC with MS. In addition, HPLC grade acetonitrile generally causes less ghost peaking than methanol when the analysis is performed in gradient elution mode.

The identification of individual phenolic compounds is generally performed on the basis of their spectroscopic and mass spectrometric spectra and comparison of retention times of the separated analytes with those of authentic standards with corresponding molecular mass

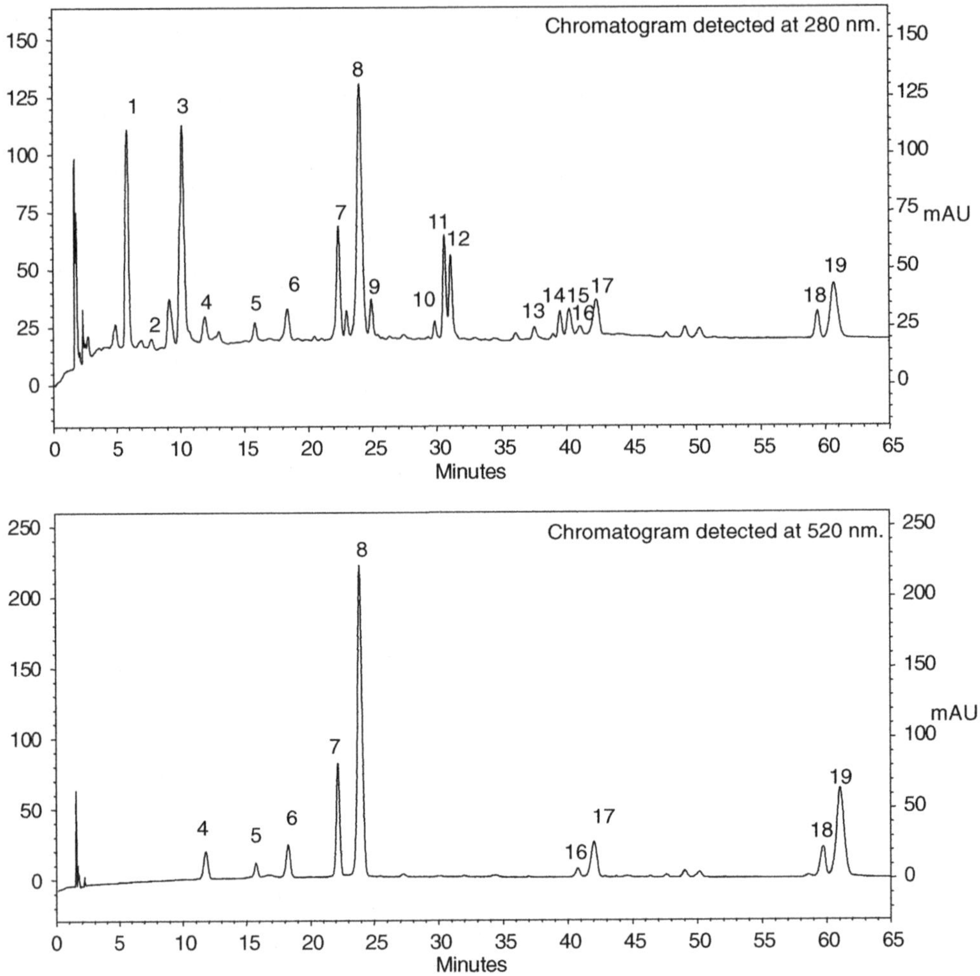

Figure 6.2 RP-HPLC separation of phenolic compounds in Aleatico red grape berries. Identity of the labelled peaks: **1** caftaric acid, **2** catechin, **3** coutaric acid, **4** delphinidin-3-*O*-glucoside, **5** cyanidin-3-*O*-glucoside, **6** petunidin-3-*O*-glucoside, **7** peonidin-3-*O*-glucoside, **8** malvidin-3-*O*-glucoside, **9** *trans*-piceid, **10** quercetin-3-*O*-glucuronide, **11** rutin, **12** quercetin-3-*O*-glucoside, **13** cis-piceid, **14** kaempferol-3-*O*-glucoside, **15** *trans*-resveratrol, **16** peonidin-3-*O*-(6-acetyl)-glucoside, **17** malvidin-3-*O*-(6-acetyl)-glucoside, **18** peonidin-3-*O*-(6-p-coumaroyl)-glucoside, **19** malvidin-3-*O*-(6-p-coumaroyl)-glucoside.

and UV-visible spectra. Libraries comprising retention times, UV-visible and mass spectra for major phenolic compounds expected in grape berry are generally made by subjecting solutions of each phenolic standard to RP-HPLC analysis with the optimized elution gradient and both spectrophotometric and MS detection. The quantification of individual phenolic compounds is usually carried out by the external standard method. However, the unavailability of authentic standards for certain phenolic compounds, and either their high cost or time-consuming purification methods, are the basis of the diffuse practice of using a single external standard for the quantification of more than one phenolic compound within a given class. For example, malvidin-3-glucoside is usually selected as the external standard for the quantification of the different anthocyanin glycosides present in a given sample detected at

520 nm, which is typically close to the wavelength of the most intense visible absorption of the individual anthocyanins.

The values of the wavelength of maximum absorbance and of the molar absorptivity reported in Table 6.1, clearly show that the capability of absorbing visible light is strongly affected by anthocyanin structural variation, including sugar substitution, and by the solvent composition. Thus it appears evident that even the quantification of closely related anthocyanins can be either underestimated or overestimated if it is carried out using the same calibration graph constructed with one standard.

6.2.2 Capillary electromigration techniques

In addition to HPLC, which still occupies the leading position in phenolic compounds analysis, capillary electromigration techniques, employing capillary tubes of typical inner diameter and lengths of 20–100 μm and 20–100 cm, respectively, are gaining increasing interest in this field. These techniques can be performed by a variety of modes, based on different separation mechanisms that can be selected by simply changing the operational conditions, in particular the composition of the electrolyte solution, which may consist of either a continuous or a discontinuous electrolyte system (Corradini, 2010).

The instrumentation employed to perform capillary electromigration techniques is basically the same for all separation modes and essentially consists of a high voltage power supply, a capillary tube and a detector. Typically, the separated sample components are revealed by on-column detection, employing a small part of the capillary as the detection cell where a property of either the analyte, such as UV-visible absorbance, or the solution, such as conductivity, is monitored. A schematic representation of a typical instrument for capillary electromigration techniques is depicted in Figure 6.3.

As in HPLC, the primarily used detectors are variable-wavelength and photodiode array spectrophotometers, whereas fluorescence detectors find narrow applications, due to the limited number of phenolic compounds that naturally fluoresce. Other detection modalities employed in capillary electromigration techniques include electrochemical detection, which is based on conductometric, amperometric, voltametric and potentiometric measurements, and on the hyphenation with spectroscopic techniques, such as MS, NMR and, more recently, NIR and MIR (Kölhed *et al.*, 2003; Corradini, 2010).

Basically, three major types of interface are employed for coupling capillary electromigration systems with MS, which are identified as coaxial sheath-flow (Smith *et al.*, 1988), sheathless (Olivares *et al.*, 1987), and liquid–junction interfaces (Lee *et al.*, 1988, 1989). The most commonly used arrangement is the coaxial sheath interface (Rentel *et al.*, 1999; Mol *et al.*, 2005). Less popular is the on-line hyphenation of capillary electromigration techniques with NMR spectroscopy (Wu *et al.*, 1994a, b).

The electromigration techniques mostly employed for the analysis of phenolic compounds are capillary zone electrophoresis (CZE) and micellar electrokinetic chromatography (MEKC), also referred to as micellar electrokinetic capillary chromatography (MECC).

A review article describing the capillary electromigration methods that are generally employed for the determination of phenolic compounds in plant food matrices has been recently published (Hurtado-Fernández *et al.*, 2010). The review summarizes and discusses a variety of relevant methods which employ capillary electromigration techniques, such as capillary zone electrophoresis and micellar electrokinetic chromatography, for the analysis of phenolic compounds in plant food matrices.

Table 6.1 Molar absorptivity of anthocyanins.

Anthocyanin	Solvent system	$\lambda_{vis\text{-}max}$ (nm)	Molar absorptivity (ε)	Reference
Delphinidin3-*O*-glucoside	1% HCl in methanol	543	29000	Asen *et al.*, 1959
	10% ethanol, pH 1.5	520	23700	Heredia *et al.*, 1998
	Water:acetonitrile: formic acid (83:12:5 v/v)	523	26092	Personal data
	Ethanol:water:HCl (70:30:1v/v), pH 1.26	544	21600	Dell'Agli *et al.*, 2005
Cyanidin3-*O*-glucoside	Aqueous buffer, pH 1.0	510	26900	Jurd and Asen,1966
	0.1N HCl	520	25740	McClure, 1968
	1% HCl in methanol	530	34300	Siegelman and Hendricks, 1958
	10% ethanol, pH 1.5	512	18800	Heredia *et al.*, 1998
	Water:acetonitrile: formic acid (83:12:5 v/v)	518	22685	Personal data
	Ethanol:water:HCl (70:30:1v/v), pH 1.26	534	26758	Dell'Agli *et al.*, 2005
Petunidin3-*O*-glucoside	0.1% HCl in methanol	546	12900	Somers, 1966
	10% ethanol, pH 1.5	520	18900	Heredia *et al.*, 1998
	Ethanol:water:HCl (70:30:1v/v), pH 1.26	543	33518	Dell'Agli *et al.*, 2005
Peonidin3-*O*-glucoside	0.1% HCl in methanol	536	11300	Asen *et al.*, 1959
	10% ethanol, pH 1.5	512	14100	Heredia *et al.*, 1998
	Water:acetonitrile: formic acid (83:12:5 v/v)	518	31073	Personal data
	Ethanol:water:HCl (70:30:1v/v), pH 1.26	531	25803	Dell'Agli *et al.*, 2005
Malvidin3-*O*-glucoside	0.1% HCl in methanol	546	13900	Asen *et al.*, 1959
	0.1% HCl in methanol	538	29500	Koeppen and Basson, 1966
	0.1N HCl	520	28000	Niketic-Aleksic and Hrazdina, 1972
	Methanol, pH 1.0	535	36400	Metivier *et al.*,1980
	10% ethanol, pH 1.5	520	20200	Heredia *et al.*, 1998
	Water:acetonitrile: formic acid (83:12:5 v/v)	526	26401	Personal data
	Ethanol:water:HCl (70:30:1v/v), pH 1.26	542	30433	Dell'Agli *et al.*, 2005

Note: For molar absorptivity calculations, the considered molecular weight comprises the weight of the counterion.

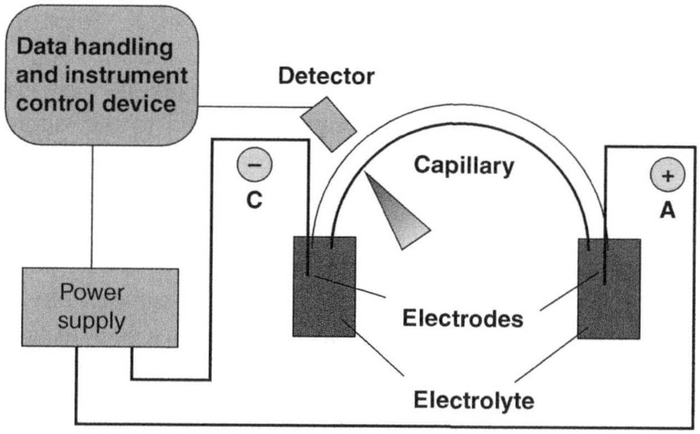

Figure 6.3 Scheme of a basic instrument for capillary electromigration separation techniques.

6.3 SIGNIFICANT VARIATIONS IN THE OCCURRENCE AND CONCENTRATION OF PHENOLIC COMPOUNDS DURING POSTHARVEST DEHYDRATION PROCESSES

The concentration of polyphenols in grape berries under dehydration treatment is strongly affected by the water loss. Thus, it is very important to specify whether the analysis is based on fresh weight (fw) or on dry weight (dw). In the former case, the result is of interest to the enologist because it gives the real situation of polyphenols content in the must, while in the latter, the value indicates whether the amount is the result of concentration or of a change in metabolism.

Increasing the content of phenolic compounds in wine is a common objective of wine producers to improve its quality, and it could be a new challenge to produce wine or grape juice with the characteristic of functional food. Postharvest grape dehydration is used to produce sweet wine in Mediterranean countries, and also dry wine such as Amarone and Sfurzat in northeast Italy. Water deficit and dehydration significantly affects primary metabolism (a shift from aerobic to anaerobic respiration) and secondary metabolisms of grape (synthesis of volatiles compounds and polyphenols) depending on the rate and the quantity of water loss (Costantini *et al.*, 2006; Moreno *et al.*, 2008; Bellincontro *et al.*, 2009).

Several studies have been carried out to investigate the influence of postharvest dehydration processes on the occurrence and concentration of phenolic compounds, either in the grape berries or in the must and wine. Some of these studies have been directed toward monitoring the variation of the total content of phenolic compounds, or their selected classes, in musts and wines produced with grapes at different stages of the postharvest dehydration process (Peinado *et al.*, 2009), as well as in grape berries during postharvest (Bellincontro *et al.*, 2004) and on-vine dehydration processes (Rolle *et al.*, 2009). In these studies the content of total phenolic compounds or of their selected classes is generally determined by either enzymatic or spectrophotometric methods.

More recent investigations have been aimed at studying the variations of specific phenolic compounds as a function of environmental conditions employed for the postharvest dehydration process and of the rate of water loss (Bellincontro *et al.*, 2009; Mencarelli *et al.*, 2010; Peinado *et al.*, 2010; Serratosa *et al.*, 2011; Marquez *et al.*, 2012). Most of these studies have been carried out employing RP-HPLC with either UV-visible spectrophotomety or mass spectrometry as the detection system.

The general enrichment in phenolic compounds during postharvest grape dehydration has been observed both in musts and berries (when referred to the single fruit and/or to fresh weight) and this can be the result of a concentration effect. As reported in Chapter 4, an activation of the phenylpropanoid pathway in dehydrated berries can be hypothesized based on the expression of 'early' genes involved in this pathway (Zamboni *et al.*, 2008; Rizzini *et al.*, 2009; Bonghi *et al.*, 2012), but the effects of withering vary according to the different classes of the considered phenolic compounds. For example, flavonoids seem to be negatively affected (with the exception of specific subcategories such as flavonols), whereas stilbenes are induced. Besides genetic differences and maturity stage of the grape berries, a main role in the management of the postharvest dehydration process under controlled conditions is played by temperature, in addition to relative humidity (RH) and air flow. The interaction of these environmental factors (combined with the duration and the intensity of the dehydration period) may induce different responses in berries in terms of phenol compound metabolism and content. This makes it extremely difficult to provide a general and comprehensive description of the effect of dehydration on the phenolic compounds.

A study carried out to investigate the effect of temperature on the content of phenolic compounds has shown that different temperatures can modulate the metabolism of these compounds in Aleatico grape berries subjected to postharvest dehydration and the consequent concentration in wine (Antelmi *et al.*, 2010). The study has evidenced that 20°C was the most effective temperature to induce phenolic compounds with antioxidant properties, such as catechin, quercetin and *trans*-resveratrol. Depending on the compounds, differences in the concentration were observed conducting the dehydration process at various values of weight loss in the range between 10 and 30%.

Significant variations in the content of phenolic compounds in grape berries from Aleatico vineyard ground without irrigation, and subjected to postharvest dehydration process under controlled conditions, have been also reported (Mencarelli *et al.*, 2010). In this study, the rise in the concentration on a dry weight basis of phenolic acids was higher at 20° than at 10°C. At 20°C and 10% weight loss, the content of stilbenes and catechins increased more than two and three times, respectively, whereas the dehydration process carried out at 10°C and at 30°C caused a lower increment of these phenolic compounds. The study has also evidenced that the concentration of anthocyanins increased by conducting the dehydration process at 10° and 20°C, whereas it diminished at 30°C. At this temperature, with an exception for the content of catechins and stilbenes, the concentration of all other phenolic compounds diminished. Significant decrease of anthocyanins at 20°C has been observed also in Montepulciano winegrape at 40% weight loss compared to 10°C treatment, whereas at 10% weight loss the concentration on a dry weight basis of anthocyanins was more or less similar to that found at harvest (Botondi *et al.*, 2008).

Noteworthy increments (mg kg^{-1} dw) in the content of the flavonols quercetin and kaempferol have been reported for Cesanese red grape berries subjected to the dehydration process under controlled conditions of RH and ventilation at either 10° or 20°C (Bellincontro *et al.*, 2009). The increase in the concentration of the above flavonols was higher at 20° than at 10°C for samples at water loss of 20%. Significant increases (fw basis) in the content of kaempferol and quercetin have been also observed during the sun-drying grape raisining of cv. Pedro Ximénez to obtain sweet wines (Serratosa *et al.*, 2008).

Under the conditions described by Bellincontro *et al.* (2009) a slight decrease in the content of myricetin was observed. The amount of the stilbene *trans*-resveratrol remained almost constant at 10°C whereas increased significantly at 20°C, whereas the amount of *trans*-piceid decreased drastically at 10°C and significantly increased at 20°C. Also observed was a slight decrease in the content of anthocyanins, which were higher at harvest and diminished during the dehydration process, either at 10° or at 20°C. This observation is dissimilar in respect to the results of a study conducted with Pinot Noir grape berries, where the content of anthocyanins remained unchanged during the dehydration process (Moreno *et al.*, 2008). However, besides the different grape cultivar employed for the two studies, data reported by Moreno *et al.* were related to a lower water loss (14–16%) than that reported by Bellincontro *et al.* (20% water loss) and a different temperature (22°C instead of 20°C).

Also investigated were the variations of occurrence and content of the phenolic compounds in red grapes of the Merlot and Tempranillo varieties during the postharvest drying process under controlled temperature and RH conditions in relation to antioxidant activity (Serratosa *et al.*, 2011). The observed increase in the content of phenolic compounds was closely related to the hydrophilic antioxidant activity, which also increased during the dehydration process.

Besides the cases reported above, other phenolic compounds have been shown to diminish their content in grape berries as a consequence of the postharvest dehydration process. Significant decrements on a dw basis in the amount of caftaric acid and catechin have been observed in Cesanese red grape berries dehydrated at 10°C (Bellincontro *et al.*, 2009). The decrements were lower in the samples dehydrated at 20°C.

The variations in the content of phenolic compounds observed during the postharvest dehydration processes of grape berries have been linked by several authors to the balance between the synthesis and oxidation processes that take place during dehydration, and might be related to temperature and rate of water loss (Serratosa *et al.*, 2008; Mencarelli *et al.*, 2010). From the data reported above, it can be inferred that the biosynthetic processes occur rapidly at high temperature (i.e. above 20°C), especially for defence phenolic compounds such as stilbenes, but immediately decline due to the physical alteration of the cells and rapid enzymatic and non-enzymatic oxidation leading to cell death, which also depends on water loss. It has been concluded (Mencarelli *et al.*, 2010) that conducting the postharvest dehydration process at lower temperatures, such as 10°C, reduces and delays the stress effect, as can be inferred from the generally lower amounts of phenolic compounds found in these samples, which might indicate a lower rate of their synthesis. On the other hand, a higher temperature (above 20°C) seems to promote the rapid response of the cells to synthesize known stress compounds such as *trans*-resveratrol or phenolic acids. Thus, 20°C appears to be the temperature that results in the most complete response in terms of an increase in phenolic compounds, especially in those having antioxidant activity, such as stilbenes, catechins and quercetin, but only for values of water loss that are not critical for cell metabolism.

It can be concluded that the large interest in phenolic compounds and the variety of studies conducted to investigate the influence of postharvest dehydration processes on their metabolism together open up good perspectives for the production of wines with high levels of nutraceutical compounds.

ACKNOWLEDGEMENT

Review paper published in the framework of the project PRIN 2008, coordinator F. Mencarelli, supported by the Italian Ministry of Education, University and Research.

REFERENCES

Allwood, J.W. and R. Goodacre. 2010. An introduction to liquid chromatography–mass spectrometry instrumentation applied in plant metabolomic analyses. *Phytochemical Analysis*, 21:33–47.

Andrade, F.D.P., L.C. Santos, M. Datchler, *et al.* 2002. Use of on-line liquid chromatography–nuclear magnetic resonance spectroscopy for the rapid investigation of flavonoids from *Sorocea bomplandii*. *Journal of Chromatography A*, 953:287–291.

Antelmi, G., A. Bellincontro, F. Mencarelli, *et al.* 2010. How dehydration temperature and weight loss affect the biosynthesis of nutritional compounds in irrigated Aleatico grape. *Acta Horticulturae*, 877:693–698.

Asen, S., N.W. Stuart, and H.W. Siegelman. 1959. Effect of various concentrations of nitrogen, phosphorus and potassium on sepal color of *Hydrangea macrophylla*. *American Society for Horticultural Science*, 73:495–502.

Bellincontro, A., D. Cozzolino, and F. Mencarelli. 2011. Application of NIR-AOTF spectroscopy to monitor Aleatico grape dehydration for Passito wine production. *American Journal of Enology and Viticulture*, 62:256–260.

Bellincontro, A., D. De Santis, R. Botondi, *et al.* 2004. Different postharvest dehydration rates affect quality characteristics and volatile compounds of Malvasia, Trebbiano and Sangiovese grapes for wine production. *Journal of the Science of Food and Agriculture*, 84:1791–1800.

Bellincontro, A., I. Nicoletti, M. Valentini, *et al.* 2009. Integration of nondestructive techniques with destructive analyses to study postharvest water stress of winegrapes. *American Journal of Enology and Viticulture*, 60:57–65.

Bellincontro, A., A. Taticchi, M. Servili, *et al.* 2012. Feasible application of a portable NIR-AOTF tool for the on-field prediction of phenolic compounds during the ripening of olives for oil production. *Journal of Agricultural and Food Chemistry*, 60:2665–2673.

Bonghi, C., F.M. Rizzini, A. Gambuti, *et al.* 2012. Phenol compound metabolism and gene expression in the skin of wine grape (*Vitis vinifera* L.) berries subjected to partial postharvest dehydration. *Postharvest Biology and Technology*, 67:102–109.

Botondi, R., M.T. Frangipane, D. Tiberi, *et al.* 2008. Influenza della temperatura di appassimento sulla qualità delle uve varietà Montepulciano. *L'Enologo*, 9:87–93.

Brandstetter, M., A. Genner, K. Anic, *et al.* 2010. Tunable external cavity quantum cascade laser for the simultaneous determination of glucose and lactate in aqueous phase. *Analyst*, 135:3260–3265.

Brkljača, R. and S. Urban. 2011. Recent advancements in HPLC-NMR and applications for natural product profiling and identification. *Journal of Liquid Chromatograghy & Related Technologies*, 34:1063–1076.

Cantos, E., J.C. Espín, and F.A. Tomás-Barberán. 2002. Varietal differences among the polyphenol profiles of seven table grape cultivars studied by LC-DAD-MS-MS. *Journal of Agricultural and Food Chemistry*, 50:5691–5696.

Castellari, M., E. Sartini, A. Fabiani, *et al.* 2002. Analysis of wine phenolics by high-performance liquid chromatography using a monolithic type column. *Journal of Chromatography A*, 973:221–227.

Castello-Muñoz, N., S. Gómez-Alonso, E. García-Romero, *et al.* 2007. Flavonol profile of *Vitis vinifera* red grapes and their single-cultivar wines. *Journal of Agricultural and Food Chemistry*, 55:992–1002.

Corradini, D. 2010. Capillary electromigration techniques. In *Handbook of HPLC*, 2nd edn (ed. D. Corradini), Taylor & Francis/CRC Press, USA, pp. 155–206.

Costantini, V., A. Bellincontro, D. De Santis, *et al.* 2006. Metabolic changes of Malvasia grapes for wine production during postharvest drying. *Journal of Agricultural and Food Chemistry*, 54:3334–3340.

Dell'Agli, M., G.V. Galli, U. Vrhovsek, *et al.* 2005. In vitro inhibition of human cGMP-specific phosphodiesterase-5 by polyphenols from red grapes. *Journal of Agricultural and Food Chemistry*, 53:1960–1965.

Downey, M.O., N.K. Dokoozlian, and M.P. Krstic. 2006. Cultural practice and environmental impacts on the flavonoid composition of grapes and wine: A review of recent research. *American Journal of Enology and Viticulture*, 57:257–268.

Esterban, M.A., M.J. Villanueva, and J.R. Lissarrague. 2001. Effect of irrigation on changes in the anthocyanin composition of the skin of cv. Tempranillo (*Vitis vinifera* L.) grape berries during ripening. *Journal of the Science of Food and Agriculture*, 81:409–420.

Exarchou, V., M. Krucker, T.A. van Beek, *et al.* 2005. LC–NMR coupling technology: recent advancements and applications in natural products analysis. *Magnetic Resonance in Chemistry*, 43:681–687.

Fresco, P., F. Borges, C. Diniz, *et al.* 2006. New insights on the anticancer properties of dietary polyphenols. *Medicinal Research Reviews*, 26:747–766.

Guilford, J.M. and J.M. Pezzuto. 2011. Wine and health: A review. *American Journal of Enology and Viticulture*, 62:471–486.

Heredia, F.J., E.M. Francia-Aricha, J.C. Rivas-Gonzalo, *et al.* 1998. Chromatic characterization of anthocyanins from red grapes. I. pH effect. *Food Chemistry*, 63:491–498.

Hermosín-Gutiérrez, I. and E. García-Romero. 2004. Anthocyanins of red wine grape cultivars grown in the Spanish region of La Mancha: characteristic cultivar patterns of grape and single-cultivar wines, and evolution during the ripening of the berry. *Alimentaria*, 41:127–139.

Huang, W., A. Xue, H. Niu, *et al.* 2009. Optimised ultrasonic-assisted extraction of flavonoids from *Folium eucommiae* and evaluation of antioxidant activity in multi-test systems in vitro. *Food Chemistry*, 114:1147–1154.

Huck, C.W., R. Maurer, M. Popp, *et al.* 1999. Quantitative Fourier transform near infrared reflectance spectroscopy (NIRS) compared to high performance liquid chromatography (HPLC) of a flavone in *Primulae veris Flos* extracts. *Pharmaceutical and Pharmacological Letters*, 9: 26–29.

Hufnagel, J.C. and T. Hofmann. 2008. Orosensory-directed identification of astringent mouthfeel and bitter-tasting compounds in red wine. *Journal of Agricultural and Food Chemistry*, 56:1376–1386.

Hurtado-Fernández E., M. Gomez-Rómero, A. Carrasco-Pancorbo, *et al.* 2010. Application and potential of capillary electroseparation methods to determine antioxidant phenolic compounds from plant food material. *Journal of Pharmaceutical and Biomedical Analysis*, 53:1130–1160.

Jurd, L. and S. Asen. 1966. The formation of metal and 'co-pigment' complexes of cyanidin 3-glucoside. *Phytochemistry*, 5:1263–1271.

Koeppen, B.H. and D.S. Basson. 1966. The anthocyanin pigments of Barlinka grapes. *Phytochemistry*, 5:183–187.

Kölhed, M., P. Hinsmann, B. Lendl, *et al.* 2003. Micellar electrokinetic chromatography with on-line Fourier transform infrared detection. *Electrophoresis*, 24:687–692.

Kühnle, M., K. Holtin, and K. Albert. 2009. Capillary NMR detection in separation science. *Journal of Separation Science*, 32:719–726.

Kuligowski, J., G. Quintás, M. de la Guardia, *et al.* 2010. Analytical potential of mid-infrared detection in capillary electrophoresis and liquid chromatography. A review. *Analytica Chimica Acta*, 679: 31–42.

Lee, E.D., W. Mück, and J.D. Henion. 1988. On-line capillary zone electrophoresis-ion spray tandem mass spectrometry for the determination of dynorphins. *Journal of Chromatography*, 458:313–321.

Lee, E.D., W. Muck, J.D. Henion, *et al.* 1989. Liquid junction coupling for capillary zone electrophoresis/ion spray mass spectrometry. *Biological Mass Spectrometry*, 18:844–850.

Lee, J., R.W. Durst, and R.E. Wrolstad. 2005. Determination of total monomeric anthocyanin pigment content of fruit juices, beverages, natural colorants, and wines by the pH differential method: Collaborative study. *Journal of AOAC International*, 88:1269–1278.

Lee, J.E., G.S. Hwang, F. Van Den Berg, *et al.* 2009. Evidence of vintage effects on grape wines using 1H NMR-based metabolomic study. *Analytica Chimica Acta*, 648:71–76.

Marquez, A., M.P. Serratosa, A. Lopez-Toledano, *et al.* 2012. Colour and phenolic compounds in sweet red wines from Merlot and Tempranillo grapes chamber-dried under controlled conditions. *Food Chemistry*, 130:111–120.

McClure, J.W. 1968. Photocontrol of *Spirodela intermedia* flavonoids. *Plant Physiology*, 43:193–200.

Mencarelli, F., A. Bellincontro, I. Nicoletti, *et al.* 2010. Chemical and biochemical change of healthy phenolic fractions in winegrape by means of postharvest dehydration. *Journal of Agricultural and Food Chemistry*, 58:7557–7564.

Metivier, R.P., F.J. Francis, and F.M. Clydesdale. 1980. Solvent extraction of anthocyanins from wine pomace. *Journal of Food Science*, 45:1099–1100.

Mol, R., G.J. de Jong, and G.W. Somsen. 2005. On-line capillary electrophoresis-mass spectrometry using dopant-assisted atmospheric pressure photoionization: Setup and system performance. *Electrophoresis*, 26:146–154.

Moreno J.J., F. Cerpa-Calderón, S.D. Cohen, *et al.* 2008. Effect of postharvest dehydration on the composition of Pinot noir grapes (*Vitis vinifera* L.) and wine. *Food Chemistry*, 109(4):755–762.

Nassiri-Asl, M. and H. Hosseinzadeh. 2009. Review of the pharmacological effects of *Vitis vinifera* (grape) and its bioactive compounds. *Phytotherapy Research*, 23:1197–1204.

Nicoletti, I., C. Bello, A. De Rossi, *et al.* 2008. Identification and quantification of phenolic compounds in grapes by HPLC-PDA-ESI-MS on a semimicro separation scale. *Journal of Agricultural and Food Chemistry*, 56:8801–8808.

Niketic-Aleksic, G.K. and G. Hrazdina. 1972. Quantitative analysis of the anthocyanin content in grape juices and wines. *Lebensmittel-Wissenschaft und-Technologie*, 5:163–165.

Ojeda, H., C. Andary, E. Kraeva, *et al.* 2002. Influence of pre- and postveraison water deficit on synthesis and concentration of skin phenolic compounds during berry growth of *Vitis vinifera* cv. Shiraz. *American Journal of Enology and Viticulture*, 53:261–267.

Olivares, J.A., N.T. Nguyen, C.R. Yonker, *et al.* 1987. On-line mass spectrometric detection for capillary zone electrophoresis. *Analytical Chemistry*, 59:1230–1232.

Peinado, J., N. López de Lerma, and R.A. Peinado. 2010. Synergistic antioxidant interaction between sugars and phenolics from a sweet wine. *European Food Research and Technology*, 231:363–370.

Peinado, J., N. López de Lerma, J. Moreno, *et al.* 2009. Antioxidant activity of different phenolics fractions isolated in must from Pedro Ximénez grapes at different stages of the off-vine drying process. *Food Chemistry*, 114:1050–1055.

Rentel, C., P. Gfrorer, and F. Baye. 1999. Coupling of capillary electrochromatography to coordination ion spray mass spectrometry, a novel detection method. *Electrophoresis*, 20:2329–2336.

Rizzini, F.M., C. Bonghi, and P. Tonutti. 2009. Postharvest water loss induces marked changes in transcript profiling in skins of wine grape berries. *Postharvest Biology and Technology*, 52:247–253.

Rodríguez-Díaz, R.C., M.P. Aguilar-Caballos, and A. Gómez-Hens. 2006. Determination of some hydroxybenzoic acids and catechins in white wine samples by liquid chromatography with luminescence detection. *Journal of Separation Science*, 29:2772–2779.

Rolle, L., F. Torchio, S. Giacosa, *et al.* 2009. Modifications of mechanical characteristics and phenolic composition in berry skins and seeds of Mondeuse winegrapes throughout the on-vine drying process. *Journal of the Science of Food and Agriculture*, 89:1973–1980.

Ryan, J.M. and E. Revilla. 2003. Anthocyanin composition of Cabernet Sauvignon and Tempranillo grapes at different stages of ripening. *Journal of Agricultural and Food Chemistry*, 51:3372–3378.

Serratosa, M.P., A. Lopez-Toledano, J. Merida, *et al.* 2008. Changes in color and phenolic compounds during the raisining of grape cv. Pedro Ximénez. *Journal of Agricultural and Food Chemistry*, 56:2810–2816.

Serratosa, M.P., A. Marquez, A. Lopez-Toledano, *et al.* 2011. Changes in hydrophilic and lipophilic antioxidant activity in relation to their phenolic composition during the chamber drying of red grapes at a controlled temperature. *Journal of Agricultural and Food Chemistry*, 59:1882–1892.

Shrikhande, A.J. 2000. Wine byproducts with health benefits. *Food Research International*, 33:469–474.

Siegelman, H.W. and S.B. Hendricks. 1958. Photocontrol of alcohol, aldehyde and anthocyanin production in apple skin. *Plant Physiology*, 33:409–413.

Singh, A., S. Holvoet, and A. Mercenier. 2011. Dietary polyphenols in the prevention and treatment of allergic diseases. *Clinical & Experimental Allergy*, 41:1346–1359.

Singleton, V.L. and J.A. Rossi Jr. 1965. Colorimetry of total phenolics with phosphomolybdic-phosphotungstic acid reagents. *American Journal of Enology and Viticulture*, 16:144–158.

Smith, R.D., J.A. Olivares, N.T. Nguyen, *et al.* 1988. Capillary zone electrophoresis mass spectrometry using an electrospray ionization interface. *Analytical Chemistry*, 60:436–441.

Somers, T.C. 1966. Grape phenolics: the anthocyanins of *Vitis vinifera*, var. Shiraz. *Journal of the Science of Food and Agriculture*, 17:215–219.

Valls, J., S. Millán, M.P. Martí, *et al.* 2009. Advanced separation methods of food anthocyanins, isoflavones and flavanols. *Journal of Chromatography A*, 1216:7143–7172.

Vauzour, D., A. Rodriguez-Mateos, G. Corona, *et al.* 2010. Polyphenols and human health: prevention of disease and mechanisms of action. *Nutrients*, 2:1106–1131.

Vian M.A., V. Tomao, S. Gallet, *et al.* 2005. Simple and rapid method for *cis*- and *trans*-resveratrol and piceid isomers determination in wine by high-performance liquid chromatography using Chromolith columns. *Journal of Chromatography A*, 1085:224–229.

Wu, N., T.L. Peck, A.G. Webb, *et al.* 1994a. 1H-NMR spectroscopy on the nanoliter scale for static and on-line measurements. *Analytical Chemistry*, 66:3849–3857.

Wu, N., T.L. Peck, A.G. Webb, *et al.* 1994b. Nanoliter volume sample cells for ^{1}H-NMR: application to on-line detection in capillary electrophoresis. *Journal of the American Chemical Society*, 116:7929–7930.

Yang, C.S., J.M. Landau, M.T. Huang, *et al.* 2001. Inhibition of carcinogenesis by dietary polyphenolic compounds. *Annual Review of Nutrition*, 21:381–406.

Zamboni, A., L. Minoia, A. Ferrarini, *et al.* 2008. Molecular analysis of post-harvest withering in grape by AFLP transcriptional profiling. *Journal of Experimental Botany*, 59(15):4145–4159.

7 Changes in Physical and Mechanical Properties of Dehydrating Berries

Luca Rolle and Vincenzo Gerbi

DIVAPRA – Food Technology Sector, University of Torino, Grugliasco, Torino, Italy

7.1 INTRODUCTION

When dealing with the consumption of fresh fruit, such as table grapes, the fruit's mechanical properties make an important contribution to consumer acceptance, with numerous instrumental measurements correlated to sensory descriptors and consumer preferences, signifying that these may be used as 'predictors' of consumer acceptability (Abbott, 1999). The instrumental assessment of the pulp compactness, crispness and berry skin consistency provides relevant information for customer acceptance of the product, and the firmness of the berry is considered to be an index for freshness (Bernstein and Lustig, 1981; Lang and During, 1990; Mencarelli *et al.*, 1994; Sato *et al.*, 1997; Sato and Yamada, 2003; Rolle *et al.*, 2011c). That is why instrumental texture parameters are also frequently used, not only to investigate the effects of vineyard practices, but also to evaluate postharvest conditions and methods (Sato *et al.*, 2004; Deng *et al.*, 2005; Porro *et al.*, 2010). Although texture studies on winegrapes were begun in the 1980s (Lee and Bourne, 1980), this research field has only recently found new impulses. In wine production, the composition of grapes, whether at the moment of picking, or after a period of withering, is an important determinant of their quality. In particular, the development of high-quality red wines requires an assessment of the grapes' phenolic maturity by determining the phenolic compound concentration and extractability during the winemaking process. In this sense, the need to investigate new parameters that allow for fast evaluation and prediction of the phenol extractability from skin and seeds, while at the same time containing costs, is most likely at the root of the recent trend towards the 'rediscovery' of instrumental texture analysis applied to the study of winegrapes (Le Moigne *et al.*, 2008; Rolle *et al.*, 2008). Several mechanical variables have been demonstrated to be eligible in estimating the anthocyanin extractability with adequate reliability (Río Segade *et al.*, 2008), as well as in evaluating the dehydration kinetics (Rolle *et al.*, 2011a) and assessing the suitability of different varieties for withering on-vine (Rolle *et al.*, 2010).

7.2 INSTRUMENTAL MEASUREMENTS OF WINEGRAPE MECHANICAL PARAMETERS

Grape texture assessment methods are performed under well-defined and controlled laboratory conditions, some of which are guaranteed by the machinery used. This type of

Sweet, Reinforced and Fortified Wines: Grape Biochemistry, Technology and Vinification, First Edition.
Edited by Fabio Mencarelli and Pietro Tonutti.

instrumentation usually goes under the heading of *Universal Testing Machines* (UTMs), or Texture Analysers, and is currently used in the texture tests applied to grape studies. Universal testing machines provide a precise evaluation of force, time, distance and/or deformation (Bourne, 2002).

Texture analysers may be fitted with different probes and fixtures, depending on the requirements dictated by the method(s). Each probe or fixture has been specifically designed for a defined group of applications and can be fitted onto the texture analyser base and/or arm. The samples are placed either on the base of the UTM, on the lower fixture, or between two fixtures. Generally, in the texture tests of table- and winegrapes, the arm of the texture analyser contains a load cell and is moved downwards to penetrate the skin and/or pulp (penetration test), to compress the whole berry and/or seed (compression test), or upwards to assess the pedicel detachment resistance (traction or tension testing). The description of a texture method necessitates knowing a lot of preliminary information, inasmuch as there are many and varied conditions involved in the texture test that may influence the final result. Among the most important and well recognized of these conditions are the effect of the berry temperature, the number of berries, or of their parts (skins, seeds, etc.) per sample, and the position of the berry on the platform. Despite this, detailed experimental conditions are often not included in the literature descriptions. Grape berries are generally analysed at environmental temperature ($+20 \pm 2°C$), and their thermal conditioning in a thermostatically controlled chamber is highly recommended before the testing phase (Rolle *et al.*, 2009a). An accurate and representative grape sampling (c.500 berries) should be carried out in the field (Carbonneau *et al.*, 1991), and for each test, a set sample number of berries should be carefully established so as to guarantee the desired degree of confidence in the results and conclusions. The analysis of too large a sample would require excessive time and resources, while, on the other hand, too small a sample would not guarantee statistically significant results. Generally, the literature places a correct sample size for a single test in the range of 20–50 berries (Maury *et al.*, 2009). The compression and penetration/puncture tests are carried out on the lateral side of the whole berry by placing it on the base of the texture analyser, with the pedicel positioned horizontally (Letaief *et al.*, 2008a). Apart from these preliminary considerations, the operative conditions of a texture test must also be defined. These include the platform and load cell to be used (from 0.5–50 kg), the test speed (expressed in mm s^{-1}, or mm min^{-1}), the probe characteristics, the frequency of acquisition carried out by UMT software (from 100–400 Hz) in the different types of curve (force/time, stress/strain, force/distance, force/deformation, pressure/deformation) and the mechanical properties required, including their correct units expressed as N, mJ, Nmm^{-1}. Traction, penetration and puncture tests are the most common tests applied when studying mechanical property changes, during field and postharvest winegrape dehydration (Rolle *et al.*, 2012a).

Traction (tension) test. This type of texture test is mainly carried out to assess the peduncle detachment resistance. When this test is used on winegrapes, the peduncle is anchored to the probe pliers (Figure 7.1a). The peduncle passes through the specific perforated UTM platform (the hole has a 5 mm diameter) during the traction phase. Whilst the berry is blocked so as to allow for the determination of the force (F_{ped}), which is expressed in N and energy (W_{ped}), expressed in mJ, of the peduncle detachment from the force–time curve (Figure 7.1b) (Rolle *et al.*, 2009a; Deng *et al.*, 2005). The traction is generally performed at 1 mm s^{-1}.

Penetration (puncture) test. Another commonly used method in the texture analysis of wine- and table grapes is the puncture test. Generally, this particular type of compression test is applied to the study of the mechanical characteristics of the grape skin. The plunger geometry (i.e. size and shape) must be taken into careful consideration in this test as it has a

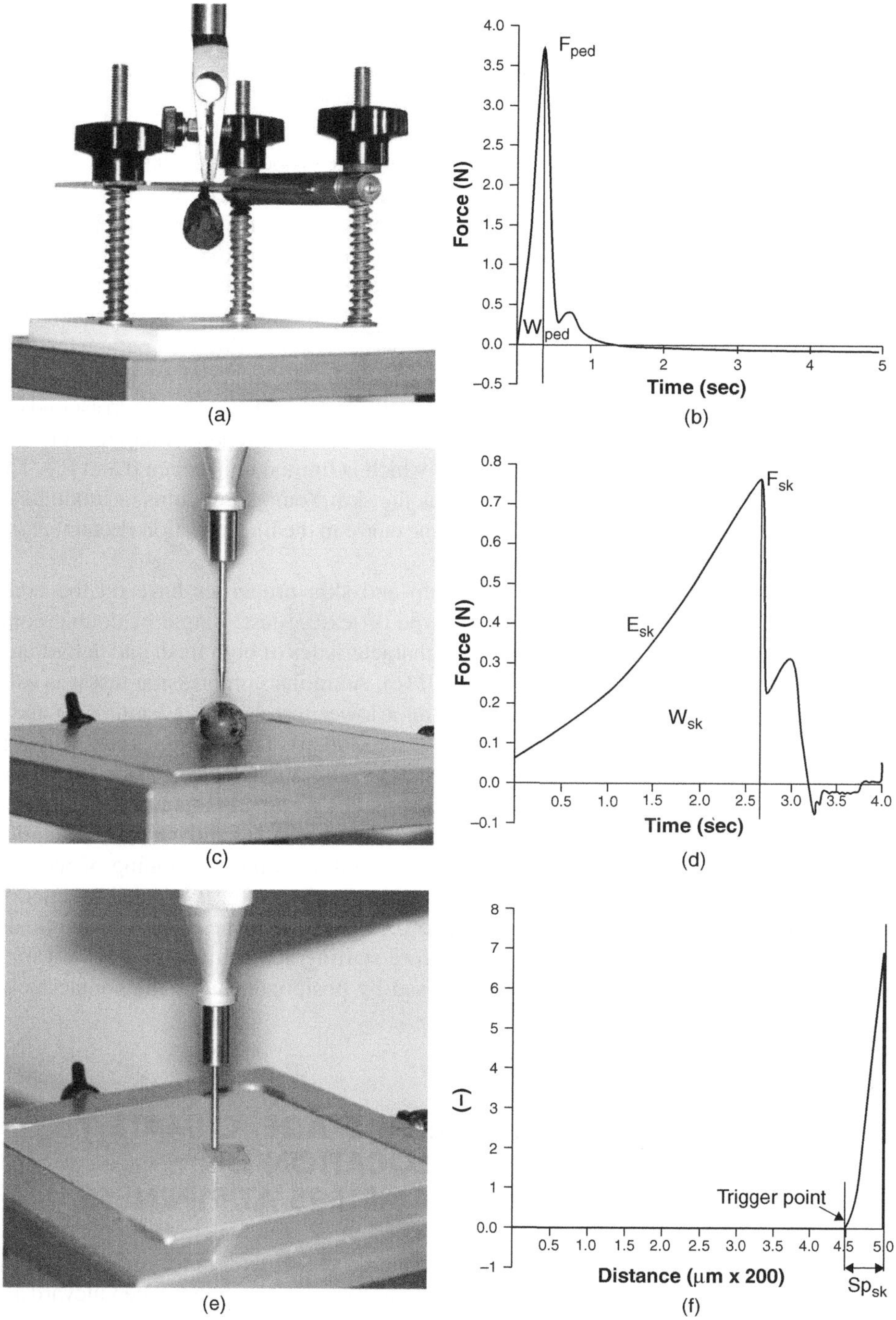

Figure 7.1 Probes and typical curves acquired by UMT during different types of texture test. (a) probe and platform used in pedicel traction test; (b) force–time curve acquired during traction test for pedicel detachment resistance assessment; (c) needle or conic probe (Ø 2 mm) used for skin hardness evaluation; (d) force–time curve acquired during puncture or penetration test of berry skins; (e) flat probe (Ø 2 mm) used for skin thickness evaluation; (f) force–distance curve acquired during compression test for skin thickness assessment.

strong influence on the results. Due to the curvature of specific probes made by some UMT companies, the 'hardness' is measured as a combination of shear and compression resistance and measurements that have been acquired with different probe forms cannot be converted, or compared one with another (Abbott, 1999; Roudot, 2006).

Common geometries for the probe used in grape-skin studies include flat probes that vary in diameter, ranging from 0.9 mm to 2 mm, needle, or conic probes with a diameter of 2 mm (Figure 7.1c), or either rounded, or hemispherical thin probes (Rolle *et al.*, 2012a). Interference caused by the pulp characteristics are minimized if the skin's mechanical properties are determined directly on the intact whole berry, with the aid of a needle probe (Rolle *et al.*, 2009b). The puncture test is also generally performed at 1 mm s^{-1}, and as a rule, a puncture of 2 mm suffices to perforate the skin. Figure 7.1d shows a force–time curve obtained by the berry skin puncture test. The berry skin hardness is assessed by different parameters: the maximum break force (F_{sk}), the break energy (W_{sk}) and by the material resistance against the axial deformation (E_{sk}). The first variable, generally expressed in N or g, corresponds to the skin's resistance to the needle probe penetration. Whilst the second variable, expressed in mJ, is represented by the area under the curve, which is limited to between 0 and F_{sk}. The third variable, expressed in N mm^{-1}, is defined as the skin Young's modulus or modulus of elasticity, and represents the slope of the force–time curve in the linear section (Letaief *et al.*, 2008b; Maury *et al.*, 2009).

Compression test. The influence that the pulp and skin properties have on the berry mechanical characteristics is aggregate in this type of texture test. Single or double compression tests are often used to evaluate the pulp characteristics of both fresh and dehydrated grapes (Bellincontro *et al.*, 2009; Rolle *et al.*, 2011c). A similar compression test was used to determine the berry skin thickness (Sp_{sk}) using a lower test speed (0.2 mm s^{-1}) and a flat 2 mm diameter probe (Figure 7.1e) (Letaief *et al.*, 2008a). The skin thickness (Sp_{sk}) is calculated as the distance (μm) between the instrument calibration distance (1 or 2 mm) and the probe trigger point (Figure 7.1f). Before measuring this parameter, manual separation of a piece of skin from the lateral side of each berry is to be carried out and particular attention is to be paid in removing the pulp from the skin, as well as in the positioning of the skin sample on the platform so as to avoid creating folds. Furthermore, it is good practice to insert an instrumental trigger threshold, equal to 0.05 N, so that the plane surface of the probe adheres completely to the skin sample before starting acquisition. This allows for a reduction, or elimination, of the 'tail' effect caused by postponement of the contact point (Río Segade *et al.*, 2011a).

7.3 EFFECT OF VARIETY, RIPENING STAGE, CLIMATE, VINTAGE AND GROWING LOCATION ON WINEGRAPE TEXTURE PARAMETERS AT RIPENESS

The textural characteristics of a grape variety depend on the structure of tissues at ripeness (Roudot, 2006). In comparative studies, carried out on grapes grown in the same vineyard and harvested at the same content of solid soluble, the skin mechanical properties, particularly skin break force (F_{sk}), are likely to be varietal markers (Figure 7.2) (Rolle *et al.*, 2011b). As can be observed, the values range from 0.964 N for Malvasia di Casorzo to 0.295 N for Pollera in black varieties and from 0.854 N for Moscato Giallo (Yellow Muscat) to 0.338 N for Cortese in white varieties. However, variations inside the same cultivar can be imputed

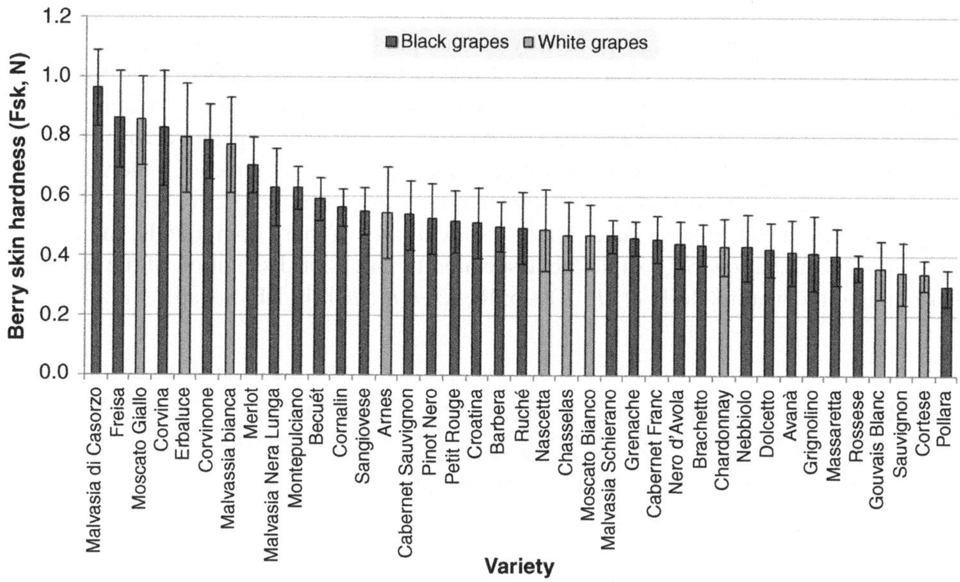

Figure 7.2 Berry skin break force (F_{sk}) of fresh grape berries for black and white winegrape varieties.

to the clonal differences (Rolle *et al.*, 2012b) and the viral status of the plants (Santini *et al.*, 2011; Giribaldi *et al.*, 2011). Moreover, the growing area heavily influences the skin break force within the same variety and provides an effective predictive tool to discriminate between vineyards located in different zones (Maury *et al.*, 2009; Torchio *et al.*, 2010; Le Moigne *et al.*, 2008; Zouid *et al.*, 2010), or even subzones (Río Segade *et al.*, 2011c). Most authors suggest that environmental factors play a dominant role more in the characterization of the skin texture attributes than in the level of sugar accumulated. Indeed, a significant increase in the F_{sk} value can be observed only in the early ripening phases after veraison, while a steady value or a slight decrease can be noted in the last 4–5 weeks preceding technological maturity (Rolle *et al.*, 2012b). This has led several authors to the conclusion that the behaviour of the berry skin break force assessed close to the harvesting period may limit the choice of this mechanical parameter as a ripeness indicator in grape berries (Maury *et al.*, 2009; Rolle *et al.*, 2012a). A renewed increase is often observed in over-ripe berries (Rolle *et al.*, 2009a).

The skin mechanical characteristics are seasonal-dependent (Letaief *et al.*, 2008a). The influence of annual variations in climate on the skin hardness has been recently investigated in coloured and white winegrape varieties, using different climatic indices (Rolle *et al.*, 2011b). Seasonal temperature parameters, particularly absolute maximum temperature, are significantly correlated to the skin break force, due to the genotype–environment interaction (Sato *et al.*, 2000; Rolle *et al.*, 2011b). Furthermore, the influence of the production area was also evaluated in the same season and it was observed that precipitation parameters were better correlated to the skin texture parameter than were others (Rolle *et al.*, 2011b). The changes observed in the skin mechanical properties during the grape-ripening process are more related to the densimetric heterogeneity of the berries harvested, than to the harvest date (Rolle *et al.*, 2011e). At technological maturity, small variations in the skin break force are imputable to different sugar contents (Torchio *et al.*, 2010; Río Segade *et al.*, 2011b; Zouid *et al.*, 2010).

Also the berry skin thickness may be considered a good varietal differentiating attribute, even if there is a generally a high variability of data present within the same grape sample (Río Segade *et al.*, 2008; Río Segade *et al.*, 2011bb; Letaief *et al.*, 2008a). The thickness and toughness of the skin are contributing factors to the resistance of grapes against fungal pathogens (Rosenquist and Morrison, 1988) as they are against handling injury during harvest and on- and off-vine withering (Kök and Çelik, 2004). Moreover, at the end of the withering process, these skin mechanical characteristics may also influence the phenol composition of the wines produced as they are related to the extractability of these compounds. In general, a high F_{sk} value and a low Sp_{sk} value are associated to an easier dissolution of the phenols, in particular anthocyanin (Rolle *et al.*, 2009b; Rolle *et al.*, 2012b; Río Segade *et al.*, 2011a).

7.4 RELATIONSHIPS BETWEEN INSTRUMENTAL TEXTURE PARAMETERS OF FRESH BERRY SKIN AND THE WINEGRAPE DEHYDRATION PROCESS

As described in the previous chapters, the complex process of withering brings about several changes in the chemical composition of winegrapes and, consequently, affects the wine quality. Knowledge of the varietal relationship between the skin hardness and the dehydration kinetics is of great relevance in the management of the withering process. In fact, the extent of these changes depends on the dehydration kinetics, which, in turn, are related to the withering climatic conditions (e.g. temperature, RH, air flow), as they are to any initial different chemical pre-treatment of the grapes, to the variety of the winegrapes themselves and the relative skin mechanical properties (Barbanti *et al.*, 2008; Gabas *et al.*, 1999; Mencarelli *et al.*, 2010; Rolle *et al.*, 2012a).

The influence of the initial skin hardness on the evolution of the berry weight due to dehydration observed during the off-vine drying process under controlled conditions, have recently been investigated (Rolle *et al.*, 2011a). Erbaluce fresh grape berries, with different break skin force values, soft (S = 0.372–0.675 N) and hard (H = 0.681–1.233 N), dehydrated in three controlled environmental conditions (A = +15°C, 55% RH; B = +18°C, 75% RH; C = +28°C, 40% RH) and under uncontrolled conditions (D) showed different kinetic behaviour. Although, significant differences in weight loss percentage (WL%) were only found among soft and hard fresh berries at withering conditions (A), the soft berries showed higher WL% values for the same time lapses, in each withering trial. This was particularly true for the soft berries (AS) which were characterized by statistic significantly higher WL% values than were hard ones (AH) from 42 days onwards and, therefore, the dehydration rates decreased as did the berry skin hardness and the number of withering days under these environmental conditions. Weight loss percentage values of about 41.3% were observed in 48 withering days for hard berries, whereas the values rose to 46.6% for soft ones at the same time lapse. An average daily WL% of about 4.1, 1.0, 1.0, 0.9 and 0.8 corresponded to withering conditions C, D, AS, AH and B, respectively. An average daily WL% of about 4.1, 1.0, 1.0, 0.9 and 0.8 corresponded to withering conditions C, D, AS, AH and B, respectively. The higher temperature may well justify the higher average daily WL% values for the dryer environmental conditions (C). While the higher RH, associated with withering conditions (B), induced lower average daily WL% values and more withering days were required to reach a WL% of 40. This behaviour is due to the smaller gradient of the water vapour pressure between grapes and air humidity. However, the decrease in the berry weight as a function of withering time was quite linear, indicating that the dehydration rates were constant for each environmental condition and berry skin hardness. This linearity is in agreement with that observed in other studies (Mencarelli *et al.*, 2010) and was, as expected, due to the fact that the dehydration rate usually slows down with time, very evident for the faster WL% (withering conditions (C)). Under the same off-vine postharvest dehydration conditions (+16°C, 60 RH%), the withering kinetics was influenced by the grape variety, inasmuch as the varieties characterized by lower berry skin break force values generally showed a quicker weight loss (Rolle *et al.*, 2011a; Rolle *et al.*, 2011d). White Muscat grapes ($F_{sk} = 0.468 \pm 0.107$ N) lost 35.9% of their weight in 21 days, whereas Erbaluce white grapes ($F_{sk} = 0.795 \pm 0.182$ N) lost 30.0% in 30 days. This means an average daily WL% of about 1.7 and 1.0, respectively. Corvinone ($F_{sk} = 0.784 \pm 0.124$ N) and Corvina ($F_{sk} = 0.827 \pm 0.194$ N) black varieties showed a WL% of 20.1 and 23.3 in 30 withering days, respectively, whereas Freisa ($F_{sk} = 0.859 \pm 0.163$ N) black grapes lost 32.3% of their weight in 27 days, daily average WL% of about 0.7, 0.8 and 1.2, respectively (Rolle *et al.*, 2011d). Therefore, the dehydration kinetics depended on the variety, as there was a faster WL% for White Muscat grapes, which have a lower skin break force than the other grapes, particularly the maximum value. However, the skin hardness can only partially explain the varietal effect on grape dehydration. In fact, although Freisa grapes are characterized by higher skin break force than are Corvinone grapes, they showed relatively faster dehydration kinetics. Other studies in the literature have already demonstrated that although the withering thermo-hygrometric conditions affect the dehydration kinetics, the same treatment does not induce the same effect on different winegrape varieties (Bellincontro *et al.*, 2004; Barbanti *et al.*, 2008; Costantini *et al.*, 2006; Chkaiban *et al.*, 2007), meaning that each grape variety probably has different response times to the dehydration process. This aspect prompts investigation into other mechanical properties, such as skin thickness, which most likely concur to determine the speed of the withering process. Indeed, berry skin is made up of an epidermis and from six to ten layers of small thick-walled cells (Lecas and Brillouet, 1994), and the number of layers in the skin of grape berries, their size and volume are cultivar-specific issues (Roudot, 2006). These different tissue characteristics may well offer a contribution to the understanding of the differences observed in the withering kinetics amongst varieties.

7.5 THE EVOLUTION OF TEXTURE PARAMETERS OF WINEGRAPES DURING ON-VINE WITHERING PROCESSES

Although few studies have been carried out on the modification of the mechanical characteristics during the grape-dehydration process, nowadays, texture parameters are accepted as efficient predictors of the winegrape suitability for on-vine dehydration. The most important grape mechanical characteristic that determines the success of the on-vine drying process is a high detachment resistance. The detachment in young berries is due to the hydrolysis of pectins in the middle lamella of the cell walls that form the separation layer at the base of the pedicel. The effects of climatic parameters and a varietal-specific sensitivity are responsible for this phenomenon during grape maturation (Ribéreau-Gayon *et al.*, 2000). This indicates that grape berries with insufficient resistance to detachment at technological maturity, should be excluded from use in Ice-wine production because they often have to withstand considerable amounts of snow. A progressive decrease of pedicel detachment force (F_{ped}) during

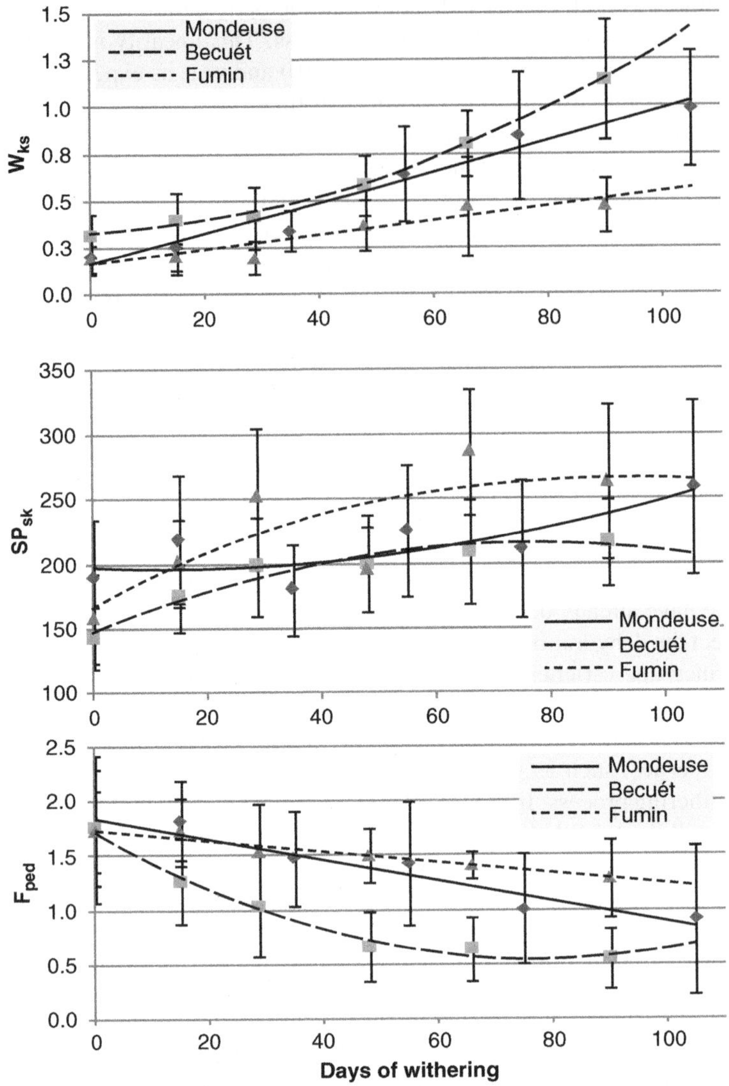

Figure 7.3 Trends of the peduncle detach resistance, the skin hardness and thickness of three black winegrape varieties for Ice wine production during on-vine dehydration.

grape dehydration on the vine for Mondeuse, Becuét and Fumin varieties used in Ice-wine production in Piedmont (northwest Italy) was noticed (Figure 7.3) (Rolle *et al.*, 2009a; Rolle *et al.*, 2010). Therefore, the pedicel detachment force is an effective parameter that should be monitored.

The evolution of the detachment force of the pedicel shows that the Becuét grapes had a greater propensity to fall throughout the whole withering period. This tendency can be noted as early as 15 days of over-ripening and lower F_{ped} values are observed when there is a strong likelihood of a snowfall. At the end of the dehydration period, the Fumin grapes had considerably higher average F_{ped} values than did Becuét grapes (+57%) and showed a more gradual decrease during the withering process. A similar constant behaviour was noted also for Mondeuse grapes (Rolle *et al.*, 2009a).

Despite the fact that the mechanical properties of berry skin vary greatly, as demonstrated by the fact that different berries often show a heterogeneous dehydration, a progressive increase in berry skin hardness was observed when assessed by break force and energy during on-vine dehydration (Rolle *et al.*, 2009a; Rolle *et al.*, 2010). However, this phenomenon was more evident for the berry skin break energy in Mondeuse and Becuét grapes. Throughout the period of dehydration on-vine the Becuét berry skin was characterized by higher values in both skin break force and skin break energy. This shows that the berry skin tends not only to harden, but also to thicken during on-vine drying. A significant increase of about 20% in the berry skin thickness was also observed during the on-vine drying process of Mondeuse and Fumin grapes. These factors play an important role as they determine grape resistance to fungal diseases and strong winds.

Both skin break force and skin thickness are important technological variables for red winegrapes as they are related to the extractability of phenolic substances from the skin (Rolle *et al.*, 2009b; Río Segade *et al.*, 2011a). Although harder and thinner skins allow for a more complete extraction of anthocyanins, the extraction kinetics are slower. Phenolic compounds are only extracted during the pressing of the frozen grapes in the production of Ice wines, due to the fact that there is no maceration during fermentation. Wines produced from hard skin varieties may have a reduced anthocyanin content, as there is only a brief period of contact between the skin and must. Therefore, this aspect has particular importance for wine produced from cultivars that are characterized by a natural low anthocyanin content, as a brief contact period may well not suffice to ensure adequate colour intensity.

ACKNOWLEDGEMENTS

The authors would like to thank Barbara Wade for her linguistic advice.

REFERENCES

Abbott, J.A. 1999. Quality measurement of fruits and vegetables. *Postharvest Biology and Technology*, 15:207–225.

Barbanti, D., B. Mora, R. Ferrarini, *et al.* 2008. Effect of various thermo-hygrometric conditions on the withering kinetics of grapes used for the production of 'Amarone' and 'Recioto' wines. *Journal of Food Engineering*, 85:350–358.

Bellincontro, A., D. De Santis, R. Botondi, *et al.* 2004. Different postharvest dehydration rates affect quality characteristics and volatile compounds of Malvasia, Trebbiano and Sangiovese grapes for wine production. *Journal of the Science of Food and Agriculture*, 84:1791–1800.

Bellincontro, A., I. Nicoletti, M. Valentini, *et al.* 2009. Integration of nondestructive techniques with destructive analyses to study postharvest water stress of wine grape. *American Journal of Enology and Viticulture*, 60:57–65.

Bernstein, Z. and I. Lustig. 1981. A new method of firmness measurement of grape berries and other juicy fruits. *Vitis* 20:15–21.

Bourne, M.C. 2002. *Food Texture and Viscosity: Concept and Measurement*, 2nd edn, Academic Press, New York.

Carbonneau, A., A. Moueix, N. Leclair, and J.L. Renoux. 1991. Proposition d'une méthode de prélèvement de raisins à partir de l'analyse de l'hétérogénéité de maturation sur un cep. *Bulletin de l'OIV*, 64:679–690.

Chkaiban, L, R. Botondi, A. Bellincontro, *et al.* 2007. Influence of postharvest water stress on lipoxygenase and alcohol dehydrogenase activities, and on the composition of some volatile compounds of

Gewürztraminer grapes dehydrated under controlled and uncontrolled thermohygrometric conditions. *Australian Journal of Grape and Wine Research*, 13:142–149.

Costantini, V., A. Bellincontro, D. De Santis, *et al.* 2006. Metabolic changes of Malvasia grapes for wine production during postharvest drying. *Journal of Agriculture and Food Chemistry*, 54:3334–3340.

Deng, Y., Y. Wu, and Y. Li. 2005. Effects of high O_2 levels on post-harvest quality and shelf life of table grapes during long-term storage. *European Food Research and Technology*, 221:392–397.

Gabas, A.L., F.C. Menegalli, and J. Telis-Romero. 1999. Effect of chemical pretreatment on the physical properties of dehydrated grapes. *Drying Technology*, 17:1215–1226.

Giribaldi, M., M. Purrotti, D. Pacifico, *et al.* 2011. A multidisciplinary study on the effects of phloem-limited viruses on the agronomical performance and berry quality of *Vitis vinifera* cv. *Nebbiolo. Journal of Proteomics*, 75:306–315.

Kök, D. and S. Çelik. 2004. Determination of characteristics of grape berry skin in some table grape cultivars (*V. vinifera* L.). *Journal of Agronomy*, 3:141–146.

Lang, A. and H. During. 1990. Grape berry splitting and some mechanical properties of the skin. *Vitis*, 29:61–70.

Lecas, M. and J.M. Brillouet. 1994. Cell wall composition of grape berry skins. *Phytochemistry*, 35:1241–1243.

Lee, C.Y. and M. Bourne. 1980. Changes in grape firmness during maturation. *Journal of Texture Studies*, 11:163–171.

Le Moigne, M., C. Maury, D. Bertrand, *et al.* 2008. Sensory and instrumental characterisation of Cabernet Franc grapes according to ripening stages and growing location. *Food Quality and Preference*, 19:220–231.

Letaief, H., L. Rolle, and V. Gerbi. 2008a. Mechanical behavior of under compression tests. *American Journal of Enology and Viticulture*, 59:323–329.

Letaief, H., L. Rolle, G. Zeppa, *et al.* 2008b. Assessment of grape skin hardness by a puncture test. *Journal of the Science of Food and Agriculture*, 88:1567–1575.

Maury, C., E. Madieta, M. Le Moigne, *et al.* 2009. Development of a mechanical texture test to evaluate the ripening process of Cabernet Franc grapes. *Journal of Texture Studies*, 40:511–535.

Mencarelli, F., A. Bellincontro, I. Nicoletti, *et al.* 2010. Chemical and biochemical change of healthy phenolic fractions in winegrape by means of postharvest dehydration. *Journal of Agriculture and Food Chemistry*, 58:7557–7564.

Mencarelli, F., R. Massantini, L. Lanzarotta, *et al.* 1994. Accurate detection of firmness and colour changes in the packing of table grapes with paper dividers. *Journal of Horticultural Science*, 69:299–304.

Porro, D., M. Ramponi, T. Tomasi, *et al.* 2010. Nutritional implications of water stress in grapevine and modifications of mechanical properties of berries. *Acta Horticulturae*, 868:73–80.

Ribéreau-Gayon, P., D. Dubourdieu, B.J. Donèche, *et al.* 2000. *Handbook of Enology. Vol. 1. The Microbiology of Wine and Vinification*, John Wiley & Sons, Ltd, Chichester, UK.

Río Segade, S., S. Giacosa, V. Gerbi, *et al.* 2011a. Berry skin thickness as main texture parameter to predict anthocyanin extractability in winegrapes. *LWT – Food Science and Technology*, 44:392–398.

Río Segade, S., I. Orriols, S. Giacosa, *et al.* 2011b. Instrumental texture analysis parameters as varietal markers and ripeness predictors. *International Journal of Food Properties*, 14:1318–1329.

Río Segade, S., L. Rolle, V. Gerbi, et al. 2008. Phenolic ripeness assessment of grape skin by texture analysis. *Journal of Food Composition and Analysis*, 21:644–649.

Río Segade, S., E. Soto Vázquez, I. Orriols, *et al.* 2011c. Possible use of texture characteristics of winegrapes as markers for zoning and their relationship with anthocyanin extractability index. *International Journal of Food Science & Technology*, 46:386–394.

Rolle, L., A. Caudana, S. Giacosa, *et al.* 2011a. Influence of skin hardness on dehydration kinetics of wine-grapes. *Journal of the Science of Food and Agriculture*, 91:505–511.

Rolle, L., V. Gerbi, A. Schneider, *et al.* 2011b. Varietal relationship between instrumental skin hardness and climate for grapevines (*Vitis vinifera* L.). *Journal of Agriculture and Food Chemistry*, 59:10624–10634.

Rolle, L., S. Giacosa, V. Gerbi, *et al.* 2011c. Comparative study of texture properties, color characteristics and chemical composition of ten white table grape varieties. *American Journal of Enology and Viticulture*, 62:49–56.

Rolle, L., S. Giacosa, F. Torchio, *et al.* 2011d. Influenza della durezza della buccia sulla cinetica di disidratazione di uve Moscato Bianco, Erbaluce e Corvina. In *Atti Territori divini – Progetti di ricerca per il settore vitivinicolo*, pp. 80–84.

Rolle, L., S. Río Segade, F. Torchio, *et al.* 2011e. Influence of grape density and harvest date on changes in phenolic composition, phenol extractability indices, and instrumental texture properties during ripening. *Journal of Agriculture and Food Chemistry*, 59:8796–8805.

Rolle, L., R. Siret, S. Río Segade, *et al.* 2012a. Instrumental texture analysis parameters as markers of table and wine-grape quality: a review. *American Journal of Enology and Viticulture*, 63:11–28.

Rolle, L., F. Torchio, E. Cagnasso, *et al.* 2010. Evolution of mechanical variables of the winegrapes for icewine production during on-vine drying. *Italian Journal of Food Science*, 22:143–149.

Rolle, L., F. Torchio, S. Giacosa, *et al.* 2009a. Modification of mechanical characteristics and phenolic composition in berry skins and seeds of Mondeuse throughout the on-vine drying process. *Journal of the Science of Food and Agriculture*, 89:1973–1980.

Rolle, L., F. Torchio, A. Ferrandino, *et al.* 2012b. Influence of wine-grape skin hardness on anthocyanin extraction kinetic. *International Journal of Food Properties*, 15:249–261.

Rolle, L., F. Torchio, G. Zeppa, *et al.* 2008. Anthocyanin extractability assessment of grape skins by texture analysis. *Journal International des Sciences de la Vigne et du Vin*, 42:157–162.

Rolle, L., F. Torchio, G. Zeppa, *et al.* 2009b. Relations between break skin force and anthocyanin extractability at different stages of ripening. *American Journal of Enology and Viticulture*, 60:93–97.

Rosenquist, J.K. and J.C. Morrison. 1988. The development of the cuticle and epicuticular wax of the grape berry. *Vitis*, 27:63–70.

Roudot, A.C. 2006. Some considerations for a theory of plant tissue mechanics. *Sciences des Aliments*, 26:409–426.

Santini, D., L. Rolle, P. Cascio, *et al.* 2011. Modifications in chemical, physical and mechanical properties of Nebbiolo (*Vitis vinifera* L.) grape berries induced by mixed virus infection. *South African Journal for Enology & Viticulture*, 32:183–189.

Sato, A. and M. Yamada. 2003. Berry texture of table, wine, and dual-purpose grape cultivars quantified. *HortScience*, 38:578–581.

Sato, A., M. Yamada, I. Hiroshi, *et al.* 2000. Optimal spatial and temporal measurement repetition for reducing environmental variation of berry traits in grape breeding. *Scientia Horticulturae*, 85:75–83.

Sato, A., M. Yamada, H. Iwanami, *et al.* 2004. Quantitative and instrumental measurements of grape flesh texture as affected by gibberellic acid application. *Journal of the Japanese Society for Horticultural Science*, 73:7–11.

Sato, A., H. Yamane, N. Hirakawa, *et al.* 1997. Varietals differences in the texture of grape berries measured by penetration tests. *Vitis*, 36:7–10.

Torchio, F., E. Cagnasso, V. Gerbi, *et al.* 2010. Mechanical properties, phenolic composition and extractability indices of Barbera grapes of different soluble solids contents from several growing areas. *Analytica Chimica Acta*, 660:183–189.

Zouid, I., R. Siret, E. Mehinagic, *et al.* 2010. Evolution of grape berries during ripening: investigations into the links between their mechanical properties and the extractability of their skin anthocyanins. *Journal International des Sciences de la Vigne et du Vin*, 44:87–99.

8 Management of Vinification and Stabilization to Preserve the Aroma Characteristic of Dehydrated Grape

Luigi Moio and Paola Piombino

Department of Food Science, University of Naples Federico II, Portici, Italy

8.1 INTRODUCTION

Wine aroma genesis largely depends on numerous and complex biochemical transformations developing during winemaking: oxidative and hydrolytic reactions arising during the pre-fermentative stages of grape pressing and maceration; primary and secondary metabolisms of micro-organisms conducting the alcoholic and malolactic fermentations; oxidation and transfer processes occurring during wine maturation in wood; and chemical and enzymatic post-fermentative reactions taking place during wine storage and bottle ageing. Alongside these sources of volatiles linked to the transformation and ageing processes, grape characteristics play a primary role in determining the quality and the evolution of wine aroma. This is true even though many wine aroma characteristics are not sensorially detectable in the grapes. Among the volatile compounds that contribute to wine aroma, grape odour active molecules play a determining role as they are responsible for the so-called 'varietal aroma', which represents an added value to wine quality. There is a substantial difference in the way varietal aroma forms in wines from 'aromatic grapes' or 'neutral grapes'. Most of the volatiles responsible for varietal notes in wines from aromatic cultivar are already present at high levels as free odour active molecules in the must, which is characterized by odours reminiscent of the grape variety. Musts from neutral grapes, however, are not recognizable, but are essentially dominated by herbaceous odours due to aldehydes and alcohols C6 derived from fatty acids. Their varietal aroma will be expressed only after alcoholic fermentation and/or ageing, as secondary metabolites of biochemical transformations or due to the release of volatiles from non-volatile aroma precursors. As a consequence of this knowledge, all factors affecting grape quality can potentially influence wine aroma characteristics.

On-plant and postharvest dehydration importantly affect grape quality because the stress due to berry water loss induces marked and variable effects on fruit physical structure, metabolism, chemical and biochemical composition. All these changes have both direct and indirect influences on the aroma of the final wine.

Sweet, Reinforced and Fortified Wines: Grape Biochemistry, Technology and Vinification, First Edition.
Edited by Fabio Mencarelli and Pietro Tonutti.

Wines from dehydrated grapes differ not only because of the grape-dehydration systems, which often reflect unique climatic and environmental conditions, but also due to the degree of grape dehydration, the employment of different grape varieties (aromatic and/or neutral; white or red) and the vinification method, which is often the consequence of traditional practices.

All this variability (from field to winemaking) in the production of wines from withered/dehydrated grapes, gives rise to different wine styles corresponding to specialized products characterized by very distinctive sensory characteristics, which have little in common other than high sugar concentration in the corresponding musts. This common feature provides unusual conditions for yeast growth with a consequent impact on flavour production during fermentation. For this reason, in wines obtained from dehydrated grapes, the equilibrium between fermentative and grape aroma shifts towards the latter, and grape varietal aroma, together with volatiles produced during dehydration, becomes even more important than in table wines. As a consequence, all winemaking practices, from field to bottle, encouraging the release and the preservation of these aromas from dehydrated grapes to wine, are suitable to make products characterized by a complex and specific aroma. This flavour complexity is needed not only because wine quality is directly correlated with its aroma richness, but also because this complexity must counter-balance the olfactory impact of acetaldehyde, acetic acid and ethyl acetate. These compounds may be present at high levels in wines obtained from dehydrated grapes, and originate from both anaerobic metabolism, to which grapes shift during water loss, and unfavourable fermentative conditions.

This chapter provides an overview of the current knowledge available to optimize the management of vinification and stabilization to preserve the aroma characteristic of dehydrated grapes. In particular, the chapter describes how the main stages of winemaking can be enhanced in order to maximize the extraction and preservation of free and bound aromas from dehydrated grapes.

8.2 VOLATILE COMPOSITION OF WINES FROM DEHYDRATED GRAPES

Although over the years, several authors have investigated the volatile fraction of different styles of dessert wines, to the best of our knowledge, the study of the volatile fraction of sound dehydrated winegrapes under different conditions, and of their musts, started less than 10 years ago (Franco *et al.*, 2004; Bellincontro *et al.*, 2004; Costantini *et al.*, 2006; Chkaiban *et al.*, 2007; Moreno *et al.*, 2008; Ruiz *et al.*, 2010; Santonico *et al.*, 2010; Piombino *et al.*, 2011; Lopez de Lerma *et al.*, 2012). The dehydration process markedly affects the primary aroma compounds, both free and glycosylated, that are present at different concentrations and ratios in the skin and in the pulp as described in Chapter 5. From the vinification point of view, it is important to stress that all free volatiles are more concentrated in the skin, with the exception of anaerobic metabolism derivatives that are mainly present in the pulp.

The aroma descriptors generally cited to describe a dessert wine are apricot, peach, coconut, exotic fruit, orange zest, candied fruit, dried fruit (apricot, fig, prune, walnut), citrus blossom, floral, marmalade, honey, caramel. The volatile compounds identified in dessert wines of different styles (from non-botrytized or partially infected grapes) belong to several chemical classes and those that have the greatest impact seem to be: ethyl acetate, acetoin, phenylacetaldehyde, benzaldehyde, 1-octen-3-ol, γ-butyrolactone, γ-hexalactone, γ-nonalactone, γ-decalactone, δ-decalactone, linalool, geraniol, nerol, 4-terpineol, β-ionone,

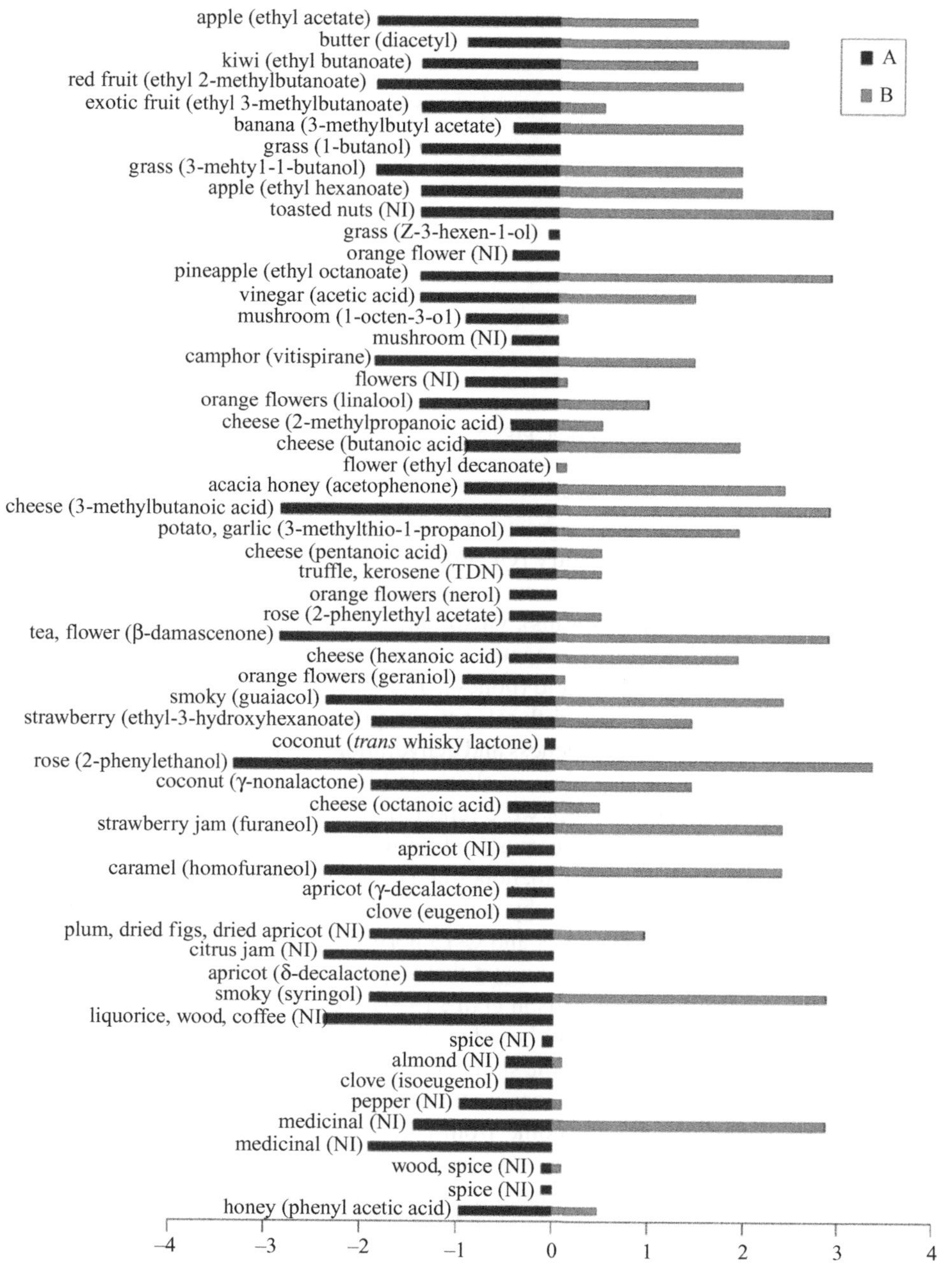

Figure 8.1 Aromagrams of sweet Fiano wine (A) and base Fiano wine (B) obtained by GC/O analysis; NI, not identified (from Genovese *et al.*, Copyright 2007, with permission from Elsevier).

β-damascenone, furfural, 5-methylfurfural, furaneol, homofuraneol, sotolon (Moreno *et al.*, 2008; Genovese *et al.*, 2007; Franco *et al.*, 2004; Castellano *et al.*, 2001). Most of these molecules are detectable also in wines obtained from fresh sound grapes, but at lower levels and consequently with less olfactory impact. Figure 8.1 shows the comparison between the aromagrams of two Fiano wines: the sweet wine (A) and the base wine (B) (Genovese *et al.*, 2007). The profiles significantly differ both in the number and in the intensity of the odour

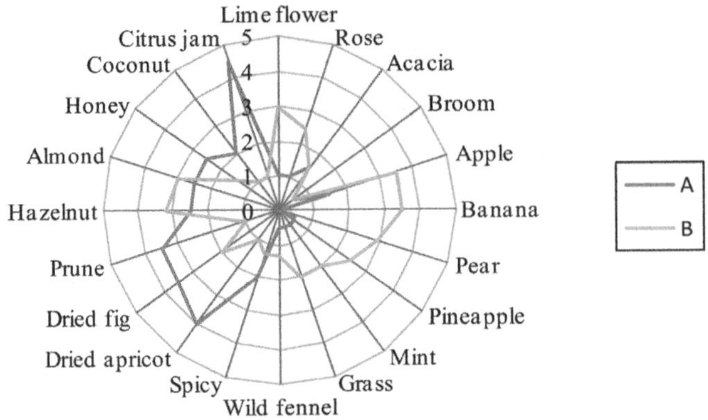

Figure 8.2 Quantitative descriptive odour profiles of sweet Fiano wine (A) and base Fiano wine (B) (adapted from Genovese *et al.*, 2007).

zones, with the sweet wine presenting a greater richness. Nineteen odour compounds detected during gas-chromatography/olfactometry (GC/O) analysis of the sweet Fiano differed from the base wine, and the most important were: ethyl 3-methylbutanoate (exotic fruits), 1-butanol (grass), 1-octen-3-ol (mushrooms), vitispirane (camphor), linalool, nerol and geraniol (orange flowers), γ-nonalactone (coconut), γ and δ-decalactone (apricot), eugenol and isoeugenol (clove) and phenyl acetic acid (honey). The other 11 odour peaks of sweet Fiano wine were not identified. They showed floral, apricot, plum, citrus jam, liquorice, almond, pepper and medicinal odours. Over-ripeness and dehydration significantly modify the flavour characteristics of the corresponding wine, as shown in Figure 8.2 where the sensory profiles of sweet (A) and base (B) Fiano wines were reported. The main influence of Fiano grape over-ripeness and dehydration on the corresponding wine appears to be the enrichment of dried fruits (apricot, fig, prune), citrus jam, honey and coconut odours. If further investigation seems necessary, these and other results by authors cited earlier show that among the compounds that derive from grapes, terpenes and C-13 norisoprenoids notably contribute to the aroma of most dessert wines. It is well known that terpenes concentration increases during ripening and after maturity fall off (Wilson *et al.*, 1984). The higher levels usually detected in wines from dehydrated grapes could be affected also by a better extraction as a consequence of the structural degradation of the skin, where most of the terpenes are located. C-13 norisoprenoids have an important role, because they are molecules characterized by very low odour thresholds. These compounds come from carotenoids, unstable compounds whose degradation can occur in the presence of oxygen, high temperatures and exposure to the sun (Rapp and Marais, 1993). Both terpenes and C-13 norisoprenoids are present in grapes also as glycoconjugated compounds, and their high levels in dessert wines may be due to major concentration in over-ripe and dehydrated grapes and to higher contents of their glycosidic precursors. The extraction and preservation of these compounds in their free and bound forms should represent one of the main targets during winemaking of dehydrated grapes.

8.3 VINIFICATION

Several styles of wines can be obtained from dehydrated grapes. They may be sweet, slightly sweet or dry, white, rosé or red and several methods are used to produce them. The different techniques often reflect environmental conditions, traditional practices, or are the evolution of fortuitous events, or are a modern development in response to consumer preferences. For this reason, there are different dessert wine styles corresponding to specialized products possessing distinctive characteristics. Variations in style depend not only on grape variety and dehydration methods, but also on the winemaking process, that is, practices applied before, during and after fermentation. Of course, it is not fair, nor yet possible, to standardize the winemaking process of this type of wine, but some physical-chemical and compositional characteristics, common to juices obtained from dehydrated grapes, universally require special attention during the pre-fermentative, fermentative and maturation stages of dessert wines production. In the following sections we mainly refer to the production of a white dessert wine.

8.3.1 Pre-fermentative phases

Managing pre-fermentative treatments is crucial for the preservation of the aromas resulting from dehydrated grapes and will indeed influence the flavour of the final wine. The crushing of dehydrated grapes represents a critical point in the production of high-quality dessert wines. Managing this step must be oriented to increase must yield and grape aroma extraction, but at the same time, prevent and minimize enzymatic oxidation due to polyphenol oxydases and increased amounts of suspended solids in the juice. As a rule, any rough manipulation (from field to cellar) of dehydrated grapes should be avoided in order to prevent the excessive release of suspended solids causing instability, difficult clarification and vegetal off-flavours. This is particularly true if part of the grapes is botrytized, because of the diffusion of glucan in the must (Ribéreau-Gayon *et al.*, 2006). This polysaccaride is located on the skin/pulp interface and, thanks to its colloid protector properties, interferes with settling of must during the pre-fermentative phases and with clarification of the future wine. The first crushing results in the extraction of most of sugars that could be enhanced by increasing the contact time between solid and liquid parts of the berries during the pre-fermentative operations. The juice recovery is favoured by the draining effect of the stems, which facilitate the passage of liquid through the pressed skins. In the specific case of dehydrated grapes, the risk of off-taste and off-odour, deriving from tannic and vegetal substances in the stems, is minimized by their lignification arising during dehydration and by the necessity for a careful pressing. Dehydrated grapes need powerful, but at the same time, slow and delicate treatments. A recent study on the modification of mechanical characteristics and phenolic composition in berry and seeds of Mondeuse wine grapes during on-vine dehydration for the production of Ice wines (Rolle *et al.*, 2009), reported that: berry skin thickness and hardness tended to increase progressively during withering, and its springiness also increased; seed hardness decreased during withering while springiness increased; at the end of the withering process, the percentage contribution of tannins extracted from seeds to total tannins was higher. These are among the dehydrated grapes characteristics that drive the pressing modality. A mighty force is necessary to break the harder withered skins; slower pressing allows a more complete extraction of flavours and anthocyanins from harder skins (Rolle *et al.*, 2008); a controlled pressing process, especially for dessert wines, minimizes tannins extraction from seeds, which significantly affect sensorial properties (astringency and bitterness) of wines (Gambuti *et al.*, 2006). Pneumatic presses give high quality musts, but due to the high density of juices from dehydrated grapes, versions with greater than 3 bars of strength may be insufficient for a proper extraction of juices with the highest sugar levels (above 22–23% volume of potential alcohol) (Ribéreau-Gayon *et al.*,

2006). Vertical presses are recommended in order to favour juice extraction, and mainly in the final step, hydraulic presses can improve the yield. For these reasons, in recent years producers of dessert wines have gone back to using vertical presses, and particularly those with low capacity (5–15 hL), as the pressing of several small aliquots, ensures a more efficient extraction of the juice. Pressing must be slow, long and constant, avoiding the shredding of dehydrated grape berries and minimizing the pressing cycles in order to control suspended solids and entrance of oxygen. The presence of oxygen during pre-fermentative operations causes the enzymatic oxidation of phenolic compounds due to endogenous tyrosinase, and a consequent browning of the juice, the extent of which also depends on the grape variety. Dehydrated grapes could be particularly susceptible to this phenomenon due to the senescence and de-structuring of berry tissue as a consequence of water loss. Even more browning should be expected in musts from botrytized grapes because of the co-presence of laccase produced by the mould, but this is not frequently observed. Often, for a better extraction of noble rot juice, two to three pressing cycles and the manual breaking up of the press cake before each pressing start could be useful. Nevertheless, this practice could be hazardous in the case of dehydrated grapes only partially infected by *Botrytis cinerea*. In totally botrytized juice, the risk of oxidation during pre-fermentative operations is lower because polyphenols were jet oxidized by laccase directly into the berry during the mould growth. However, in sound dehydrated grapes only partially infected by noble rot, the substrate for laccase activity is available, so that in this kind of juice the entrance of oxygen increases the risk of enzymatic oxidation. Laccase is a more general and active enzyme than tyrosinase and other phenol oxidases. This enzyme is particularly dangerous because it is active and stable at must pH values; it is not inhibited by SO_2 at the levels used in winemaking, is able to consume dissolved oxygen, and is reactive towards the quinone-glutathione browning protective complex (Ugliano, 2009). The most effective way to control its activity is by decreasing juice temperature and aeration (Boulton *et al.*, 1999). According to the results reported by Selli *et al.* (2011), pressing uniformly increased the levels of aromatic constituents, but this treatment lowered the grape juice quality for winemaking by increasing the total phenolic compounds, browning index, and C6-alcohol levels (green-herbaceous odour). Moreover, even if results reported by Moio *et al.* (2004b) were on wines from sound fresh grapes, these authors demonstrated that the protection of the must against oxidation gives wines with higher levels of volatiles responsible for fruity aromas. On the contrary, hyperoxygenation provides wines with increased colour stability (Schneider, 1998), but it seems to have a negative effect on the varietal character of wine (Dubourdieu and Lavigne, 1990) becoming undesirable in wines that emphasize the contribution of the grape (Boulton *et al.*, 1999), as is the case in wines from dehydrated grapes.

In order to aid the production of high-quality musts from dehydrated grapes, minimize some risks linked to the pre-fermentative steps, maximize juice yield, and favour free and bound aromas extraction, it could be very useful to add enzymes in the production process of dessert wines. Enzymes are able to help the grape must yield during pressing as they favour the settling of musts, and improve clarification and filtration. Different types of enzymes are used in winemaking (pectinase, cellulase, hemicellulase, oxidoreductase, protease, β-glycosidase) and their principal and secondary activities significantly affect the quality of the final wine (Guérin *et al.*, 2009). For this reason, a careful selection of commercial enzymes is necessary and knowledge of their specific effect on different grape varieties, which may vary according to the specific condition of transformation. A recent paper (Espejo and Armada, 2010) evaluated the effect of pectolytic enzymes addition in the making of Pedro Ximénez sweet wines, using dynamic pre-fermentative maceration (room temperature,

3 hours). Treated wines varied significantly with respect to the control sample in total soluble solids (> °Brix), total juice (> yield) and final sensory characteristics (≫ flavour intensity, flavour quality, aroma intensity, equilibrium; < astringency; = herbaceous); on the contrary, no effects on the content of total polyphenols and other chemical characteristics of the musts were observed. The employment of selected pectolytic enzymes represents an effective means of improving aroma and taste quality of dessert wines. The enzyme solution must be added at the beginning of the pre-fermentative operations, when the skins and pulps of crushed dehydrated grapes are in contact with their own juice prior to pressing: 1–2 g hL^{-1} directly added during the grape must and homogenized in a rotating macerator to achieve the best distribution.

Together with pectolytic enzymes, in the winemaking of some dessert wines, the pre-fermentative maceration also constitutes a valid means to obtain high-quality products from dehydrated grapes. On the one hand, the pre-fermentative maceration enhances the extraction of desirable compounds from grapes (sugars, free and bound aromas), and on the other hand, in the absence of ethanol, there is a relatively low extraction of tannin compounds and flavan-3-ols (Salinas *et al.*, 2003; Pinelo *et al.*, 2006). Pre-fermentative maceration may be carried out at room temperature, but in modern enology, the pre-fermentative maceration at low (~5°C) or very low (≤0°C) temperatures is here to stay, and Amarone is a prestigious example of dessert wine that is produced using cryomaceration (Paronetto and Dellaglio, 2011). Regarding another renowned Italian wine produced from sun-dried grapes, Piombino *et al.* (2010) reported that as an effect of cold-maceration (6°C, 12 hours), both total free and glycoconjugated aromas of Malvasia delle Lipari wines increased by 45% and 36%, respectively. Among free volatiles, alcohols increased by 79% (mainly due to 3-methyl-1-butanol and 2-phenylethanol) while esters were 14% lower, probably due to grape solids in the must, which limit esters and enhance alcohols production during alcoholic fermentation (Moio *et al.*, 2004a). The cold-macerated Malvasia was richer in the following free terpenols: β-citronellol (96%), farnesol (69%), and 3,7-dimethyl-1,5-octadien-3,7-diol (21%). The latter compound is considered to be one of the key aromas of Malvasia delle Lipari passito wine (Guarrera *et al.*, 2005). With regard to the increased amount of glycoconjugated volatiles, bound terpenoids were the most affected by cold skin contact increasing by 44% and becoming 78% of the total bound compounds detected in the cryomacerated sample. Four terpenols significantly increased: linalool (67%), α-terpineol (37%), geraniol (33%), nerol (21%); only epoxylinalool decreases by 52%. The potential varietal aroma of the Malvasia delle Lipari was increased by cryomaceration, which augmented the concentration of bound terpenoids. Sensory analysis showed that cryomaceration of Malvasia grapes affected the aroma of the corresponding wines and perceptible odour modifications occurred. These results show that cryomaceration is an interesting means of managing the aroma profile of sweet wines, offering the advantage of preserving grapes' varietal compounds.

At this stage of winemaking, a light juice sulfiting is recommended. A concentration ranging from 30 to 50 mg L^{-1}, depending on grape health, generally represents a good equilibrium between enological requirements and legal total SO_2 limits. Sulfiting juice from dehydrated grapes leads to high levels of bound SO_2 in the final wine, due to interaction with higher presence of acetaldehyde (from both grape and unfavourable fermentation conditions). In grapes with noble rot, some compounds produced by the infection (γ- and δ-gluconolactone produced by gluconic acid, combined with the bacterial byproducts 5-oxofructone and dihydroxyacetone), act as the principal SO_2-binding compounds in must and wine (Barbe *et al.*, 2002). Moreover, thiamine deficiency due to *Botrytis* infection causes accumulation of pyruvic acid, which is responsible for much of the non-acetaldehyde sulphite binding pool (Boulton *et al.*, 1999). Fortunately, only a portion of SO_2 added to juice is found in the wine as bound form, because generally at least 50–60% is in the free form. Juice sulfiting has more crucial effects, some of which become even more important during vinification of dehydrated grapes: protection against excessive oxidation of grape phenolic substrates; inhibition of the proliferation of spoilage micro-organisms, all forming byproducts responsible for wine off-flavours and quality depletion, as well as of the development of acetic and lactic bacteria; blockage of fermentation for several hours, enhancing clarification by natural sedimentation.

The spontaneous settling of musts from dehydrated grapes is not easy. The level of suspended solids is high as a consequence of berry tissues degradation due to over-ripening and water loss, and also by the presence of glucan in the case of *Botrytis*-infected grapes. Moreover, the high sugars concentration increases must viscosity. Therefore, natural settling is difficult as the suspended solids and high density must present comparable specific weights. The addition of selected pectolytic enzymes at the beginning of pre-fermentative operations when skins and pulps are crushed, accelerates must clarification. This, together with controlled temperature (8–12°C) as an accelerator of phases separation, leads to the recovery of spontaneous clarified musts within 24–72 hours. Must turbidity will not be as low as in dry white wine production (80–120 NTU), but a value between 500–600 NTU is absolutely appropriate. The improvement of clarification efficiency by adding fining agents is not necessary, and on the contrary, it could be detrimental. Indeed, from the sensory point of view, dessert wines are less subject to development of reduction and vegetal off-odours deriving from insufficient clarification, due to the masking effect of the aroma richness. Moreover, Moio *et al.* (2004a) evaluated the impact of several clarification treatments on the aroma of Falanghina must and wine. Results showed that glycosylated precursors of linalool, geraniol, benzyl alcohol, 2-phenylethanol and eugenol were significantly decreased by must clarification with fining agents, which also affected flavour production during fermentation. The authors concluded that pre-fermentative clarification might influence both expression of varietal character and ageing potential of the treated wine. Moreover, as mentioned by the same authors, clarification by fining agents could significantly affect the composition of must, reducing the concentration of some constituents important for yeast growth, such as long-chain fatty acids and sterols, and reducing the level of amino nitrogen. It could also affect nutrient depletion. For all these reasons, the best way to preserve grape aroma is to clarify musts from dehydrated grapes by the spontaneous settling that could be favoured by adding pectolytic enzymes at the beginning of the pre-fermentative stage.

8.3.2 Fermentation

After decantation (24–72 hours at 8–12°C), juice obtained from dehydrated grapes (500–600 NTU) is on its way to becoming wine through the alcoholic fermentation. The fermentation of dehydrated grape juices is generally slower than fermentation of musts from sound grapes vinified at their technological maturity. This is due to the sluggish rate of yeast growth as a consequence of multiple inhibitions of the substrate. The high sugar content is the principal limiting factor, but the nutrient deficit, especially in the case of botrytized grapes, also contributes to slowing alcoholic fermentation. At this stage of dessert wine production, the main target in order to preserve the aroma quality should be the contraction of the yeasts growth latency. In dessert wines the so-called 'fermentation aroma' plays second fiddle to aromas originating from dehydrated grapes, because their production is not supported by the physical-chemical and chemical characteristics of the must. For this reason, it is very important to manage the alcoholic fermentation in order to control the spread of off-flavours, and prevent too great a rise in acetaldehyde, acetic acid and ethyl acetate. This is possible by refining fermentation conditions by inoculating with selected yeast strains, adding nutrients and controlling temperature.

As long ago as 1935, Hopkins and Roberts (1935a, b), showed that at the sugar concentration normally found in late-harvest musts (250–300 g L^{-1}), the fermentation rate decreases at least by 20% only because of sugar levels. It has been largely reported that increasing sugar levels in fermentative media reduces yeast cell size, cell growth, viable cell concentration, and fermentation activities (Charoenchai *et al.*, 1998). During yeast growth in dehydrated grape musts, they are exposed to a strong osmotic stress. Erasmus *et al.* (2003) reported that the transcription of 589 genes in wine yeast was at least twofold lower in must containing 40% (w/v) sugar (water activity $\sim$ 0.939) when compared to 22% (w/v) sugar (water activity $\sim$ 0.981). As a consequence of osmotic stress due to high sugar levels, the structural genes concerned with the synthesis of acetic acid from acetaldehyde and of glycerol from dihydroxyacetone phosphate is up-regulated. Yeast cells produce more glycerol as a response to osmotic stress, as it limits water loss from the cytoplasm preventing dehydration of the yeast. The production of acetic acid has been correlated by different authors (Remize *et al.*, 1999) to the over-production of glycerol. The shift in redox balance ($NADH/NAD^+$) caused by the rise of glycerol is compensated by using acetic acid to transform NAD^+ back to NADH. The acetic acid is produced by yeasts through the oxidation of acetaldehyde to acetate. A recent study on Ice wines (Erasmus *et al.*, 2004) has shown that different yeast strains in the same condition of osmotic pressure, respond differently to stress, producing different concentrations of glycerol and acetic acid. The same paper also shows that, the higher the sugar concentration is (high osmotic stress), the more acetic acid and glycerol are produced. They also found a positive correlation between acetic acid and ethanol formation, indicating that as more sugar is consumed, the yeast could potentially form more acetic acid. Therefore, the choice of yeast strain may strongly affect the overall sensory quality of dessert wines, mainly as a function of the initial juice sugar content. This is true for different reasons: first, because an overproduction of acetic acid is obviously detrimental for a dessert wine; second, because too high a level of ethanol reduces the headspace concentration of some volatile aroma compounds, likely suppressing their perception during wine tasting (Robinson *et al.*, 2009). In descending order, the importance of each wine matrix-volatile interaction is ethanol > glucose > glycerol > catechin. For 20 volatiles, also including some esters (isoamyl acetate, ethyl esters), terpenes (linalool, nerol) and norisoprenoids (β-damascenone, α- and β-ionone), increasing ethanol in the matrix was negatively correlated with the analyte peak area and was linear over the range 10–18% v/v. In the light of this, from an aromatic point of view, it is very important that the fermentation of high sugar musts from dehydrated grapes should be stopped as soon as sufficient ethanol (13.5–14.5 % v/v) is produced.

The winemaking of most wines produced from dehydrated grapes is usually strictly linked to traditional procedures that in practice bring a faint control of fermentation parameters, such as microbial population and temperature. In spite of this, inoculation of the musts with selected yeasts is strongly recommended. The principal intent of this operation is a better control of the fermentative process in order to control the volatile acidity and optimize the wines' quality level. The criteria followed to choose a good selected yeast strain for the production of a dessert wine are: high tolerance to sugar and ethanol; low production of acetic acid in difficult fermentation conditions; low production of SO_2-binding compounds such as acetaldehyde; low demand of nutrients (nitrogen, vitamins) mainly if grapes are partially infected by *B. cinerea*; sensibility to SO_2 at levels necessary to stop the alcoholic fermentation (~100 mg L^{-1}) in order to allow easy *mutage*.

Inoculation modalities also have a significant impact on the fermentation of highly sugar concentrated juice obtained from dehydrated grapes. Kontkanen *et al.* (2004) studied the effect of yeast inoculation rate, acclimatization, and nutrient addition on Ice-wine fermentation. The authors showed that fermentation at the lower inoculation rate (0.2 g L^{-1}) consumed less sugar. In these conditions, cells converted a higher proportion of sugar consumed to acetic acid and glycerol in comparison to cells at a higher (0.5 g L^{-1}) inoculation rate. The main impact of acclimatization with respect to the direct inoculation of rehydrated yeast into the juice was higher biomass and viable cell concentration, which allowed for more sugar to be consumed in a shorter time and higher ethanol production. The addition of the yeast nutrient during yeast rehydration increased the rate of biomass accumulation, reduced the fermentation time, reduced the ethanol concentration in the Ice wines, and reduced the rate of acetic acid produced as a function of sugar consumed. Such nutrient supplementation (vitamins and minerals) at the yeast rehydration stages improved cell viability, reducing volatile acidity and off-odours. When necessary, the addition of nitrogen to fermentations also reduces sulfur-like off-odour due to high hydrogen sulphide, production of which is associated with sluggish or stuck fermentations caused by a lack of nitrogen in the must. Low must yeast assimilable nitrogen (YAN) leads to higher alcohols, and low production of esters and long-chain volatile fatty acids. High must YAN leads to a rise of ethyl acetate, acetic acid and volatile acidity due to increased biomass and higher fermentation temperature as a consequence of the vigour. Too high YAN increases the concentration of proteins inducing haze, urea, ethyl carbamate, biogenic amines, and the risk of microbial instability, potential taint from botrytized bunches and atypical ageing character (Bell and Henschke, 2005). Intermediate must YAN favours the best balance between desirable and undesirable chemical and sensory wine attributes, and in harmony with Bely *et al.* (2003), in high Brix musts a correct level of nitrogen could be 160–200 mg L^{-1}. Another survival factor is oxygen. Ribéreau-Gayon *et al.* (2006) suggest that in sweet wine production, a late aeration prevents increases in volatile acidity, and for this reason oxygen should be introduced during the stationary phase of the yeast growth, rather than during the growth phase. This is particularly true when sweet must is not fermented in wood, but when fermentation occurs in tanks in stricter anaerobic conditions.

Finally, temperature control is also a crucial factor in managing fermentation of dehydrated grapes juice. As reported earlier, the composition of this kind of juice is characterized by different inhibitory factors affecting alcoholic fermentation. As a consequence, the setting of a constant temperature (18–20°C) represents a means to allow a good rate of alcoholic fermentation in musts from dehydrated grapes. Extreme temperatures during fermentation can severely affect yeast growth and metabolism, also affecting sensory characteristics of the final wine. Too low a temperature prolongs yeast growth latency causing a rise in volatile acidity, ethyl acetate and acetaldehyde, and affecting yeast population dynamics, mainly during spontaneous fermentation. Too high a temperature during fermentation, influences yeast ethanol resistance because the cell membrane fluidity increases and ethanol can enter the cell more readily, adversely affecting metabolism and cell viability. Moreover, tolerance to both ethanol and temperature is also very strain-dependent (Bisson, 1999). At this point, as soon as sufficient ethanol (13.5–14.5% v/v) is produced, fermentation must be stopped (*mutage*) by adding SO_2 (100–200 mg L^{-1}, depending on pH). The strong addition of SO_2 is necessary to block yeast activity as rapidly as possible, in order to limit the acetaldehyde production, which strongly combines with SO_2. Yeasts are more sensitive to SO_2 at higher temperature, but at the same time low temperatures negatively affect yeast activity. For these reasons, after SO_2 addition, a gradual diminution of the temperature until 4–6°C is useful, but this operation should be carried out at least 2–3 hours after sulfiting.

8.3.3 Maturation

The overall sensory quality, and principally the aroma of dessert wines, improve with maturation. This is possible largely due to their richness in bound volatiles arising from dehydrated grapes, which represent a real 'aroma tank' throughout maturation and bottle ageing of dessert wines. The structure of the whole volatile fraction of this kind of wine benefits a long maturation in wood barrels more than dry wines. Wines from dehydrated grapes are sufficiently rich in aromas, so that volatiles arising from the wood barrel (e.g. whisky lactone, eugenol, vanillin) do not create a disharmony, but rather, add complexity to the final wine. The wood barrel adds woody, spicy and toasted notes to dried fruit, honey, citrus and orange blossom flavours resulting from dehydrated grapes. This union enhances the quality of the wine, determining a significant increase in the overall aroma complexity. This is true throughout the time because wines from dehydrated grapes are richest in flavourless precursors (Genovese *et al.*, 2007) which could be transformed during maturation and ageing into flavour active compounds. Table 8.1 shows the concentrations of bound aroma compounds

Table 8.1 Quantitative data of main glycosylated aroma compounds in sweet Fiano wine (A) and base Fiano wine (B) (adapted from Genovese *et al.*, 2007).

	Concentration (μg L^{-1})[a]		
Compound	**A**	**B**	**%**
3-methyl-1-butanol	160 ± 12	38 ± 1	+321
2-octanone	92 ± 6	11 ± 0.3	+736
1-hexanol	116 ± 9	18 ± 0.8	+544
cis-3hexen-1-ol	10 ± 1	3 ± 0.2	+233
trans-3-hexen-1-ol	1 ± 0.1	2 ± 0.1	−100
cis-linalool oxide	17 ± 3	3 ± 0.2	+467
1-octen-3-ol	11 ± 2	4 ± 0.1	+175
trans-linalool oxide	3 ± 0.2	2 ± 0.1	+50
benzaldehyde	5 ± 0.1	4 ± 0.2	+25
linalool	50 ± 3	2 ± 0.1	+2400
1-octanol	12 ± 1	5 ± 0.2	+140
α-terpineol	19 ± 1	5 ± 0.1	+280
epoxylinalool	25 ± 2	6 ± 0.2	+317
nerol	29 ± 2	13 ± 0.4	+123
β-damascenone	5 ± 0.2	1 ± 0.1	+400
hexanoic acid	8 ± 0.3	2 ± 0.1	+300
geraniol	34 ± 1	9 ± 0.3	+278
esohydroxycineol	17 ± 1	3 ± 0.3	+467
benzyl alcohol	193 ± 13	51 ± 3	+278
2-phenylethanol	167 ± 10	61 ± 3	+174
3-phenylpropanol	13 ± 0.5	1 ± 0.1	+1200
octanoic acid	136 ± 8	31 ± 1	+339

[a]: means of triplicate analysis.
%: change in the concentration of compounds in sweet Fiano wine (A) expressed as percentages with respect to base Fiano wine (B).

measured in the volatile fraction of a sweet (A) and base (B) Fiano wine obtained from the homonym grapes, a neutral variety from southern Italy. Interestingly, the sweet wine showed the highest levels (four times greater) of total bound compounds with respect to the base Fiano wine. The main increments have been observed for compounds that have an active role in wine aroma: terpenes, β-damascenone, benzyl alcohol and 2-phenylethanol. Ultimately, correct barrel maturation has an important fining action on sweet wines. Wood tannins stabilize the proteinaceous colloids, thus minimizing the employment of fining agents such as bentonite, which unfortunately impoverishes the wine of some aromas (Lisanti *et al.*, 2009), and small-sized barrels favour the spontaneous process of clarification, therefore they are to be preferred in the case of dessert wines.

Finally, a microbiological stabilization of these wines is necessary before bottling as, due to high sugar concentration, fermentation re-start is possible with consequent alteration of the wine quality. The free SO_2 concentration should be monitored and the level corrected in order to obtain the right stability during ageing. For example, for a safe ageing of a wine with 14% (v/v) of ethanol and a pH of 3.4, an approximate level of 1.5–2 mg L^{-1} of molecular SO_2 corresponding to 40–50 mg L^{-1} of free SO_2 is suitable before bottling. It is important that bottling is carried out by sterile microfiltration and periodically monitored through microbiological stability tests.

8.4 CONCLUSIONS

Modern enology is not based on a single approach, but on the adaptation of winemaking practices to the compositional peculiarity of a grape variety and to the 'wine design' intended by the winemaker. This chapter is an example of a proper interpretation of this concept of enology, where the main goal is to produce a wine that best expresses the sensory qualities of a particular grape variety.

ACKNOWLEDGEMENTS

Review paper published in the framework of the project PRIN 2008, coordinator F. Mencarelli, supported by the Italian Ministry of Education, University and Research. The authors would like to thank Dr Alessandro Genovese for his kind collaboration.

REFERENCES

Barbe, J.C., G. De Revel, and A. Bertrand. 2002. Gluconic acid, its lactones, and SO_2 binding phenomena in musts from botrytized grapes. *Journal of Agriculture and Food Chemistry*, 50:6408–6412.

Bell, S.J. and P.A. Henschke. 2005. Implications of nitrogen nutrition for grapes, fermentation and wine. *Australian Journal of Grape and Wine Research*, 11:242–295.

Bely, M., A. Rinaldi, and D. Dubourdieu. 2003. Influence of assimilable nitrogen on volatile acidity production by *Saccharomyces cerevisiae* during high sugar fermentation. *Journal of Biosciences and Bioengineering*, 96:507–512.

Bisson, L.F. 1999. Stuck and sluggish fermentations. *American Journal of Enology and Viticulture*, 150:1–13.

Boulton, R.B., V.L. Singleton, L.F. Bisson, *et al.* 1999. Preparation of musts and juice. In *Principles and Practices of Winemaking*, Springer, New York, pp. 65–101.

Castellano, L., A. Genovese, L. Moio, *et al.* 2001. Identificazione del sotolone: molecola di impatto dell'aroma del Vin Santo Toscano. *L'Enologo*, 10:95–102.

Charoenchai, C., G.H. Fleet, and P.A. Henschke. 1998. Effects of temperature, pH, and sugar concentration on the growth rates and cell biomass of wine yeasts. *American Journal of Enology and Viticulture*, 49:283–288.

Chkaiban, L., R. Botondi, A. Bellincontro, *et al.* 2007. Influence of postharvest water stress on lipoxygenase and alcohol dehydrogenase activities, and on the composition of some volatile compounds of Gewürztraminer grapes dehydrated under controlled and uncontrolled thermohygrometric conditions. *Australian Journal of Grape and Wine Research*, 13:142–149.

Costantini, V., A. Bellincontro, D. De Santis, *et al.* 2006. Metabolic changes of Malvasia grapes for wine production during postharvest drying. *Journal of Agriculture and Food Chemistry*, 54:3334–3340.

Dubourdieu, D. and V. Lavigne. 1990. Incidence de l'hyperoxygénation sur la composition chimique et les qualities organoleptiques des vins blanc secs du Bordelais. *Revue Française d'Œnologie*, 124:58–61.

Erasmus, D.J., M. Cliff, and H.J.J. Van Vuuren. 2004. Impact of yeast strain on the production of acetic acid, glycerol, and the sensory attributes of icewine. *American Journal of Enology and Viticulture*, 55:371–378.

Erasmus, D.J., G.K. van der Merwe, and H.J.J. van Vuuren. 2003. Genome-wide expression analyses: Metabolic adaptation of *Saccharomyces cerevisiae* to high sugar stress. *FEMS Yeast Research*, 3:375–399.

Espejo, F. and S. Armada. 2010. Effect of enzyme addition in the making of Pedro Ximénez sweet wines using dynamic pre-fermentative maceration. *South African Journal for Enology & Viticulture*, 31:133–142.

Franco, M., R.A. Peinado, M. Medina, *et al.* 2004. Off-vine grape drying effect on volatile compounds and aromatic series in must from Pedro Ximénez grape variety. *Journal of Agriculture and Food Chemistry*, 52:3905–3910.

Gambuti, A., A. Rinaldi, R. Pessina, *et al.* 2006. Evaluation of Aglianico grape skin and seed polyphenol astringency by SDS-PAGE electrophoresis of salivary proteins after the binding reaction. *Food Chemistry*, 97: 614–620.

Genovese, A., A. Gambuti, P. Piombino, *et al.* 2007. Sensory properties and aroma compounds of sweet Fiano wine. *Food Chemistry*, 103:1228–1236.

Guarrera, N., S. Campisi, and C. Nicolosi Asmundo. 2005. Identification of the odorants of two passito wines by gas chromatographyolfactometry and sensory analysis. *American Journal of Enology and Viticulture*, 56:394–399.

Guérin, L., D.H. Sutter, A. Demois, *et al.* 2009. Determination of activity profiles of the main commercial enzyme preparations used in winemaking. *American Journal of Enology and Viticulture*, 60:322–331.

Hopkins, R.H. and R.H. Roberts. 1935a. The kinetics of alcoholic fermentation by brewers yeast. 1. Effect of concentration of yeast and sugar. *Biochemical Journal*, 29:919–930.

Hopkins, R.H. and R.H. Roberts. 1935b. The kinetics of alcoholic fermentation by brewers yeast. 2. The relative rates of fermentation glucose and fructose. *Biochemical Journal*, 29:931–936.

Kontkanen, D., D.L. Inglis, G.J. Pickering, *et al.* 2004. Effect of yeast inoculation rate, acclimatization, and nutrient addition on ice wine fermentation. *American Journal of Enology and Viticulture*, 55:363–370.

Lisanti, M.T., P. Piombino, A. Gambuti, *et al.* 2009. Oenological treatments for the removal of geosmin, responsible for earthy off-flavour, in wine. *Bulletin de l'OIV*, 82:71–82.

Lopez De Lerma, N., A. Bellincontro, F. Mencarelli, *et al.* 2012. Use of electronic nose, validated by GC-MS, to establish the optimum off-vine dehydration time of wine grapes. *Food Chemistry*, 130:447–452.

Moio, L., M. Ugliano, A. Gambuti, *et al.* 2004a. Influence of clarification treatment on concentrations of selected free varietal aroma compounds and glycoconjugates in Falanghina (*Vitis vinifera* L.) must and wine. *American Journal of Enology and Viticulture*, 55:7–12.

Moio, L., M. Ugliano, A. Genovese, *et al.* 2004b. Effect of antioxidant protection of must on volatile compounds and aroma shelf life of Falanghina (*Vitis vinifera* L.) wine. *Journal of Agricultural and Food Chemistry*, 52:891–897.

Moreno, J.J., F. Cerpa-Calderon, S.D. Cohen, *et al.* 2008. Effect of postharvest dehydration on the composition of pinot noir grapes (*Vitis vinifera* L.) and wine. *Food Chemistry*, 109: 755–762.

Paronetto, L. and F. Dellaglio. 2011. Amarone a modern wine coming from an ancient production technology. *Advances in Food Nutrition Research, Speciality Wines*, 63:285–306.

Pinelo, M., A. Arnous, and A.S. Meyer. 2006. Upgrading of grape skins: significance of plant cell-wall structural components and extraction techniques for phenol release. *Trends in Food Science and Technology*, 17:579–590.

Piombino, P., A. Genovese, A. Gambuti, *et al.* 2010. Effects of off-vine bunches shading and cryomaceration on free and glycosylated flavours of Malvasia delle Lipari wine. *International Journal of Food Science and Technology*, 45:234–244.

Piombino, P., A. Genovese, M.T. Lisanti, *et al.* 2011. Effect of different conditions of dehydration on the volatile composition of Nebbiolo grapes (*Vitis vinifera* L.). 'ŒNO 2011' Ninth International Symposium of Oenology, 15–17 June, Bordeaux, France.

Rapp, A. and J. Marais. 1993. The shelf life of wine: Changes in aroma substances during storage and ageing of white wines. In *Shelf Life Studies of Food and Beverages. Chemical, Biological, Physical and Nutritional Aspects* (ed. G. Charalambous), Elsevier Science, Amsterdam, pp. 891–921.

Remize, F., J.L. Roustan, J.M. Sablayrolles, *et al.* 1999. Glycerol verproduction by engineered *Saccharomyces cerevisiae* wine yeast strains leads to substantial changes in byproduct formation and to a stimulation of fermentation rate in stationary phase. *Applied and Environmental Microbiology*, 65:143–149.

Ribéreau-Gayon, P., D. Dubourdieu, B. Donèche, *et al.* 2006. Other winemaking methods. In *Handbook of Enology*, Vol. 1, John Wiley & Sons, Ltd, Chichester, pp. 445–480.

Robinson, A.L., S.E. Ebeler, H. Heymann, *et al.* 2009. Interactions between wine volatile compounds and grape and wine matrix components influence aroma compound headspace partitioning. *Journal of Agricultural and Food Chemistry*, 57:10313–10322.

Rolle, L., F. Torchio, S. Giacosa, *et al.* 2009. Modifications of mechanical characteristics and phenolic composition in berry skins and seeds of Mondeuse winegrapes throughout the on-vine drying process. *Journal of the Science of Food and Agriculture*, 89:1973–1980.

Rolle, L., F. Torchio, G. Zeppa, *et al.* 2008. Anthocyanin extractability assessment of grape skins by texture analysis. *Journal International des Sciences de la Vigne et du Vin*, 42:157–162.

Ruiz, M.J., L. Zea, L. Moyano, *et al.* 2010. Aroma active compounds during the drying of grapes cv. Pedro Ximenez destined to the production of sweet Sherry wine. *European Food Research and Technology*, 230:429–435.

Salinas, M.S., J. Garijo, F. Pardo, *et al.* 2003. Color, polyphenol, and aroma compounds in rosé wines after prefermentative maceration and enzymatic treatments. *American Journal of Enology and Viticulture*, 54:195–202.

Santonico, M., A. Bellincontro, D. De Santis, *et al.* 2010. Electronic nose to study postharvest dehydration of wine grapes. *Food Chemistry*, 121:789–796.

Schneider, V. 1998. Must hyperoxidation: A review. *American Journal of Enology and Viticulture*, 49:65–73.

Selli, S., B. Bagatarb, K. Sena, *et al.* 2011. Evaluation of differences in the aroma composition of free-run and pressed neutral grape juices obtained from Emir (*Vitis vinifera* L.). *Chemistry and Biodiversity*, 8:1776–1782.

Ugliano, M. 2009. Enzymes in winemaking. In *Wine Chemistry and Biochemistry*, Springer, New York, pp. 103–126.

Wilson, B., C.R. Strass, and P.J. Williams. 1984. Changes in free and glycosidically bound monoterpenes in developing Muscat grapes. *Journal of Agriculture and Food Chemistry*, 32: 919–924.

9 Methods of Vinification to Preserve the Colour in Red Grape Passiti Wines. Aleatico: a Case Study

Riccardo Cotarella

Falesco Winery, Montecchio, Italy

9.1 INTRODUCTION

This chapter describes the techniques used in the production of red Passiti wines with the aim of achieving optimum results with respect to colour and varietal aroma expression. The work begins in the vineyard using various agronomic and agricultural techniques to arrive at technical solutions for the dehydration and processing of the grapes. The need to evaluate the process in its entirety is fundamental to achieving the best results; therefore the work needed in the vineyard cannot be separated from the skilful techniques employed during dehydration and vinification.

9.2 GENERAL INFORMATION

In the following sections, we describe as a case study the experimental work carried out on Aleatico grapes, to obtain the maximum varietal and chromatic expression of the wine. Aleatico is a red berry variety with a medium-sized, tri-lobed pentagonal dark-green leaf; the grape bunch is medium, elongated with a wing; the grape is medium in size, of an irregular shape and is blue-vermilion in colour. The berry has a very characteristic flavour similar to that of Muscat with high terpene alcohol, mainly nerol and geraniol, and a bright red colour where malvidin-3-monoglucoside represents almost 90% of the total anthocyanin monoglucosides (Bellincontro *et al.*, 2006). The plant prefers a warm climate and hilly ground with good sun exposure.

There are conflicting opinions as to its origin: some believe it originated in Tuscany (where it goes by the name of Liatico) as a mutation of Muscat, and others that it was a vine introduced from Greece from where it spread to Puglia and central Italy. The Aleatico variety is characterized by a very heterogeneous clone population which, over time, has adapted itself to different microclimates and *terroir*. A particular and optimal microclimate has been found along the north-western shore of Lake Bolsena. This area includes the cities of Gradoli, Grotte di Castro, San Lorenzo Nuovo and Latera, all within the Province of Viterbo, where the historical DOC 'Aleatico di Gradoli' is produced as a fortified wine.

Sweet, Reinforced and Fortified Wines: Grape Biochemistry, Technology and Vinification, First Edition.
Edited by Fabio Mencarelli and Pietro Tonutti.

9.3 CHARACTERISTICS OF THE VINEYARD AND ITS MANAGEMENT

Management of the vine canopy is carried out to achieve a correct balance between leaves and bunches through careful agronomic practices with the aim of obtaining healthy and ripe grapes with their skins intact. These are the essential elements that will permit an optimum dehydration process and subsequent vinification.

With respect to soil management, spontaneous grassing on alternate rows is maintained and the grass is cut periodically; this improves the soil structure, reduces soil erosion and enables machines to enter the vineyard during unfavourable weather conditions, allowing pesticide treatment and harvesting. The natural availability of exchangeable potassium in the soil should be checked because together with irrigation, this can control potassium concentration. When vines manifest water stress they tend to absorb potassium from the soil in order to boost internal osmosis and avoid losing water. This phenomenon must be limited especially for the dehydration process, as potassium-rich grapes undergo a marked fall in tartaric acid, and a subsequent rise in pH; this condition is unfavourable to obtaining maximum chromatic expression in the wine.

Management of the vine canopy is very important, especially green pruning. Eliminating excess buds regulates production, enhances the development and sun exposure of the grapes, and helps to prevent attack by fungi, such as *Botrytis*. Two buds per spur are left, eliminating the sterile ones and, at the same time, leaving useless buds for the next pruning. These techniques enable an equilibrium of production to be reached; overcrowding of vegetation is avoided, thus allowing for an efficient aeration of the foliage and preventing excessive shading of the grapes, which causes uneven ripening. This canopy management is worthwhile to achieve healthy grapes that reach optimal phenolic ripening.

During the last period of ripening special attention must be paid to *Botrytis* control. The factors that favour this disease are high temperatures and high levels of relative humidity, grape morphology (sparse grape bunches are less susceptible than compact ones), damage from hail and attack by moths. A particular characteristic of the clone present in the vineyard is that it has a distinct aroma and grape bunches that tend naturally to lock the individual berries tightly making a very compact bunch. In the case of compact bunches, using copper salt treatment during flowering can result in bunches with fewer berries. We carried out an experiment using two treatments of copper sulphate during flowering with the aim of artificially inducing the flowers to fall, and thereby reducing the probability of fungal attack. Of the 3.5 ha set aside for the experiment, 3 ha were treated, while the remaining half hectare was used as a control and managed conventionally with two systemic treatments during flowering.

The effect was evident right from the pre-closure phase. The treated bunches were sparser and had fewer berries than those that were untreated (Figure 9.1). At the end of ripening the treated bunches were smaller, lighter in weight, more uniform in colour, with a thicker skin, and disease-free (Table 9.1).

Table 9.1 Influence of copper sulfate and conventional treatment (untreated) on bunch features. Data are the mean of 100 bunches ± SD.

	Bunch weight (g)	Berries number	Berry weight (g)	Sugar content at harvest (g L^{-1})
Copper sulfate	180 ± 10	91 ± 6	1.97 ± 0.03	229.5 ± 4.1
Conventional	300 ± 13	140 ± 10	2.10 ± 0.05	204.8 ± 5.3

Figure 9.1 Bunches of grapes untreated (left) and treated with copper sulfate (right). (For a colour version of this figure, please see the colour plate section.)

During the veraison period, the bunches were thinned out, leaving 5–6 per stem and 600 g of grape per vine, resulting in a production at harvest of 30 q ha^{-1}. Another important operation involved removing the leaves covering the bunches a few days before harvest to enhance aeration and sun exposure.

9.4 HARVEST

In our case study, harvesting was done when the grapes had reached a technological maturity of 229.5 g L^{-1}, a total acidity of 8.20 g L^{-1} and a pH of 3.3. In particular, it was observed that the sparse bunches of copper-treated vines were riper and healthier than the compact ones of untreated vines, which began to show the first signs of rot in the internal part of the bunch, with skins split, and leaking of the must. The bunches were harvested manually and placed in special perforated plastic boxes, specific for the dehydration process.

9.5 DEHYDRATION TECHNIQUES AND VINIFICATION

With respect to grape-drying techniques, the use of a controlled environment for dehydration is now normal practice for wineries that intend to produce wines with a high enological value. A purpose-built specially insulated room was used where the environmental conditions, temperature and relative humidity were fixed initially and then maintained automatically. A fixed air flow was passed through the stacks of perforated boxes containing the grapes.

Table 9.2 Kinetics of dehydration; 1, 2, 3 and 4 are the number of plastic boxes weighed. The % refers to the variation of plastic boxes weight.

Day	1 (kg)	2 (kg)	3 (kg)	4 (kg)	Relative humidity (%)	Temperature (°C)	Weight loss (%)
20/9	4650	5610	4320	5060	72	16.0	0.0
21/9	4300	5480	4210	4950	70	16.0	3.6
22/9	4230	5410	4120	4900	70	17.0	4.9
23/9	4190	5310	4050	4870	70	16.5	6.2
24/9	4110	5240	3960	4670	67	17.0	8.4
26/9	4050	5120	3900	4590	68	17.5	10.1
27/9	4000	5060	3860	4550	67	17.0	11.0
29/9	3940	5010	3820	4520	67	16.5	11.9
1/10	3900	4990	3800	4480	68	16.0	12.6
3/10	3840	4860	3710	4320	68	16.0	14.8
6/10	3749	4655	3586	4153	68	16.0	17.8
10/10	3508	4148	3248	3863	67	16.0	24.8
13/10	3456	3942	3127	3765	66	17.0	27.2
17/10	3394	3932	3085	3682	67	17.0	28.2
21/10	3256	3765	2942	3546	64	17.0	31.2
24/10	3188	3628	2822	3450	65	17.0	33.4

A very important aspect of the dehydration technique is the rate of dehydration (% of weight loss per day); if this is done too quickly, the skins of the berries may split. Such a phenomenon can be seen above all in varieties with a large berry size and thin skin, such as Aleatico. Injury to the skin carries a high risk of oxidation and mould development, mainly *Botrytis* and *Aspergillus*, which are responsible for increasing the oxidation of phenolic compounds, and can even cause the formation of ochratoxin.

Dehydration was performed with relative humidity below 70% and temperatures between 15 and 18°C; a lower temperature of dehydration (10°C) for Aleatico is beneficial to maintain varietal aroma and favour the increase of some polyphenols (Mencarelli *et al.*, 2010), but would entail a greater energy cost at this time. Table 9.2 shows the evolution of weight loss obtained during the dehydration process of Aleatico grapes; it was noted that weight loss was constant and the environmental conditions were within acceptable limits throughout the working period. The condition of the grapes was monitored daily, and measurements of weight loss taken every 1–2 days by weighing sample crates from the dehydration chamber.

Every 4–5 days samples were taken for analysis especially of total anthocyanins and polyphenols (Table 9.3).

Table 9.3 Values of the various grape features during dehydration.

Date	Total anthocyanins mg L^{-1} malvidin	Total polyphenols mg L^{-1} gallic acid	°Brix	Total acidity g L^{-1}	pH
20/09	485	2206	20.5	8.40	3.22
26/09	515	2300	23.6	8.50	3.24
02/10	586	2567	26.7	8.20	3.25
12/10	611	2742	28.8	8.00	3.25
19/10	635	2791	31.8	7.80	3.31
24/10	666	2903	33.1	7.60	3.30

Vinification of the dehydrated grapes was carried out with the aim of maintaining the phenolic patrimony and varietal aroma of the grape. Grapes were removed from the dehydration chamber, mechanically destalked and pressed; during this phase the skins are broken by allowing the must to flow out, and to avoid uncontrollable oxidative phenomena, micronized dry ice was added to the grapes directly on the press. The dry ice cools the product and produces CO_2 which preserves it from contact with oxygen, limiting oxidation to a minimum. Metabisulphate was spread directly onto the grapes in the amount of 60 ppm. The must together with the skins was stored in thermo-regulated tanks where pre-fermentation maceration allowed the release of anthocyanins to be controlled separately from that of tannins, which will be extracted later during alcoholic fermentation.

In order to obtain good results from this process, it is essential that the grapes are in a good chemical and physical condition at the start of the process because the must has to reach the maceration stage in a perfect state.

Pre-fermentation maceration consists of lowering the temperature of the pressed grapes to 3–4°C, uniformly within the whole mass. Good results also depend on low potassium content of the berries, because as stated earlier, low temperatures cause tartaric precipitation and subsequently a fall in total acidity and a rise in pH. Therefore it is extremely important that potassium concentration is accurately measured.

Maceration can last from a few hours to several days; the length of this process depends on the condition of the grapes and the solubility of the anthocyanins, which need to be monitored frequently. The mass was stirred during maceration, paying special attention to avoiding excessive oxygenation. When the pre-fermentation maceration was over, the mass was heated to a fermentation temperature of 20–22°C.

In yeast inoculation, choosing the right dose and type are important to guarantee regular fermentation for must with very high sugar concentration. High concentration of yeasts is not advisable, since the capacity of the yeast membrane to absorb anthocyanins, particularly during long fermentations where contact with the yeast can be excessive, results in a greater loss of colour, which is removed with the lees.

Pumping over and plunging the cap down was carried out to extract as much colour as possible without removing excessive quantities of tannins, which tends to harden the wine. For this reason the majority of pumping over and cap plunging took place in the first 3–4 days of fermentation, decreasing radically towards the end of the fermentation process. The final treatment of the cap was simply a light dampening to prevent excessive superficial drying and volatile acidity. As an alternative, a 'sweet drawing off' can be carried out, by separating the must from the skins once the colour release has ceased and before the release of tannins from grape seeds begins. This process entails risk to the progress of fermentation and can require a second inoculum of yeasts, which raises the absorption problem described earlier.

Table 9.4 details the chemical analysis of the dehydrating Aleatico grapes during the maceration and fermentation processes.

9.6 AGEING AND STORAGE

As soon as the desired alcohol content was reached, fermentation was blocked by rapidly lowering the temperature to 0°C and adding 30–40 ppm of SO_2, to protect the product from oxidation. This last operation can make all the difference and must be done accurately. Indeed, SO_2 can provoke decolouration of the anthocyanins because the cationic nature of anthocyanins makes them susceptible to attack from any agent with a nucleophile character,

Table 9.4 Chemical analyses during maceration and fermentation.

	Days of maceration				Days of fermentation							
	1	**3**	**4**	**5**	**6**	**8**	**10**	**12**	**14**	**16**	**18**	**20**
Tank number	20	20	20	20	20	20	20	20	20	20	20	20
Temperature °C	4	4	4	12	22	22	22	22	22	21	22	22
Sugars g L^{-1}	340	340	340	339	302	267	234	208	177	152	136	128
Distilled alcohol % v/v	0.00	0.00	0.00	0.20	2.40	4.56	6.56	7.89	9.83	11.50	12.56	12.80
Total alcohol % v/v	20.40	20.40	20.40	20.54	20.52	20.58	20.60	20.40	20.45	20.62	20.72	20.48
Total acidity g L^{-1}	7.70	7.40	7.35	7.45	7.45	7.50	7.45	7.40	7.40	7.40	7.45	7.40
Volatile acidity g L^{-1}	0.06	0.06	0.07	0.22	0.25	0.26	0.28	0.34	0.38	0.44	0.51	0.57
L-malic acid g L^{-1}	3.12	3.09	3.00	3.00	3.00	3.00	2.88	2.78	2.69	2.56	2.50	2.50
pH	3.24	3.35	3.38	3.39	3.40	3.40	3.40	3.40	3.42	3.40	3.40	3.42
Free SO_2 mg L^{-1}	12	12	12	14	14	14	16	16	16	16	16	16
Total SO_2 mg L^{-1}	32	32	32	34	34	34	38	38	38	40	38	40
Total anthocyanins mg L^{-1}	150	278	390	450	509	556	602	600	598	572	569	550
Colour index	3.20	4.40	6.40	7.58	8.90	10.24	11.86	11.54	11.40	11.32	11.30	1.25
Total polyphenols mg L^{-1}	761	842	912	1.012	1.122	1.455	1.671	1.876	1.996	2.034	2.134	2.250

such as the bisulfite ion (HSO_3^-). For this reason, the use of SO_2 must be cleverly managed to avoid an excessive loss of colour, but at the same time guarantee protection of the product. If the product presents particular chemical characteristics, principally due to its pH value, the use of SO_2 can be limited. The level of molecular sulfur must be analysed daily because it varies greatly depending on the concentration of hydrogen, in other words, a wine with a free sulfur level of 30 mg L^{-1} will have a molecular SO_2 that is notably different if the pH is 3.3 rather than 3.8.

After sulfuring, we proceeded to filtration of the product; this is necessary to remove the lees. The presence of lees can produce various undesirable events: first, if the product is not kept cool, the risk of fermentation is high and can produce wines with an unpleasant aroma. Furthermore, as mentioned earlier, the lees absorb colour and therefore prolonged contact of the lees with wine is inadvisable especially if the production of polysaccharides and mannoproteins from yeast lysis is of no interest for the wine stabilization.

Filtration was carried out using a membrane filtration system, starting at a porosity of 20 μ and taking it down to 0.45 μ. The use of cardboard filtration, although rapid and cheap, has the inconvenience of reducing the chromatic level of the wines. In fact, cardboard, even when washed and rinsed with wine, can withhold a significant aliquot of anthocyanins.

ACKNOWLEDGEMENT

The author would like to thank all the technical staff at Riccardo Cotarella Consulting srl.

REFERENCES

Bellincontro A., A. Fardelli, D. De Santis, *et al.* 2006. Postharvest ethylene and 1-MCP treatments both affect phenols, anthocyanins, and aromatic quality of Aleatico grapes and wine. *Australian Journal of Grape and Wine Research*, 12:141–149.

Mencarelli F., A. Bellincontro, I. Nicoletti, *et al.* 2010. Chemical and biochemical changes of healthy phenolic fractions in winegrape by means of postharvest dehydration. *Journal of Agricultural and Food Chemistry*, 58:7557–7564.

10 Role of Yeasts in Sweet Wines

Juan C. García-Mauricio and Teresa García-Martínez

Department of Microbiology, University of Córdoba, Córdoba, Spain

10.1 YEAST BIODIVERSITY IN HIGH SUGAR CONCENTRATION GRAPE MUSTS

The yeasts effecting the fermentation of grape musts come from the grapes and the cellar ambient. A number of non-*Saccharomyces* species (particularly *Hanseniaspora*, *Candida*, *Pichia* and *Metschnikowia*) spontaneously trigger alcoholic fermentation in grape must but are soon surpassed in growth by strains of *Saccharomyces cerevisiae*, which governs the process from middle to end and is frequently the sole species ultimately remaining in the medium, and perhaps strains of closely related species such as *Saccharomyces bayanus* (Fleet, 2008; Pretorius *et al.*, 1999). Although the presence of yeasts and other microbes in high-sugar musts has scarcely been studied, a number of *S. cerevisiae* strains are known to be highly resistant to osmotic stress and capable of synthesizing ethanol under these conditions (Malacrinò *et al.*, 2005).

Traditionally, Andalusia (south Spain) has produced special sweet wines with must from sundried Pedro Ximénez or Muscat grapes, which can easily contain about 50% (w/v) reducing sugars. García-Martínez *et al.* (2008) found the contents in moulds and lactic and acetic bacteria of high-sugar must to depend on the health status of the starting grapes and on the grape-drying conditions; moulds to disappear during fermentation and bacteria to drop in number. These authors (García-Martínez *et al.*, 2007) found partially fermented Pedro Ximénez musts with an initial sugar content of 400 g L^{-1} to contain strains of the species *S. cerevisiae* alone. Research by Urso *et al.* (2008) into the biodiversity and dynamics of yeast populations during production of sweet wine from raisins revealed the presence of a high diversity of moulds on the grapes; by contrast, yeasts occurred virtually exclusively in the must and were limited to non-*Saccharomyces* genera (*Hanseniaspora*, *Metschnikowia*, *Pichia*, *Candida*, *Torulaspora* and *Debaryomyces*) and the species *Saccharomyces cerevisiae*. The last species was that contributing to the greatest extent to the properties of the resulting wines. Recently, Tofalo *et al.* (2009) detected various osmotolerant yeast species including *Candida zemplinina*, *Candida apicola* and *Zygosaccharomyces bailii* throughout the fermentation of high-sugar musts, and found *S. cerevisiae* to prevail after 15 days of fermentation. Finally, the role of the non-*Saccharomyces* yeasts in wine fermentation is receiving increasing attention by wine microbiologists (Comitini *et al.*, 2011), and some authors have suggested that non-*Saccharomyces* species such as *Kloeckera apiculata*, *Candida stellata* and

Sweet, Reinforced and Fortified Wines: Grape Biochemistry, Technology and Vinification, First Edition.
Edited by Fabio Mencarelli and Pietro Tonutti.

Torulaspora delbrueckii might grow more efficiently than *S. cerevisiae* during the fermentation of high-sugar musts and lead to sweet wines with improved properties as a result (Bely *et al.*, 2008; Tofalo *et al.*, 2009).

10.2 MOLECULAR RESPONSE OF *S. CEREVISIAE* TO HIGH SUGAR STRESS CONDITIONS

In their book entitled *Yeast Stress Responses*, Hohman and Mager (2003) defined stress as follows: 'Environmental conditions that threaten the survival of a cell, or at least prevent it from performing optimally, are commonly referred to as cell stress'. During wine production, yeast cells are affected by several conditions that are adverse to growth and viability (oxidative, osmotic and ethanol stress among others) and they should detect and respond to these conditions, otherwise alcoholic fermentation can be negatively affected. So, one of the stress conditions that can affect *S. cerevisiae* cells during their growth in high sugar grape juices is osmotic stress. Yeast cells in high glucose concentration have a decreased growth rate during the initial phase of fermentation and even this condition can contribute to stuck fermentations (Guidi *et al.*, 2010). *S. cerevisiae* adapts to increased osmotic stress by enhanced production of intracellular glycerol as the main compatible solute to counter-balance the osmotic pressure, and some yeasts even have active glycerol uptake pumps (Hohmann, 2002). However, osmostress not only induces osmolyte accumulation, but also has a great impact on cellular physiology, such as cytoskeleton reorganization, changes in cell-wall dynamics, alteration of ion homeostasis, metabolic adjustments and cell-cycle arrest, as well as a very notable effect on gene expression (Hohmann, 2002; Nadal and Posas, 2008). The osmoregulatory molecular response in *S. cerevisiae* has been well characterized and it is complex and influenced by a number of factors (Erasmus *et al.*, 2003; Hohmann and Mager, 2003; Jiménez-Martí *et al.*, 2011a). The yeast high-osmolarity glycerol (HOG) mitogen-activated protein kinase (MAPK) pathway is the most prominent signalling pathway that allows *S. cerevisiae* cells to perceive and to quickly respond to altered osmolarity (Capaldi *et al.*, 2008).

The key step of glycerol synthesis is catalysed by an NADH-dependent cytosolic glycerol-3-phosphate dehydrogenase that converts dihydroxyacetone phosphate to glycerol-3-phosphate with the production of NAD^+. Two isoforms of this enzyme are encoded by the *GPD1* and *GPD2* genes. *GPD1* is strongly induced by osmotic stress (Erasmus *et al.*, 2003; Zuzuarregui *et al.*, 2005), expression of *GPD2* is not up-regulated in high sugar fermentations and does not reflect the elevated levels of glycerol found in these wines; it is reported to have a role in maintaining redox balance under anaerobic conditions (such as fermentation) and is not induced by osmotic stress (Pigeau and Inglis, 2007). An increase in glycerol production requires an equimolar increase of cytoplasmic NADH formation. Under high osmotic stress, this requirement seems to be partially met by decreased reduction of acetaldehyde to ethanol on the one hand, and an increased oxidation to acetic acid on the other. Under conditions of stress, acetic acid formation, therefore, plays an important role in maintaining the redox balance in yeast cells (Erasmus *et al.*, 2003). This may occur through the action of $NAD(P)^+$-dependent aldehyde dehydrogenase activity which reduces $NAD(P)^+$ to NAD(P)H while oxidizing acetaldehyde to acetic acid. Pigeau and Inglis (2007) reported that the aldehyde dehydrogenase encoded by *ALD3* but not *ALD2*, *-4*, *-5* and *-6* is responsible, in part, for the increased acetic acid found in Ice wines based on the expression profile of these genes during fermentation. Additional responses of yeast under osmotic stress are required to increase acetic acid in sweet wines.

Erasmus *et al.* (2003) have studied the effect of 40% (w/v) sugars in Riesling grape juice on the transcriptional response of a polyploidy industrial wine yeast strain. Genome-wide expression analyses revealed that the transcription of 589 genes was affected by more than twofold. They found that genes involved in the glycolytic and pentose phosphate pathways were up-regulated. Gene expression profiles indicate that the oxidative and non-oxidative branches of the pentose phosphate pathway were up-regulated and might be used to shunt more glucose-6-phosphate and fructose-6-phosphate, respectively, from the glycolytic pathway into the pentose phosphate pathway. It has been suggested that the yeast cell prevents the accumulation of fructose-1,6-biphosphate: by controlling the influx of glucose by inhibition of hexose transporters (Salmon and Mauricio, 1994); by feedback inhibition of hexokinase by trehalose-6-phosphate limiting phosphorylation of glucose; or by creating a demand for ATP by activating the glycerol, trehalose and glycogen futile cycles which act as safety valves to prevent substrate-accelerated death (Erasmus *et al.*, 2003; Jiménez-Martí *et al.*, 2011b). Furthermore, the genes involved in production of acetic acid from acetaldehyde, and succinate from glutamate, were also up-regulated. By contrast, genes involved in *de novo* synthesis of purines, pyrimidines, histidine and lysine were down-regulated by sugar stress. Proteomic characterization of *S. cerevisiae* response to high glucose concentrations also has revealed that most of the proteins involved in glycolysis and pentose phosphate pathways are up-regulated under these conditions (Pham and Wright, 2008). Recently, a similar approach has shown the over-expression of peroxiredoxin up 20% w/v glucose, as being the protein involved in protection against oxidative stress insult (Guidi *et al.*, 2010). The *YHR087W* gene (also known as *RTC3*) has been reported to be induced by several conditions of high sugar stress (Erasmus *et al.*, 2003; Jiménez-Martí *et al.*, 2011b).

Jiménez-Martí *et al.* (2011a) consider the possibility that the response to osmotic stress found at the beginning of vinification could be an important trait for the adaptation of wine yeasts. They demonstrate that the pre-adaptation of the wine yeast strains showing growth problems at the beginning of vinification in a rehydration medium containing 2% or 5% glucose (depending on the yeast strain) may increase their vitality when inoculated into high sugar media. Pham and Wright (2008) also reported that enhanced ethanol fermentation under very high glucose conditions can be achieved with the aid of amino acid supplementation.

10.3 INFLUENCE OF *S. CEREVISIAE* ON BYPRODUCT FORMATION IN SWEET WINES

Yeast influence on wine composition and quality is well known. Besides ethanol and CO_2, the metabolism of yeasts yields a great number of byproducts, for example glycerol, acetaldehyde, acetic acid, succinic acid and lactic acid. Fermentation conditions (e.g. high sugar concentration grape musts) affect concentrations of these compounds. As stated earlier, *S. cerevisiae* adapts to increased osmotic stress by enhanced production of intracellular glycerol as the main compatible solute to counter-balance the osmotic pressure. For this reason, glyceropyruvic fermentation is very active in these conditions. Due to sugar stress by yeast, pyruvate concentration is greatly increased and to maintain the redox potential balance other mechanisms in addition to the acetic acid formation from acetaldehyde are necessary. Succinic acid may also be formed from pyruvate to maintain the redox balance. In conclusion, the high contents in glycerol and acetic acid found in the partially fermented musts relative to dry wines obtained from ripe and undried grapes of the same variety are consistent with the above-described effect of osmotic stress on yeasts under such drastic conditions. In addition, as a consequence of previous acetaldehyde, acetoin, succinic acid and 2,3-butanediol also show increased concentrations in these special wines (García-Martínez *et al.*, 2011; López de Lerma and Peinado, 2011).

ACKNOWLEDGEMENT

This work was part of a research project funded by Spain's Ministry of Economy and Competitiveness (INIA-CCAA) and FEDER (RTA2011-00020-C02-02).

REFERENCES

Bely, M., P. Stoeckle, I. Masneuf-Pomarède, *et al.* 2008. Impact of mixed *Torulaspora delbrueckii–Saccharomyces cerevisiae* culture on high-sugar fermentation. *International Journal of Food Microbiology*, 122:312–320.

Capaldi, A.P., T. Kaplan, Y. Liu, *et al.* 2008. Structure and function of a transcriptional network activated by the MAPK Hog1. *Nature Genetics*, 40:1300–1306.

Comitini, F., M. Gobbi, P. Domizio, *et al.* 2011. Selected non-*Saccharomyces* wine yeasts in controlled multistarter fermentations with *Saccharomyces cerevisiae*. *Food Microbiology*, 28:873–882.

Erasmus, D.J., G.K. van der Merwe, and H.J.J. van Vuuren. 2003. Genome-wide analyses, metabolic adaptation of *Saccharomyces cerevisiae* to high sugar stress. *FEMS Yeast Research*, 3:375–399.

Fleet, G.H. 2008. Wine yeasts for the future. *FEMS Yeast Research*, 8:979–995.

García-Martinez, T., A. Bellincontro, M.N. López de Lerma, *et al.* 2011. Discrimination of sweet wines partially fermented by two osmo-ethanol-tolerant yeasts by gas chromatographic analysis and electronic nose. *Food Chemistry*, 127:1391–1396.

García-Martínez, T., O. Maestre, R.A. Peinado, *et al.* 2007. Aislamiento y caracterización de levaduras tolerantes a altas concentraciones de azúcares. In *Avances en Ciencias y Técnicas Enológicas*, Editorial: Junta de Extremadura, pp. 102–104.

García-Martínez, T., R.A. Peinado, O. Maestre, *et al.* 2008. Fermentación de mostos con elevado contenido en azúcares mediante bioinmovilización de levaduras. *Bulletin de l'OIV*, 81:559–568.

Guidi, F., F. Magherini, T. Gamberi, *et al.* 2010. Effect of different glucose concentrations on proteome of Saccharomyces cerevisiae. *Biochimica et Biophysica Acta*, 1804:1516–1525.

Hohmann, S. 2002. Osmotic stress signaling and osmoadaptation in yeasts. *Microbiology and Molecular Biology Reviews*, 66:300–372.

Hohmann, S. and W.H. Mager. 2003. *Yeast Stress Responses*, Springer-Verlag, Berlin/Heidelberg.

Jiménez-Martí, E., M. Gomar-Alba, A. Palacios, *et al.* 2011a. Towards an understanding of the adaptation of wine yeasts to must: relevance of the osmotic stress response. *Applied Microbiology and Biotechnology*, 89:1551–1561.

Jiménez-Martí, E., A. Zuzuarregui, M. Gomar-Alba, *et al.* 2011b. Molecular response of *Saccharomyces cerevisiae* wine and laboratory strains to high sugar stress conditions. *International Journal of Food Microbiology*, 145:211–220.

López de Lerma, N. and R.A. Peinado. 2011. Use of two osmoethanol tolerant yeast strains to ferment must from Tempranillo dried grapes. Effect on wine composition. *International Journal of Food Microbiology*, 145:342–348.

Malacrinò, P., E. Tosi, C. Caramia, *et al.* 2005. The vinification of partially dried grapes: a comparative fermentation study of *Saccharomyces cerevisiae* strains under high sugar stress. *Letters of Applied Microbiology*, 40:466–472.

Nadal, E. and F. Posas. 2008. Regulation of gene expression in response to osmostress by the yeast stress-activated protein kinase Hog1. In *Stress-activated Protein Kinases* (eds F. Posas and A.R. Nebreda), Springer-Verlag, Berlin/Heidelberg.

Pham, T.K. and P.C. Wright. 2008. The proteomic response of *Saccharomyces cerevisiae* in very high glucose conditions with amino acid supplementation. *Journal of Proteome Research*, 7:4766–4774.

Pigeau, G.M. and D.L. Inglis. 2007. Response of wine yeast (*Saccharomyces cerevisiae*) aldehyde dehydrogenases to acetaldehyde stress during Icewine fermentation. *Journal of Applied Microbiology*, 103:1576–1586.

Pretorius, I.S., T.J. van der Westhuizen, and O.P.H. Augustyn. 1999. Yeast biodiversity in vineyards and wineries and its importance to the South African wine industry. A review. *South African Journal for Enology & Viticulture*, 20:61–74.

Salmon, J.M. and J.C. Mauricio. 1994. Relationship between sugar uptake kinetics and total sugar consumption in different industrial *Saccharomyces cerevisiae* strains during alcoholic fermentation. *Biotechnology Letters*, 16:89–94.

Tofalo, R., C. Chaves-López, F. Di Fabio, *et al.* 2009. Molecular identification and osmotolerant profile of wine yeasts that ferment a high sugar grape must. *International Journal of Food Microbiology*, 130:179–187.

Urso, R., K. Rantsiou, P. Dolci, *et al.* 2008. Yeast biodiversity and dynamics during sweet wine production as determined by molecular methods. *FEMS Yeast Research*, 8:1053–1062.

Zuzuarregui, A., P. Carrasco, A. Palacios, *et al.* 2005. Analysis of the expression of some stress-induced genes in several commercial wine yeast strains at the beginning of vinification. *Journal of Applied Microbiology*, 98:299–307.

11 *Botrytis* Infection: Grey Mould and Noble Rot

Andrea Vannini and Gabriele Chilosi

DIBAF, University of Tuscia, Viterbo, Italy

11.1 INTRODUCTION

The aim of this short review is to provide an overview of the *Botrytis cinerea* disease cycle on grape, and the conditions inducing 'bunch rot' or 'noble rot' development. The complex factors governing host–parasite–environment interactions and the degree of susceptibility of grape to *B. cinerea* are described. In particular, the process of noble rot and the site, climatic and host factors driving the process are explored. In addition, a section is dedicated to new technologies available to induce noble rot development under postharvest conditions.

Botrytis cinerea Pers. ex Fr. (teleomorph: *Botryotinia fuckeliana* (de Bary) Whetzel), is the cause of grey mould disease in the flowers, fruits, leaves, shoots and soil storage organs of more than 200 plant hosts worldwide. This fungus is ubiquitous in dehiscent or dead plant material of any origin and in soil (Jarvis, 1977). The fungus is considered a necrotroph; however, it can also be considered a true plant pathogen due to the severe pre- and postharvest losses it inflicts on crops worldwide, including grape, kiwifruit, strawberry and ornamentals. Over the last three decades, a large number of studies on *Botrytis*-grape and *Botrytis–Arabidopsis* interactions have been conducted for the purpose of elucidating various stages of development of this fungus and to understand the mechanisms involved in pathogenesis and resistance. The genomes of two strains of *B. cinerea* have been sequenced and compared to each other and to the genome of the related fungal pathogen *Sclerotinia sclerotiorum*. These resources will facilitate the functional studies designed to better understand what makes these fungi such successful and persistent pathogens of agronomic crops (Amselem *et al.*, 2011). However, despite the large amount of research performed on these diseases, many crops are dramatically damaged in terms of direct product losses and indirect economic and environmental impact in the form of pesticides applied to prevent this disease.

Grape is one of the most economically valuable hosts of *B. cinerea*. The fungus can be latent in different tissues and organs, although most of the damage is caused on bunches. The disease occurs on ripe berries, the impact being highest in association with heavy rainfall or long periods of high humidity. Under these conditions, the colonized berries rot completely (bunch rot) and reduce the quality of wine by interfering with fermentation and affecting the flavour and colour (Elad *et al.*, 2004). By contrast, under specific climatic conditions involving moist nights, foggy mornings and dry days, *B. cinerea* infection is limited to the outermost layers of epidermis (Ribéreau-Gayon *et al.*, 1980), promoting dehydration of

Sweet, Reinforced and Fortified Wines: Grape Biochemistry, Technology and Vinification, First Edition.
Edited by Fabio Mencarelli and Pietro Tonutti.

berries, increased sugar concentration and the occurrence of specific sensory compounds. This form of *B. cinerea* infection results in an improvement of grape quality (termed 'noble rot') which can then be used for the production of sweet, smooth, full-bodied wines, the so-called 'botrytized wines'.

11.2 THE INFECTION PROCESS OF *B. CINEREA* AND 'BUNCH ROT' DEVELOPMENT

Spores produced in late winter and early spring on over-wintering mycelium and/or sclerotia are considered the most important source of *B. cinerea* infection (Elmer and Michailides, 2004). The inoculum can potentially originate from a large number of host plants. The rate and extent of an epidemic depends on a dynamic interaction between the pathogen, the host and the environment (Deytieux-Belleau *et al.*, 2009). Key factors are represented by the genetic structure of the pathogen population (Fournier *et al.*, 2005; Martinez *et al.*, 2003, 2005, 2008), the climatic conditions (Fermaud *et al.*, 2001), the grapevine canopy development (Valdés-Gòmez *et al.*, 2008), the bunch architecture and compactness (Vail and Marois, 1991), the thickness and anatomy of the berry cuticle (Bernard, 1976; Commenil *et al.*, 1997), and the stomata changes on the berry epidermis (Bessis, 1972). Infection of grape occurs during blooming, with the flower receptacle area being the most receptive site for infection. However, a minor portion of infections may also occur through the stigma and style (Viret *et al.*, 2004).

Unripened berries are resistant to *B. cinerea* bunch rot and the pathogen commonly remains latent in the infection sites during berry growth (Keller *et al.*, 2003). Various constitutive anti-fungal compounds may play an active role in preventing *B. cinerea* pathogenesis of unripened berries (Pezet *et al.*, 2004). For example, constitutive phenolic compounds and polymeric proanthocyanidins from berry skin tissues are toxic to *B. cinerea* and show inhibition of polygalacturonases (PGs) and laccase-like stilbene oxidase activity (Pezet *et al.*, 2003, 2004). More recently, Deytieux-Belleau *et al.* (2009) showed that berry susceptibility to *B. cinerea* is negatively correlated with the total skin tannin content.

Different active resistance mechanisms are also induced following the challenge of unripe berries with *B. cinerea* such as the accumulation of phytoalexins, specifically the stilbene resveratrol (Keller *et al.*, 2003).

Pathogenesis-related (PR) proteins are inducible proteins implicated in active defence against disease, restricting pathogen development and spread in the plant (van Loon *et al.*, 2006). Prominent PRs in grape berries are chitinase and thaumatin-like proteins (osmotin) (Robinson and Davies, 2000; Monteiro *et al.*, 2003a, b). These proteins exert *in vitro* anti-fungal activity against *B. cinerea* and accumulate during ripening (Salzman *et al.*, 1998). Moreover, PRs are induced during pre-ripening in response to abiotic stress and attack by pathogens (Jacobs *et al.*, 1999; Robinson *et al.*, 1997; Monteiro *et al.*, 2003a, b). By contrast, *B. cinerea* infection results in decreased levels of PR proteins in berry juice and in a total protein extract from infected berries compared to uninfected grapes (Marchal *et al.*, 1998; Girbau *et al.*, 2004). It was suggested that in the grape environment characterized by an abundance of polyphenols, *B. cinerea* laccase not only detoxifies the *trans*-resveratrol phytoalexin but also modifies the solubility of grape proteins (Favaron *et al.*, 2009).

The cuticle also represents a potential source of signalling for both pathogen and plant, leading to the onset of defence responses against *B. cinerea* (Bessire *et al.*, 2007; Chassot *et al.*, 2007; Mang *et al.*, 2009). While cutin components may induce cutinase secretion

(Lin and Kolattukudy, 1978; Woloshuk and Kolattukudy, 1986) and appressorium formation (Gilbert *et al.*, 1996; Dickman *et al.*, 2003) in fungal pathogens, cutin monomers and cuticular waxes can also function as elicitors of plant defence responses (Fauth *et al.*, 1998; Chassot and Métraux, 2005; Kim *et al.*, 2008; Park *et al.*, 2008). When cuticle is permeabilized, upon degradation by either enzymes secreted by pathogens or mechanical action, elicitors may have a facilitated access to the cell, being recognized and, eventually, leading to a rapid release of reactive oxygen species (ROS) and subsequent defence reactions (L'Haridon *et al.*, 2011).

It is known that grape berries respond to *B. cinerea* infection by up-regulating a number of defence-related genes such as the PGIP gene that expresses the polygalacturonase-inhibiting proteins (PGIPs) located in the plant cell wall (Bezier *et al.*, 2002). The role of PGIP in limiting *B. cinerea* infection of grape has been described in heterologous systems (Agüero *et al.*, 2005; Joubert *et al.*, 2006, 2007). Heterologous host over-expression of the grape PGIP encoding gene, *Vvpgip1* was found to confer a reduced susceptibility to infection and to infiltration of *B. cinerea* PGs (Joubert *et al.*, 2006, 2007). Furthermore the inhibition of fungal PGs facilitates the release of oligalacturonides from plant cell walls, which serve as elicitors for further defence responses (Mattei *et al.*, 2005).

At the onset of ripening ('veraison'), grape berries become more susceptible to *B. cinerea*. Maturation of grape berries leads to a favourable nutritive environment for seed germination and, subsequently during this process there is an increase in berry susceptibility which could be considered a necessary step in the life cycle of the plant (Pezet *et al.*, 2003). Different structural and biochemical changes that occur during berry ripening may act simultaneously to affect the susceptibility of berries to *B. cinerea* (Deytieux-Belleau *et al.*, 2009).

Colonization of berries may be related to the presence of inhibitory substances produced during the early stages of development, but which are ineffective after ripening (McClellan and Hewitt, 1973; Nyerges *et al.*, 1975). It has been shown that following the onset of ripening, the production of sugars provides more favourable carbon sources for the fungus, enabling the achievement of higher specific growth rates, greater hyphal extension, and hence, greater colonizing potential (Vercesi *et al.*, 1997). Furthermore, both water availability at the berry surface and the total tannin content in the skin are considered potential major determinants affecting both fungal growth and berry colonization (Deytieux-Belleau *et al.*, 2009). Water availability seems to represent a critical factor in *B. cinerea* pathogenesis on berries. It has been found that water availability values higher than 0.95 are highly favourable to *B. cinerea* growth and metabolism (Deytieux-Belleau *et al.*, 2009); however, it has also been shown that water availability is negatively correlated to susceptibility during ripening. The explanation can be found in the increase of exosmosis during ripening and the release of available nutrients for *B. cinerea* metabolism which negatively affects water availability values (Donèche, 1986; Padgett and Morrison, 1990). However, high osmotic potential values, followed by low water availability, are believed to represent the key factors limiting *B. cinerea* metabolism and growth on dehydrated berries and favouring the development of noble rot (Magyar, 2011).

The mode of penetration is a crucial step in determining the result of the interaction. *Botrytis cinerea* has a limited ability to penetrate intact skins, although there is some evidence presented to the contrary by Coertze and Holz (1999), Coertze *et al.* (2001), and Williamson *et al.* (2007). More often penetration has been described through stigmata (McClellan and Hewitt, 1973), pedicels (Holz *et al.*, 2003; Pucheu-Planté and Mercier, 1983), and natural openings (Pucheu-Planté and Mercier, 1983). However, the presence of wounds of exogenous origin, such as powdery mildew attack or injuries provoked by insects, has been regarded as a major pathway for the ingress of the pathogen (Elmer and Michailides, 2004). Wounds causing microfissures in mature berry skin and stomata are another important route for penetration (Holz *et al.*, 2003). It has been shown that increased density of *B. cinerea* conidia forms hyphal mats leading to infection through microfissures that develop in the cuticle under the droplet (Coertze *et al.*, 2001).

Once *B. cinerea* has penetrated the cuticle, it secretes a group of hydrolases that target the plant cell wall, such as PGs (Kars and van Kan, 2004), pectin methylesterases (Reignault *et al.*, 1994), and proteases (Mohavedi and Heale, 1990). *B. cinerea* acidifies the cellular environment via secretion of various organic acids such as oxalic acid generating a more suitable pH for enzymatic degradation of plant cell walls and plant chemical defences by PGs, proteases and laccases (Vercesi *et al.*, 1997; Manteau *et al.*, 2003). PGs are specialized isoenzymes considered virulent factors, which are sequentially expressed during penetration and colonization, depending on the stage of infection as well as plant species and its development phase (Leone and van den Heuvel 1987; Mohavedi and Heale, 1990; Chilosi and Magro, 1997; Ten Have *et al.*, 2001; Kars *et al.*, 2004). Moreover, oxalic acid is involved in pathogenicity not only by suppressing oxidative burst as described above, but also by chelating the calcium ions embedded in pectin rendering more efficient the activity of PGs (Bateman and Beer, 1965), by weakening the plant cells, and facilitating invasion due to its direct toxic effect (Noyes and Hancock, 1981; Marciano *et al.*, 1983).

Upon colonization of berries by the pathogen, typical bunch-rot symptoms occur accompanied by significant yield losses. Wine made from such grapes is of very poor quality, particularly red wines, which often have a substandard colour, texture and off-odours (Hornsey, 2007). Laccases produced by *B. cinerea* are involved in the oxidation of the anthocyanins and flavonoids to brown oxidation products. These enzymes are very stable and can pass to the must and wine resulting in wine becoming brownish; red wines in particular lose their characteristic red colour (Kassemeyer and Berkelmann-Löhnertz, 2009).

11.3 THE NOBLE ROT

Under particular climatic (temperature, water relations) and edaphic (texture, chemistry, nutrients cycle, water availability) conditions, typical of certain wine regions, grapes infected by *B. cinerea* can develop the noble rot (Figure 11.1) The shift from 'bunch rot' to 'noble rot' is determined by a delicate equilibrium among the cited factors; as a consequence both rots can be present in the same vineyard in a balance determined by site characteristics, including the host. It has been assumed that noble rot is favoured in areas with nutrient-poor (especially N), well-drained limestone soils, where plants are deep-rooted and have a constant water supply (Ribéreau-Gayon *et al.*, 1980). These conditions limit splitting of berries reducing the chance of penetration by *B. cinerea* through mechanical wounds. Grape varieties with sparse bunches and thick berry skins also reduce the likelihood of mechanical damage to berries and negatively affect *B. cinerea* penetration and infection potential. Climate strongly influences post-infection disease progression. In general, 10–15 days of cyclical meteorological conditions, late in the season, are crucial for noble rot development. Night-time humidity, dew and frequent morning mists in the valleys, stimulate *B. cinerea* infection and development, whereas warm and sunny windy afternoons facilitate water evaporation, increase osmotic potential of berries and thus limit the fungal growth.

The disease cycle during noble rot development is characterized by penetration of the pathogen and is favoured by dew and moist mornings. Infection occurs through microfissures that occur naturally during berry growth when the cuticle is not too thick. This is

Figure 11.1 Natural development of noble rot on vine. (For a colour version of this figure, please see the colour plate section.)

followed by the development of an intercellular mycelium limited to the outermost layers of epidermis (Ribéreau-Gayon *et al.*, 1980; Magyar and Bene, 2006). In this phase of *Botrytis* colonization, the berry still appears turgid and the colour becomes chocolate-brown ('pourri plein'). The colonized berry skin becomes permeable and gradually shrivels ('pourri rôti'). The loss of water from the berries, during sunny and windy afternoons, leads to the concentration of sugars and other constituents. The increase in sugar concentration reduces the water availability values and retards and modifies the metabolism of *B. cinerea*, limiting its colonization (Ribéreau-Gayon *et al.*, 1980; Antunovics *et al.*, 2003; Magyar, 2011). Finally, drying limits secondary invasion by bacteria and fungi, such as *Penicillium*, *Aspergillus*, and *Mucor* spp. responsible for most of the mouldy off-odours and tastes associated with bunch rot (Hornsey, 2007). By contrast, grape-colonizing yeasts, such as *Metschnikovia pulcherrima*, which commonly colonizes grapes during maturation, may positively influence the onset of noble rot. *M. pulcherrima* develops more frequently on noble-rotted grapes (Antunovics *et al.*, 2003). Pigment-producing *Metschnikowia* strains from noble-rotted grapes were found to possess antagonistic activity against yeasts, bacteria and filamentous fungi, including *B. cinerea* (Sipiczki, 2006). Therefore, one of the conditions predisposing the development of noble rot could be the onset of a microbial population on berries during *B. cinerea* infection capable of antagonizing and limiting the spread of the pathogen.

Another factor that stimulates noble rot development is maturation precocity (Ribéreau-Gayon *et al.*, 2006). The infected grapes are over-ripened, dehydrated and the sugars are more concentrated than the acids. The grape acquires the characteristic aromas that permit the production of renowned sweet white wines such as Sauternes-Barsac, Côteaux du Layon (France), Tokaji (Hungary), Trockenbeerenauslese (Germany and Austria) (Ribéreau-Gayon *et al.*, 2006; Tosi *et al.*, 2009). Botrytized wines are produced throughout wine regions in Europe. Most recently noble rot was utilized to produce sweet wines in Australia, New Zealand, South Africa and the United States. While most of the botrytized wines are produced from white varieties, the utilization of noble rot recently has been considered to improve the characteristics of valuable red wines such as Amarone (Valpolicella, Italy) (Tosi *et al.*, 2009).

11.4 INDUCED BOTRYTIZATION

The production of sweet botrytized wines is considered risky and expensive since it depends on the proper combination of climatic and site factors, particular vine management and grape harvest. For instance, grapes left on the vine after ripening increases the risk of damage by animals, other rot fungi, and meteorological adverse conditions. The selection of noble-rotted bunches requires time and experienced operators. Finally, the scenario of global climatic change puts uncertainty on the regular development of the noble rot for the near future even in those regions considered climatically ideal. The latter aspect is of particular relevance. Several studies have been undertaken in order to forecast the effect of global climatic changes on vine cultivation and wine quality (Schultz, 2000; Jones *et al.*, 2005; Ohmart, 2007). Climatic scenarios for the next 50 years forecast a predictable increase in average temperatures, changes in precipitation regimes, and increase of climatic anomalies (prolonged drought periods, heavy flooding). The potential consequences may be changes in the duration and quality of the growing season which could have a strong impact on grape phenology and exposure to adverse stress factors, including pests and diseases. Within this scenario, the delicate equilibrium determining the process of noble rot can be severely impaired. In particular, increases in the frequency of climatic anomalies could result in longer and more intense drought or humid periods that might impair the synchronism of alternate mists, dews and dry afternoons, crucial for noble rot development.

Thus, the possibility of obtaining controlled botrytization in postharvest conditions, represents an option for producing sweet botrytized wines independently from climatic and edaphic conditions in the vineyard and, possibly, from the grape varieties. The earliest records in the literature are by Nelson and Amerine (1956, 1957) and Nelson (1958). In their experiments, the grapes were sprayed with spores of *B. cinerea* and kept at high humidity for 24–26 hours at 20–25°C, to induce spore germination and penetration. Afterward, the grapes were exposed to dry air to promote berries dehydration and to slow down the colonization process. After 8–13 days, Nelson and Amerine obtained good artificial noble rot. The process was not developed further due to the high costs at that time.

Watanabe and Shimazu (1976) used a different approach by inoculating *B. cinerea* mycelia and spores directly into the juice.

More recently, an Italian enterprise registered two international patents, Sordato Kinensis™ and Sordato *Botrytis* control™, offering an automated system combining grape withering and induction of development of latent *B. cinerea* inoculum present in over-ripened bunches. The grape treatment, carried out in a drying room, includes over-ripening and dehydration at 76% and 11°C for 30 days. This is followed by exposure to 100% RH for 7 days and a new cycle of dehydration of 76% RH at 8–10°C (Ferrarini *et al.*, 2009). Using this method, it has been possible to successfully induce *B. cinerea*, and the resulting wine had analytical values typical of botrytized wines.

Figure 11.2 Trays with bunches in the VIMACO withering chamber dedicated to postharvest botrytization and noble rot development. (For a colour version of this figure, please see the colour plate section.)

An alternative automated process of post-harvest development of noble rot has been investigated in the Italian project VIMACO (Vannini *et al.*, 2004). The postharvest grapes are exposed simultaneously to dehydration and artificial botrytization in a dedicated withering chamber (Figure 11.2) The system is completely automated: the grapes are inoculated with *B. cinerea* at high RH and 21°C through a dispenser inside the chamber, utilizing a commercial product made with selected strains of the fungus (Freschi *et al.*, 2011). After incubation in the presence of induced mist for 3–4 days at 21°C, the grapes are subjected to ventilation at 50% RH in order to simulate the process, similar to grapes in the field. The botrytization/withering process is completed within 15–25 days depending on the variety used (Figure 11.3). Five

Figure 11.3 Different phases of noble rot development on berry after *Botrytis* infection under controlled conditions in the VIMACO withering chamber. (For a colour version of this figure, please see the colour plate section.)

varieties have been tested, including Chardonnay, Sauvignon Blanc, Trebbiano, Malvasia Puntinata and Grechetto. Botrytization varied between 25% for Chardonnay to 75% for Grechetto with no symptoms of sour rot by other organisms. Dehydration ranged between 40 and 60% (over 32° Brix). Sensorial and analytical characteristics of the resulting wines were compatible with botrytized wines (Tiberi *et al.*, 2008; Guenzi *et al.*, 2011). Furthermore, presence of ochratoxin A and aflatoxin (B1, B2, G1, G2) was approximately 20 times lower than the EU prescribed limits (2 and 4 ppm, respectively). In similar vein, several strains of *B. cinerea* have been selected during dehydration of Amarone grapes; strains which can be used for inducing botrytization to produce Passito wines (Tosi *et al.*, 2012).

Postharvest botrytization technologies offer different opportunities. They allow the production of sweet *Botrytis* wines from grapes cultivated in areas that are not suitable for noble rot development. Furthermore, these technologies can compensate for the yield losses common in traditional noble rot areas during climatically adverse seasons. A large number of grape varieties can be successfully botrytized, giving winemakers an opportunity to make different products. The proposed technologies are flexible, as it is possible to regulate dehydration and botrytization levels. Finally, the management of damaging micro-organisms, such as those causing sour rot or producing mycotoxins, will contribute to consumer protection.

REFERENCES

Agüero, C.B., S.L. Uratsu, L.C. Greve, *et al.* 2005. Evaluation of tolerance to Pierce's disease and *Botrytis* in transgenic plants of *Vitis vinifera* L. expressing the pear PGIP gene. *Molecular Plant Pathology*, 6:43–51.

Amselem, J., C.A. Cuomo, J.A.L. van Kan, *et al.* 2011. Genomic analysis of the necrotrophic fungal pathogens *Sclerotinia sclerotiorum* and *Botrytis cinerea*. *PLoS Genetics*, 7:e1002230.

Antunovics, Z., H. Csoma, and M. Sipiczki. 2003. Molecular and genetic analysis of the yeast flora of botrytized Tokaj wines. *Bulletin de l'OIV*, 76:380–397.

Bateman, D.F. and S.V. Beer. 1965. Simultaneous production and synergistic action of oxalic acid and polygalacturonase during pathogenesis by *Sclerotium rolfsii*. *Phytopathology*, 55:204–211.

Bernard, A.C. 1976. Resistance mécanique des baies de *Vitis vinifera* au *Botrytis cinerea* Pers. *France Viticole*, 8:301–307.

Bessire, M., C. Chassot, A-C. Jacquat, *et al.* 2007. A permeable cuticle in *Arabidopsis* leads to a strong resistance to *Botrytis cinerea*. *EMBO Journal*, 26:2158–2168.

Bessis, R. 1972. Etude en microscopie électronique à balayage des rapports entre l'hôte et le parasite dans le cas de la pourriture grise. *Comptes Rendus de l'Academie des sciences de Paris, Serie D*, 274:2991–2994.

Bezier, A., B. Lambert, and F. Baillieul. 2002. Study of defence-related gene expression in grapevines leaves and berries infected with *Botrytis cinerea*. *European Journal of Plant Pathology*, 108:111–120.

Chassot, C. and J.P. Métraux. 2005. The cuticle as source of signal for plant defense. *Plant Biosystems*, 139:28–31.

Chassot, C., C. Nawrath, and J.P. Métraux. 2007. Cuticular defects lead to full immunity to a major plant pathogen. *Plant Journal*, 49:972–980.

Chilosi, G. and P. Magro. 1997. Pectin lyase and polygalacturonase isoenzymes production by *Botrytis cinerea* during the early stages of infection on different host plants. *Journal of Plant Pathology*, 78:61–69.

Coertze, S. and G. Holz. 1999. Surface colonisation, penetration and lesion formation on grapes inoculated fresh and after cold storage with single airborne conidia of *Botrytis cinerea*. *Plant Disease*, 83:917–924.

Coertze, S., G. Holz, and A. Sadie. 2001. Germination and establishment of infection on grape berries by single airborne conidia of *Botrytis cinerea*. *Plant Disease*, 85:668–677.

Commenil, P., L. Brunet, and J-C. Audran. 1997. The development of the grape berry cuticle in relation to susceptibility to bunch rot disease. *Journal of Experimental Botany*, 48:1599–1607.

Deytieux- Belleau, C., L. Geny, J. Roudet, *et al.* 2009. Grape berry skin features related to ontogenic resistance to *Botrytis cinerea*. *European Journal of Plant Pathology*, 125:551–563.

Dickman, M.B., Y.S. Ha, Z. Yang, *et al.* 2003. A protein kinase from *Colletotrichum trifolii* is induced by plant cutin and is required for appressorium formation. *Molecular Plant Microbe Interaction*, 16:411–421.

Donèche, B. 1986. La nature des exsudats de raisin et leur rôle dans la germination des conidies de *Botrytis cinerea*. *Agronomie*, 6:67–73.

Elad,Y., B. Williamson, P. Tudzynski, *et al.* 2004. *Botrytis* spp. and diseases they cause in agricultural systems: an introduction. In Botrytis: *Biology, Pathology and Control* (eds Y. Elad, B. Williamson, P. Tudzynski, N. Delen), Kluwer Academic Publishers, Dordrecht, pp. 1–6.

Elmer, P.A.G. and T.M. Michailides. 2004. Epidemiology of *Botrytis cinerea* in orchard and vine crops. In Botrytis: *Biology, Pathology and Control* (eds Y. Elad, B. Williamson, P. Tudzynski, N. Delen), Kluwer Academic Publishers, Dordrecht, pp. 243–272.

Fauth, M., P. Schweizer, A. Buchala, *et al.* 1998. Cutin monomers and surface wax constituents elicit H_2O_2 in conditioned cucumber hypocotyl segments and enhance the activity of other H_2O_2 elicitors. *Plant Physiology*, 117:1373–1380.

Favaron, F., M. Lucchetta, S. Odorizzi, *et al.* 2009. The role of grape polyphenols on trans-resveratrol activity against *Botrytis cinerea* and of fungal laccase on the solubility of putative grape PR proteins. *Journal of Plant Pathology*, 91:579–588.

Fermaud, M., J.M. Limiñana, G. Froidefond, *et al.* 2001. Grape cluster microclimate and architecture affect severity of Botrytis rot of ripening berries. *IOBC/WPRS Bulletin*, 24:7–10.

Ferrarini, R., E.M. Casarotti, G. Zanella, *et al.* 2009. Botrytis cinerea noble form induction on grapes during withering. *American Journal of Enology and Viticulture*, 60:400A.

Fournier, E., T. Giraud, C. Albertini, *et al.* 2005. Partition of the *Botrytis cinerea* complex in France using multiple gene genealogies. *Mycologia*, 97:1251–1267.

Freschi, G., S. Coranelli, A. Vannini, *et al.* 2011. Compositions suitable for botrytization. Tipology: Industrial invention. European patent number 09425093.3-2114.

Gilbert, R.D., A.M. Johnson, and R.A. Dean. 1996. Chemical signals responsible for appressorium formation in the rice blast fungus *Magnaporthe grisea*. *Physiological and Molecular Plant Pathology*, 48:335–346.

Girbau, T., B.E. Stummer, K.F. Pocock, *et al.* 2004. The effect of *Uncinula necator* (powdery mildew) and *Botrytis cinerea* infection of grapes on the levels of haze-forming pathogenesis-related proteins in grape juice and wine. *Australian Journal of Grape and Wine Research*, 10:125–133.

Guenzi, S., F. Mencarelli, A. Bellincontro, *et al.* 2011. Botritizzazione controllata. *VQ*, 1:46–49.

Holz, G., M. Gütschow, S. Coertze, *et al.* 2003. Occurrence of *Botrytis cinerea* and subsequent disease expression at different positions on leaves and bunches of grape. *Plant Disease*, 87:351–358.

Hornsey, I. 2007. *The Chemistry and Biology of Winemaking*, Royal Society of Chemistry, Cambridge, UK.

Jacobs, A., I. Dry, and S. Robinson. 1999. Induction of different pathogenesis-related cDNAs in grapevine infected with powdery mildew and treated with ethephon. *Plant Pathology*, 48:325–336.

Jarvis, W.R. 1977. Botryotinia and Botrytis species; taxonomy, physiology and pathogenicity. Monograph No. 15, Canadian Department of Agriculture, Ottawa.

Jones, G.V., M.A. White, O.R. Cooper, *et al.* 2005. Climate change and global wine quality. *Climatic Change*, 73:319–343.

Joubert, D.A., I. Kars, L. Wagemakers, *et al.* 2007. A polygalacturonase-inhibiting protein from grapevine reduces the symptoms of the endopolygalacturonase BcPG2 from *Botrytis cinerea* in *Nicotiana benthamiana* leaves without any evidence for *in vitro* interaction. *Molecular Plant Microbe Interactions*, 20:392–402.

Joubert. D.A., A.R. Slaughter, G. Kemp, *et al.* 2006. The grapevine polygalacturonase-inhibiting protein (VvPGIP1) reduces *Botrytis cinerea* susceptibility in transgenic tobacco and differentially inhibits fungal polygalacturonases. *Transgenic Research*, 15:687–702.

Kars, I. and J.A. van Kan. 2004. Extracellular enzymes and metabolites involved in pathogenesis of *Botrytis*. In *Botrytis: Biology, Pathology and Control* (eds Y. Elad, B. Williamson, P. Tudzynski, N. Delen), Kluwer Academic Publishers, Dordrecht, pp. 99–118.

Kassemeyer, H.H. and B. Berkelmann-Löhnertz. Fungi of grapes. In *Biology of Microorganisms on Grapes, in Must and in Wine* (eds H. Konig, G. Unden, J. Frohlich), Springer-Verlag, Berlin/Heidelberg.

Keller, M., O. Viret, and F.M. Cole. 2003. *Botrytis cinerea* infection in grape flowers: Defense reaction, latency, and disease expression. *Phytopathology*, 93:316–322.

Kim, T.H., J.H. Park, M.C. Kim, *et al.* 2008. Cutin monomer induces expression of the rice OsLTP5 lipid transfer protein gene. *Journal of Plant Physiology*, 165:345–349.

L'Haridon, F., A. Besson-Bard, M. Binda, *et al.* 2011. A permeable cuticle is associated with the release of reactive oxygen species and induction of innate immunity. *PLoS Pathogens*, 7:e1002148.

Lin, T.S. and P.E. Kolattukudy. 1978. Induction of a polyester hydrolase (cutinase) by low levels of cutin monomers in *Fusarium solani* f.sp. pisi. *Journal of Bacteriology*, 133:942–951.

Leone, G. and J. van den Heuvel. 1987. Regulation by carbohydrates of the sequential in vitro production of pectic enzymes by *Botrytis cinerea*. *Canadian Journal of Botany*, 65:2133–2141.

van Loon, L.C., M. Rep, and C.M.J. Pieterse. 2006. Significance of inducible defense-related proteins in infected plants. *Annual Review of Phytopathology*, 44:35–62.

Magyar, I. 2011. Botrytized wines. *Advances in Food and Nutrition Research*, 63:147–206.

Magyar, I. and Z.S. Bene. 2006. Morphological and taxonomic study on mycobiota of noble rotted grapes in the Tokaj wine district. *Acta Alimentaria*, 35:237–246.

Mang, H.G., K.A. Laluk, E.P. Parsons, *et al.* 2009. The *Arabidopsis* RESURRECTION1 gene regulates a novel antagonistic interaction in plant defense to biotrophs and necrotrophs. *Plant Physiology*, 151:290–305.

Manteau, S., S. Abouna, B. Lambert, *et al.* 2003. Differential regulation by ambient pH of putative virulence factor secretion by the phytopathogenic fungus *Botrytis cinerea*. *FEMS Microbiology Ecology*, 43:359–366.

Marchal, R., L. Berthier, L. Legendre, *et al.* 1998. Effects of *Botrytis cinerea* infection on the must protein electrophoretic characteristics. *Journal of Agricultural and Food Chemistry*, 46:4945–4949.

Marciano, P., P. DiLennam, and P. Magro. 1983. Oxalic acid, cell wall degrading enzymes and pH in pathogenesis and their significance in the virulence of two *Sclerotinia sclerotiorum* isolates on sunflower. *Physiological Plant Pathology*, 22:339–345.

Martinez, F., B. Dubos, and M. Fermaud. 2005. The role of saprotrophy and virulence in the population dynamics of *Botrytis cinerea* in the vineyards. *Phytopathology*, 95:692–700.

Martinez, F., D. Blancard, P. Lecomte, *et al.* 2003. Phenotypic differences between vacuma and transposa subpopulations of *Botrytis cinerea*. *European Journal of Plant Pathology*, 109:479–488.

Martinez, F., M.F. Corio-Costet, C. Levis, *et al.* 2008. New PCR primers applied to characterize distribution of *Botrytis cinerea* populations in French vineyards. *Vitis*, 47:217–226.

Mattei, B., R. Galletti, C. Manfredini, *et al.* 2005. Recognition and signalling in the cell wall: the case of endopolygalacturonase, PGIP and oligogalacturonides. *Plant Biosystems*, 139:24–27.

McClellan, W.D. and B. Hewitt. 1973. Early botrytis rot of grapes: Time of infection and latency of *Botrytis cinerea* Pers. in *Vitis vinifera* L. *Phytopathology*, 63:1151–1157.

Mohavedi, S. and J.B. Heale. 1990. The roles of aspartic proteinase and endopectin lyase enzymes in the primary stages of infection and pathogenesis of various host tissues by different isolates of *Botrytis cinerea* Pers. ex. Pers. *Physiological and Molecular Plant Pathology*, 36:303–324.

Monteiro, S., M. Barakat, M. Piçarra-Pereira, *et al.* 2003a. Osmotin and thaumatin from grape: a putative general defense mechanism against pathogenic fungi. *Phytopathology*, 93:1505–1512.

Monteiro, S., M.A. Piçarra-Pereira, A. Teixeira, *et al.* 2003b. Environmental conditions during vegetative growth determine the major proteins that accumulate in mature grapes. *Journal of Agricultural and Food Chemistry*, 51:4046–4053.

Nelson, K.E. 1958. Some studies on the action of sulphur dioxide in the control of botrytis rot of Tokay grapes. *Proceedings of the American Society of Horticultural Science*, 71:183–189.

Nelson, K.E. and M.A. Amerine. 1956. Use of *Botrytis cinerea* for the production of sweet table wines. *American Journal of Enology and Viticulture*, 7:131–136.

Nelson, K.E. and M.A. Amerine. 1957. Further studies on the production of natural, sweet table wines from botrytized grapes. *American Journal of Enology and Viticulture*. 8:127–134.

Noyes, R.D. and J.G. Hancock. 1981. Role of oxalic acid in the sclerotinia wilt of sunflower. *Physiological Plant Pathology*, 18:123–132.

Nyerges, P., E. Szabo, and E. Donko. 1975. The role of anthocyanin and phenol compounds in the resistance of grapes against *Botrytis* infection. *Acta Phytopathologica Academiae Scientarium Hungaricae*, 10:21–32.

Ohmart, C. 2007. Climate change and viticulture. *Wines & Vines*, 88:88–91.

Padgett, M. and J.C. Morrison. 1990. Changes in grape berry exudates during fruit development and their effect on mycelial growth of *Botrytis cinerea*. *Journal of the American Society for Horticultural Science*, 115:269–273.

Park, J.H., M.C. Suh, T.H. Kim, *et al.* 2008. Expression of glycine-rich protein genes, AtGRP5 and AtGRP23, induced by the cutin monomer 16-hydroxypalmitic acid in *Arabidopsis thaliana*. *Plant Physiology and Biochemestry*, 46:1015–1018.

Pezet, R., O. Viret, and K. Gindro. 2004. Plant microbe interaction: The *Botrytis* gray mold of grapes. In *Advances in Plant Physiology*, Vol. 7 (ed. A. Hemantaranjan), Varanasi, India, pp. 75–120.

Pezet, R., O. Viret, C. Perret, *et al.* 2003. Latency of *Botrytis cinerea* Pers.: Fr. and biochemical studies during growth and ripening of two grape berry cultivars, respectively susceptible and resistant to grey mould. *Journal of Phytopathology*, 151:208–214.

Pucheu-Planté, B. and M. Mercier. 1983. Ultrastructural study on the host-parasite relationship between the grape and the fungus *Botrytis cinerea*: model of the noble rot in Sauternais. *Canadian Journal of Botany*, 61:1785–1797.

Reignault, P., M. Mercier, G. Bompeix, *et al.* 1994. Pectin methyl esterase from *Botrytis cinerea*: physiological, biochemical and immunochemical studies. *Microbiology*, 140:3249–3255.

Ribéreau-Gayon, J., P. Ribéreau-Gayon, and G. Segun. 1980. *Botrytis cinerea* in enology. In *The Biology of Botrytis* (eds J.R. Coley-Smith, K. Verhoeff and W.R. Jarvis), Academic Press, London, UK, pp. 251–274.

Ribéreau-Gayon, P., D. Dubourdieu, B. Donèche, *et al.* 2006. The grape and its maturation. In *Handbook of Enology: The Microbiology of Wine and Vinifications*, Vol. 1, 2nd edn, John Wiley & Sons, Ltd, Chichester, UK.

Robinson, S.P. and C. Davies. 2000. Molecular biology of grape berry ripening. *Australian Journal of Grape and Wine Research*, 6:175–188.

Robinson, S.P., A.K. Jacobs, and I.B. Dry. 1997. A class IV chitinase is highly expressed in grape berries during ripening. *Plant Physiology*, 114:771–778.

Salzman, R.A., I. Tikhonova, B.P. Bordelon, *et al.* 1998. Coordinate accumulation of antifungal proteins and hexoses constitutes a developmentally controlled defense response during fruit ripening in grape. *Plant Physiology*, 117:465–472.

Schultz, H. 2000. Climate change and viticulture: A European perspective on climatology, carbon dioxide and UV-B effects. *Australian Journal of Grape and Wine Research*, 6:2–12.

Sipiczki, M. 2006. *Metschnikowia* strains isolated from botrytized grapes antagonize fungal and bacterial growth by iron depletion. *Applied and Environmental Microbiology*, 72:6716–6724.

Ten Have, A., W. Oude Breuil, J.P. Wubben, *et al.* 2001. *Botrytis cinerea* endopolygalacturonase genes are differentially expressed in various plant tissues. *Fungal Genetics and Biology*, 33:97–105.

Tiberi, D., P. Pietromarchi, G. Vuono, *et al.* 2008. Botrytization in a controlled environment and in field of Malvasia Puntinata: effects on grapes and wines. *Rivista di Viticoltura ed Enologia*, 61:255–260.

Tosi E., M. Azzolini, M. Lorenzini, *et al.* 2012. Caratterizzazione di ceppi di *Botrytis cinerea* da utilizzare per la produzione di vini passiti. *L'Enologo*, 9:85–90.

Tosi, E., M. Azzolini, G. Zapparoli, *et al.* 2009. Incidenza della muffa nobile sulle caratteristiche chimiche e sensoriali del vino amarone. *L'Enologo*, 10:81–87.

Vail, M.E. and J.J. Marois. 1991. Grape cluster architecture and the susceptibility of berries to *Botrytis cinerea*. *Phytopathology*, 81:188–191.

Valdés-Gómez, H., M. Fermaud, J. Roudet, *et al.* 2008. Grey mould incidence is reduced on grapevines with lower vegetative and reproductive growth. *Crop Protection*, 27:1174–1186.

Vannini, A., F. Mencarelli, G. Chilosi, *et al.* 2004. Procedimento per disidratare acini d'uva. Tipology: Industrial invention. Deposit N. VR2004A000071 del 23.04.2004.

Vercesi, A, R. Locci, and J.I. Prosser. 1997. Growth kinetics of *Botrytis cinerea* on organic acids and sugars in relation to colonization of grape berries. *Mycological Research*, 101:139–142.

Viret, O., M. Keller, V.G. Jaudzems, *et al.* 2004. *Botrytis cinerea* infection of grape flowers: light and electron microscopical studies of infection sites. *Phytopathology*, 94:850–857.

Watanabe, M. and Y. Shimazu. 1976. Application of *Botrytis cinerea* for wine making. *Journal of Fermentation Technology*, 54:471–478.

Williamson, B., B. Tudzinsky, and J.A.N. van Kan. 2007. *Botrytis cinerea*: the cause of grey mould disease. *Molecular Plant Pathology*, 8:561–580.

Woloshuk, C.P. and P.E. Kolattukudy. 1986. Mechanism by which contact with plant cuticle triggers cutinase gene expression in the spores of *Fusarium solani* f. sp. *pisi*. *Proceedings of the National Academy of Sciences U.S.A.*, 83:1704–1708.

12 Vinification and Aroma Characteristic of Botrytized Grape

Pierre Louis Teissedre and Bernard Donèche

Faculty of Oenology, Université Bordeaux Segalen, Institut des Sciences de la Vigne et du Vin, Villenave d'Ornon, France

12.1 INTRODUCTION

Making wine from botrytized grapes presents a real challenge as the grapes have a particular composition and the various steps in the winemaking process can modify the wine's organoleptic quality. The loss of aroma components that occurs during noble rot development is typical of Muscat varieties and can be explained by the destruction of terpenes, and the formation of other odorous compounds which give the wine its characteristic *Botrytis* aroma. Several aromatic compounds, such as linalool, geraniol and nerol, can be metabolized by *Botrytis cinerea* in compounds with less volatility (α-terpineol, ß-pinene, furan and pyran oxides) (Bock *et al.*, 1986, 1988). Phenolic and iodine-like odours have been reported in some botrytized wines (Boidron *et al.*, 1978). *Botrytis* can also produce esterases which degrade the fruit esters that give these wines their typical fruity character (Dubourdieu *et al.*, 1983). The losses and their significance are directly linked to the importance of the destroyed compounds to the varietal aroma of the wine. For example, Riesling and Semillon generally gain more aromatic complexity than they lose in varietal distinctiveness, but Muscat varieties lose more aromas than they gain. Thus, on the one hand, fungal metabolism destroys certain aromatic compounds, but conversely, it also synthesizes others.

One of the distinctive *Botrytis* flavours that can appear is sotolon [4,5-dimethyl-3-hydroxy-2(5H)-furanone]. Sotolon assayed at concentrations above its perception threshold has been shown to contribute to botrytized aroma; in particular, when combined with other volatile aroma compounds found or produced in botrytized wines, it helps to give a honey-like fragrance to the wine. However, Sponholz and Huehn (1994) claimed that sotolon was not necessarily linked with *B. cinerea* infection but was formed instead by Maillard reactions during ripening.

Botrytized grapes can also contain the typical mushroom alcohol compound, 1-octen-3-ol, which provides another characteristic aroma. Bock *et al.* (1985) isolated more than 20 derivatives from terpenes in *Botrytis*-infected grapes. In 2000, Tominaga *et al.* identified volatile thiols in botrytized wines and found surprisingly high concentrations of 3-mercaptohexan-1-ol. Genovese *et al.* (2002) analysed botrytized wines and found that lactones, for example γ-nonalactone, γ-decalactone, and δ-decalactone, had a considerable impact on aroma. A combination of *B. cinerea* and over-ripening may be used or adopted to modify grape

Sweet, Reinforced and Fortified Wines: Grape Biochemistry, Technology and Vinification, First Edition.
Edited by Fabio Mencarelli and Pietro Tonutti.

composition, creating conditions similar to those in oxidized wines, and leading to the formation of phenylacetaldehyde, methional and sotolon.

12.2 MANAGING THE VINIFICATION OF BOTRYTIZED GRAPE

Grape berries used in the production of botrytized wines are generally pressed several times rather than crushed. This is particularly important for grape harvests that have been invaded by noble rot. Indeed, the roasted grape juices can only be satisfactorily extracted if a light pressing opens the film. De-stalking is not recommended for these grape harvests and is only very rarely implemented. In view of the low volumes of juice removed, due to the evaporation of the water in berries under the influence of the *Botrytis*, the presence of rachis favours the extraction of juices. However, in certain regions, it is sometimes recommended that the rachis is removed from the marc to avoid tannic juices from stem particles entering the must. The tannic juices can be released by the often heavy pressure needed to extract the juice of grapes affected by noble rot.

The need to carry out several successive presses, interrupted by crumbling the marc, increases the risk of cutting the rachis and crushing pips and thus the release of undesirable tannins. Three types of presses can be used:

- *Vertical presses.* Vertical presses have the advantage of enabling good extraction of the juice, and low deposits of thick sludge. Regrettably, the duration of a pressing cycle is long and the operation requires considerable manual handling during crumbling of the marc.
- *Horizontal presses.* Horizontal presses are frequently used in medium-sized wineries because they are easy to control and yield sufficient quantities; however, the musts obtained are rather muddy due to the mashing of the grapes by chains during crumbling. It is necessary to work at low speed in order to achieve a more progressive extraction of the must.
- *Pneumatic horizontal presses.* These presses are more powerful than vertical or horizontal presses. The grapes are introduced into the press between membranes, reaching 3 bars of pressure, enabling good volume of extraction.

The selection of juices from the different presses can create a delicate problem; in the case of grapes affected by noble rot, it is the juice of the last pressing that is of the best quality, because it is the richest in sugars.

It is possible to use a combination of presses, employing a pneumatic press for the first pressing and then vertical basket presses for the second and third pressings, to obtain the richest and sweetest juice. In order to make a top-class botrytized wine, it is necessary to obtain potential alcohol by volume of 20–22%.

The must extracted from the grape harvest has to undergo various treatments to prevent oxidation; the most important of these is the addition of SO_2 and the elimination of the lees, by clarifying the must. The 'lees' are the deposits of residual particles that result from the 'fining' process and descend to the bottom of the wine vat. The addition of sulphite is straightforward: SO_2 is added to the extracted musts, at a rate of 2–5 g hL^{-1}, and this should be done in the optimum conditions during the winemaking process. The treatment is carried out in temperature-controlled vats at a low temperature (between 6 and 15°C) for 12 to 24 hours (static operation). It greatly inhibits oxidation of the must before fermentation begins, and it enables clarification of the unrefined must, by eliminating the heaviest suspended particles. A turbidity of 500 NTU is acceptable in the clarified must. This extremely delicate operation is realized by natural sedimentation. Indeed, on the one hand, the difference in the specific weights of the particles in suspension and the must is low due to the high density of the must, but by contrast, the viscosity of the liquid is great, not only because of the raised sugars content, but also due to the presence of protective colloids (gums, mucilages and dextrans) formed by *B. cinerea*. In practice, a partial 'de-sludge' is generally sufficient to remove the largest and heaviest particles. The operation takes on average about 20 hours.

It can be useful to make limited chemical corrections to the clarified must to ensure a balanced composition of the final wine. Thus, to regulate the nitrogen content of musts, the addition of 10–15 g hL^{-1} (25–40 mg L^{-1}) of ammonium ions (NH_4^+) is required. Normally, for each gram of sugar that has to be transformed into alcohol, 1 mg of nitrogen is needed. In addition, it is often beneficial to add thiamine to musts that derive from botrytized clusters; this not only has the effect of accelerating the alcoholic fermentation, but also limits the combination of sulphurous anhydride during the arrest of the fermentation.

Because grapes have a low moisture content, pressing is a long process, the rises in pressure are slow, and few crumblings are needed. The pneumatic press is generally used to apply pressures from 2 to 3 bars, but pressure can be raised up to 6 bars. Some châteaux use the vertical press with maximum pressures of 9 bars. Others may use a system of selective pressing by cryoextraction. This process is inspired by the method for producing Ice wines. The grapes are cooled to a negative temperature (between -4 and -10°C), which freezes the water on the surface of the berries. Only the grapes that are richest in sugars will release some juice during the pressing, the others will remain solid (Chauvet *et al.*, 1986). This method of concentrating musts is the only one authorized under the appellation regulations. However, attempts have been made to use reverse osmosis. De-sludging musts, a classic operation during the making of white wines, is a very delicate process for achieving viscosity, due to the presence of glycerol and ß-glucans from *B. cinerea* (Robertson *et al.*, 1980). It is carried out at a temperature of between 4 and 10°C, but it is not applied in all the châteaux.

After pressing, the must is poured into barrels, where the alcoholic fermentation can begin. At this stage the winemakers have two choices: (i) fermentation by native yeasts, carried out in classified vineyards, to strengthen the characteristics of the wines; (ii) the use of commercially prepared or 'active dry' yeasts to avoid fermentation problems. In particular, because of the high concentration of sugars in musts and the lack of essential elements, such as nitrogen and vitamins, the alcoholic fermentation is slow and the volatile acidity increases considerably. A maximum content of 25 meq L^{-1} is authorized for this type of wine under an official EU directive. Sometimes ammonium sulfate or phosphate are added (up to a cumulative maximum of 250 mg L^{-1}) to help the fermentation process. However, it is more prudent to add ammonium in steps of 25 mg L^{-1} (corresponding to 10 g hL^{-1}), as too great a level of ammonium can also increase ethyl acetate synthesis by yeasts and, in case of a residual ammonium content in the finished wine, it can favour the development of *Brettanomyces* (a contamination yeast). *Brettanomyces* is a micro-organism that is related to wine spoilage; most notably, contamination can result in phenolic off-flavours (Chatonnet *et al.*, 1992) or Brett character. The flavour induced can be described as horsey, leathery, medicinal, sticky plaster, smoky or savoury (Chatonnet *et al.*, 1992; Licker *et al.*, 1999; Wirz *et al.*, 2004; Norris, 2004; Saurez *et al.*, 2007; Romano *et al.*, 2009). The two main spoilage compounds traditionally associated with *Brettanomyces* are 4-ethylphenol and 4-ethylguaiacol. The aroma of 4-ethylphenol is associated with leather and sticky plaster, while 4-ethylguaiacol is associated with a medicinal or spicy smell (Saurez *et al.*, 2007). *Brettanomyces bruxellensis* has been found on grape skins (Renouf *et al.*, 2007) and this is considered to be an important source of contamination. A correlation between the presence of *Brettanomyces* and *Botrytis* on grape skins has also been noted. This may be because excessive heat and moisture favour both *Brettanomyces* and *Botrytis*, rather than due to a direct interaction between the two species. Damaged grapes can enhance the development of *Brettanomyces* on berries, as nutrients previously trapped in the berries are liberated (Renouf *et al.*, 2007). The microbiological origin of ethylphenols in wines has been reported. *Saccharomyces cerivisae* (Chatonnet *et al.*, 1993), other yeast species and *Oenococcus oeni* can produce 4-vinylphenol and 4-vinylguaiacol from ferulic and p-coumaric acid, through the action of hydrocinnamate decarboxylase (Chatonnet *et al.*, 1995), but only *Lactobacillus plantarum* and *Dekkera/Brettanomyces* possess the enzyme vinylphenol reductase which converts the vinylphenols to their respective ethylphenols (Chatonnet *et al.*, 1995). Only *Dekkera/Brettanomyces* can produce ethylphenols at the levels found in wines, and they have a 50–60% conversion rate of the available substrate (Chatonnet *et al.*, 1992). In a model solution, perception thresholds were found to be 440 $\mu g\ L^{-1}$ and 47 $\mu g\ L^{-1}$ for 4-ethylphenol and 4-ethylguaiacol, respectively, whereas in water they were found to be 130 $\mu g\ L^{-1}$ and 35 $\mu g\ L^{-1}$, respectively. Another volatile phenol reported in *Brettanomyces*-infected wines is 4-ethylcatechol (Hesford *et al.*, 2004), which has a horsey flavour and is formed from caffeic acid in an analogous manner to 4-ethylphenol and 4-ethylguaiacol from p-coumaric and p-ferulic acid, respectively. The detection threshold of 4-ethylcatechol is under debate: 60 $\mu g\ L^{-1}$ (Hesford and Schneider, 2004); 100–400 $\mu g\ L^{-1}$ and 774 $\mu g\ L^{-1}$ (Larcher *et al.*, 2008). Recent studies have led to the authorization of a new practice by the European Union (EC 53/2011), after the works of Bornet and Teissedre (2005, 2007, 2008). A product of fungal origin, called chitosan, can now be used as a tool to control *Brettanomyces*. Chitosan is a polysaccharide derived (via a patented process; Bornet *et al.*, 2011) from chitin, a natural polymer produced by a specific strain of *Aspergillus niger*. Recently, the use of chitosan, as a new natural enological treatment to prevent the growth of *Brettanomyces* during winemaking processes, was reported to be successful. The antimicrobial activity of chitosan is effective from between 5 to 10 days on different strains of *Brettanomyces bruxellensis* with a dose of 4 g hL^{-1} (Renou *et al.*, 2010). This new enological practice could also be interesting as an alternative to DMDC (dimethyl dicarbonate) in preventing potential *Brettanomyces* off-flavours in botrytized wines.

Thiamine can also be added up to 0.6 mg L^{-1} to facilitate the start of fermentation, to reduce volatile acidity creation and intermediary compounds, such as acetaldehyde, and to help preserve the wine aroma's qualities (Lafon-Lafourcade *et al.*, 1981). The use of active dry yeasts is sometimes useful to prevent the arrest of fermentation.

The presence of inhibitors produced by *B. cinerea*, known as botryticine, which are rich in the heteropolysaccharides, rhamnose and mannose, increases the formation of glycerol and acetic acid by yeasts via stimulation during glycerol-pyruvic fermentation. Fermentation takes place at between 17 and 23°C in air-cooled cellars for 18–30 days depending on the batch. Fermentation can take place in tank or in finely grained oak barrels (at least 30% of which are renewed each year). When a certain balance is achieved between the content of formed alcohol and residual sugars, the alcoholic fermentation is stopped by the addition of SO_2 in a single massive dose. The precise moment at which to stop the alcoholic fermentation is left to the expertise of the winemaker. However, it is often dictated by yeasts, which sometimes stop fermenting because of a raised content in ethanol (12.5% vol. minimum according to the appellation regulations). The operation to arrest fermentation (called 'mutage'), may or may not be preceded by a cooling of the wine using static clarification in a tank (de-sludge). The two essential operations of mutage (i.e. cooling and the large single-dose addition of SO_2 after de-sludge) are recommended notably by Peynaud and Lafourcade (1952) for arresting fermentation and eliminating a maximum of yeasts. An indicative reduction of 0.1 unit of the density is generally desirable before carrying out mutage so as to establish a good organoleptic balance in the wine. The added dose of SO_2 is generally around 250 mg L^{-1}. Botrytized wines in the Sauternes area can be aged from between 6 and 36 months in barrels, with a regular monthly topping up and quarterly rackings. At each racking (every 3 months subject to tasting), the SO_2 content is checked and maintained at around 50 mg L^{-1}.

Mutage is a key stage in making botrytized wines. In the production of certain natural sweet wines, a neutral alcohol or brandy is added to stop the alcoholic fermentation. In wines such as Sauternes, the addition of alcohol would upset the precarious organoleptic balance established by the winemaker. For this reason, mutage is provoked by a massive addition of SO_2. The successive but indispensable addition of SO_2 during fermentation has led to the introduction of legislation authorizing doses of 400 mg L^{-1} of total SO_2 in wines produced from botrytized grapes (Resolution OENO 9/98 (OIV)).

The antioxidant activity and the anti-oxidase and anti-microbial power of SO_2 make its use in enology indispensable, particularly in the making of botrytized wines. Botrytized wines are characterized above all by their pH, which is exceptionally raised in comparison to other wines (often close to 3.90). The concentrations of glycerol (produced by *B. cinerea* on the grapes and by yeasts during the alcoholic fermentation) and residual sugars are so high that they exercise a strong osmotic pressure on the wine. Furthermore, these wines do not go through malolactic fermentation, because they contain residual sugars, which do not facilitate microbiological stability. However, very few ecological studies on the microflora of botrytized grapes have been carried out. There is a difference between the microflora of healthy grapes and that of grapes infected by *B. cinerea*. The latter accommodate preferentially species such as *Kloeckera apiculata* or *Candida stellata*. Generally, on healthy berries, yeasts such as *Rhodotorula glutinis*, *Cryptococcus humicolus*, *Hanseniaspora sp.* and *Metschnikowia sp.* can be found (Rementaria *et al.*, 2003; Fleet, 2003). *Botrytis cinerea* is also accompanied by other fungi on the grape berries. Fleet (2003) reported the joint presence of *Aspergillus* and *Penicillium*, and *Cladosporium sp.* was noted by Loureiro and Malfeito-Ferreira (2003). On damaged botrytized grapes, the species *Hanseniaspora*, *Candida*, *Metschnikowia*, *Saccharomyces* and *Zygosaccharomyces* (although rare, according to Loureiro and Malfeito-Ferreira (2003)) seem to be more dominant. The presence of *Hanseniaspora uvarum* and *Candida stellata* has been noted as a major species accompanying *B. cinerea* infection (Donèche *et al.*, 1992; Mills *et al.*, 2002; Loureiro and Malfeito-Ferreira *et al.*, 2003). Although the microflora of botrytized grapes is still underestimated, the presence of yeasts such as *Hanseniaspora uvarum* (an anamorphic of *Kloeckera apiculata*) and *Candida stellata* in the must obtained from botrytized grapes has been confirmed for some time (Donèche, 1992). These two species of yeasts could play a role in the sensorial aspect of the wine. As secondary products of the alcoholic fermentation, they possess a similar profile, characterized by a strong production of acetoin and ethyl acetate, while the species *S. cerevisiae* produces more amylic alcohol and butane-2,3-diol. The interaction between these species and *S. cerevisiae* has been reported to have a positive effect on the organoleptic quality of the wine (Jolly *et al.*, 2003).

Peynaud and Domercq (1959) showed that *Candida* represents only 17.3% of the total yeasts present on botrytized grapes. It is found to a great extent in the must of grapes from

Ontario, the big Ice-wine producing region of Canada (Holloway *et al.*, 1992), because of its tolerance to free SO_2 (up to 50 mg L^{-1}) (Moore *et al.*, 1988), to ethanol (up to 10% vol.) (Benda, 1982), and the high production of glycerol (Ciani *et al.*, 2000). During alcoholic fermentation, *Candida stellata* seems to decrease the concentrations of superior alcohols, isoamyl acetate, ethyl caproate, ethyl caprylate and ethyl caprate (Ferraro *et al.*, 2000). In botrytized musts, the presence of botryticine favours the growth of *S. cerevisiae* instead of *S. cerevisiae Var. Oviformis* (also called in the past *Saccharomyces bayanus*). Due to its tolerance towards ethanol, and its good fermentation capacity in strong sugars concentration, *S. cerevisiae* quickly dominates and is considered to be responsible to accomplish the alcoholic fermentation (Ribéreau-Gayon, 1985). However, *S. bayanus* has also been observed in botrytized wines (Naumov *et al.*, 2002), and other species of yeasts have been reported: apiculated yeasts were found at the beginning of fermentation, but ceded their place to *Candida* during the alcoholic fermentation (Fleet *et al.*, 1984; Mills *et al.*, 2002); these included: *Schizosaccharomyces pombe*, *Zygosaccharomyces bailii* or *Zygosaccharomyces florentinus* (Fleet *et al.*, 2003).

The amount of volatile acidity (mainly acetate) plays a significant role in wine aroma and excessive concentrations of this alcoholic fermentation byproduct are highly detrimental to wine quality. The quantity of volatile acidity produced by *S. cerevisiae* is usually low (0.25–0.50 g L^{-1} expressed in acetic acid), but may be higher under certain fermentation conditions. In particular, during fermentation of high-sugar media, such as botrytized musts, the volatile acidity content may be greater than 1.8 g L^{-1}, that is, above the EU legal limit of 1.5 g L^{-1}. *S. cerevisiae* produces acetic acid as a byproduct of the hyperosmotic stress response to the high-sugar concentrations in grape must (Blomberg and Alder, 1992). Recently, to reduce the volatile acidity of botrytized wines, a specific yeast, *Torulaspora delbrueckii*, has been studied and developed. The objective is to achieve an aromatic diversity and intensity, a positive impact on the length in the mouth, as well as an aromatic purity with a low production of acetaldehyde, acetoin and diacetyl (Bely *et al.*, 2008). *Torulaspora delbrueckii* has been used in pure cultures and mixed with *S. cerevisiae* yeast to ferment botrytized musts. Fermentation rate, biomass growth, and the formation of volatile acidity, acetaldehyde and glycerol were considered. Unlike *S. cerevisiae*, this species did not respond to the hyper-osmotic medium by increasing acetic production as soon as it is inoculated into the must. Nevertheless, this yeast produces low ethanol and biomass yields, and *Torulaspora delbrueckii* fermentations do not reach the required ethanol content (14% vol.), although the species can survive at this concentration. A mixed culture of *Torulaspora delbrueckii* and *S. cerevisiae* was demonstrated to be the best combination for improving the analytical profile of sweet wine, particularly volatile acidity and acetaldehyde production. A mixed *Torulaspora delbrueckii/S. cerevisiae* culture at a 20 : 1 ratio produced 53% less volatile acidity and 60% less acetaldehyde than a pure culture of *S. cerevisiae*. Inoculating *S. cerevisiae* with *Torulaspora delbrueckii* after 5 days' fermentation, however, had less effect on volatile acidity and acetaldehyde production, and resulted in a stuck fermentation.

12.3 MANAGEMENT OF STABILIZATION OF BOTRYTIZED WINE

Before bottling, the wine is fined by adding a fining agent, such as bentonite or an animal-based substance, such as gelatine, to bind suspended particles and to remove soluble substances that may cause cloudiness in the finished wine. The wine is then ready to be filtered by passing it through a porous medium (generally with a porosity of about 5–6 μm). Some châteaux use tangential (or cross-flow) filtration. All these processes have been established over time through experimentation and every châteaux has its own way of making wine. Tartrate crystals can occur naturally in wine bottles, and although their presence neither affects taste nor represents a health concern, they should be avoided both for aesthetic and commercial reasons, as consumers normally consider them unacceptable. To eliminate tartrate, the wine is chilled and introduced via the bottom of a tank at a temperature of −5°C. This operation takes from 24 to 96 hours, and means that the wine that exits from the top of the vat is stabilized, and 95% or more of the generated bitartrate has been removed. Any remaining suspended microcrystals of potassium bitartrate can easily be separated later by using a simple candle-type filter. A recent alternative practice for the tartaric stabilization of white wines is the use of carboxymethylcellulose (CMC), which has been authorized by the European Union since August 2009 at the dose of 10 g hL^{-1}. CMC functions in a similar way to mannoproteins (Bowyer and Moine-Ledoux, 2007; Bowyer, 2009). The possibility of degradation of sorbic acid (possibly added to the wine for its action against yeasts and moulds) by certain lactic bacteria in malodorous compounds was also described in this type of wine (Crowell and Guyman, 1975). Sorbic acid can be added to wines in amounts up to 200 mg L^{-1} as a preserving agent.

The refermentation is a new alcoholic fermentation that occurs after adding SO_2. This unwanted phenomenon can constitute the most serious microbiological problem of botrytized wines during ageing or conservation. Ribéreau-Gayon and Peynaud (1960) suggested classifying the full phenomenon of diseases in the wine, because of the disastrous impact on the organoleptic quality of the wine, with its inherent commercial and financial consequences. These refermentations can occur during ageing in barrel (most often), or in bottle after racking. The winemaker is alerted by a gaseous release or more exactly by a crackling perceived when topping up the wine. Aromatic modifications are noted, accompanied by the smell of rotten apple, yeast, and a certain bitterness in the mouth. If this is noticed quickly in barrel, the winemaker will generally filter the wine to eliminate the yeasts, and adjust the free SO_2 content. A non-systematic filtration is carried out using open-weave filters (6–14 μm) Correcting the level of free SO_2 is often a delicate process, because the wine's binding power is suddenly increased. If refermentation arises after bottling and if the bottles are still in the wine storehouse, they are opened and filtered and the free SO_2 is corrected as described earlier. The wine is not unfit for consumption from a health point of view, but its organoleptic quality is considerably depreciated.

The accidental refermentation of a sweet wine is generally a pure fermentation led by a single species of yeast which was selected by the conditions of preservation of the wine, contrary to the first alcoholic fermentation of the must provoked by a mixture of yeasts species, such as *S. cerevisiae*, *Zygosaccharomyces bailii* and *Saccharomycodes ludwigii*, the first of these being particularly responsible for cases of refermentation in wines with high alcohol levels. Refermentation usually occurs when the temperatures are clement, in spring or in autumn, generally a few months after the initial alcoholic fermentation, and in wines that are difficult to stabilize (e.g. wines with a strong rate of binding, a high pH, from lees, or a low concentration of residual sugars). When refermentation occurs, the concentrations of acetaldehyde are always very high (generally, greater than 100 mg L^{-1}) and the free content of SO_2 is very low (close to 10 mg L^{-1} or even less). The wine's binding power is usually well advanced, and the volatile acidity has often increased.

12.4 CHARACTERISTIC AROMA OF BOTRYTIZED GRAPE

Beerenauslese wines are produced from grapes that have been selected and harvested by hand. They are similar to Sauternes, in that they have typically been exposed to *B. cinerea* (noble rot), but tend to be lower in alcohol and have greater acidity. They have a great ageing potential, and their richness of honey, caramel and tropical fruits flavours make them sought after by wine collectors. A Beerenauslese can age for 10–35 years on average and up to 50 years for exceptional wines.

Trockenbeerenauslese is the richest of the German dessert wines. In the best years, the grapes shrivel up like raisins and are overtaken by *Botrytis*. Trockenbeerenauslese displays an overwhelming intensity and complexity of flavours. These wines have the potential to age for up to 80 years and improve with age.

The region of Sauternes produces incomparable wines that differ strongly from other sweet wines. Great wines from Sauternes develop a strong aromatic intensity in the nose and mouth, which is balanced by a beautiful unctuousness, heightened by a certain liveliness. Five types of aromas can be revealed in these wines: plant aromas (green tea, mint, lime tree, verbena tea, hawthorn, eucalyptus, honey of forest, undergrowth), fruity (citrus fruits, candied fruits, white fruits, yellow fruits, dried fruits, exotic fruits), floral (honeysuckle, locust tree, rose, violet), spiced (vanilla, paprika, cumin, saffron, cinnamon, clove, nutmeg), empyreumatics (coffee, toffee, cocoa, vanilla, toast overtones) and woody (cedar, oak, sandalwood, and lime tree) (descriptions inspired by Cobbold (2001)). When they are young, these wines display, rather, aromas of pineapple, lemon, grapefruit, pear, locust tree, honey and vanilla. With age, the crystallized characters increase: apricot candies, orange, quince, cinnamon and clove. It should be noted that great botrytized wines have the highest life expectancy in the world, for the best vintage wines.

However, phenolic acids have been reported in botrytized wines from the Sauternes area for seven vintages; total phenols content has been found to be between 552 and 773 mg L^{-1} and between 22.34 and 30.1 mg L^{-1} for total catechins (catechin, epicatechin, procyanidins B_1, B_2, B_3, B_4). These results show that total phenols are 1.6 times higher in botrytized wines than in dry white wines (Landrault *et al.*, 2001). White botrytized Sauternes wines from Semillon have also been shown to have a 1.7 times greater antioxidant capacity (3.2 mmol L^{-1}) than traditional dry white wines (1.91 mmol L^{-1}) (Landrault *et al.*, 2001). A particular aspect concerns the pressing and crushing carried out on noble rotted grapes to obtain botrytized wines. A progressive crushing method (using vertical and pneumatic presses) for botrytized berries can better protect typical volatile thiols compounds, which convert to new additional products, such as 3-sulfanylhexan-1-ol (3SH) and (+)-catechin, (–)-epicatechin, and caftaric acid. One caftaric acid/3SH adduct, three (+)-catechin/3SH adducts and three (–)-epicatechin/3SH adducts have been recently characterized; and in a white must matrix caftaric acid/3SH and catechin/3SH adducts provided evidence of 3SH loss in must during alcoholic fermentation and in wine. The concentration of the (+)-catechin/3SH adduct was found to be higher than that of the caftaric acid/3SH adduct (Nikolantonaki *et al.*, 2012). In a previous study, Nikolantonaki *et al.* (2010) established, under wine oxidation conditions, a clear difference between (+)-catechin and (–)-epicatechin in their reactivity toward aromatic volatile thiols such as 3SH, 4MSP (4-methyl-4-sulfanyl-pentan-2-one), and 2-furanmethanethiol. In this work, the presence of ferric iron was also found to accelerate the decrease of the volatile thiols concentrations, as these metallic species increased the rate of both (+)-catechin and (−)-epicatechin oxidation. The same study showed that SO_2 modulated in a compound-dependant manner the oxidation of flavan-3-ols with a concomitant direct effect on volatile thiols instability, manifested by their concentration decrease in the model medium. If oxidized (+)-catechin and other potential catechins compounds can react with 3SH and 4MSP, this should result in a loss of the fruity varietal character of the wines, and during winemaking, prevention of oxidation should protect aromas of botrytized grapes.

12.4.1 Semillon and Sauvignon Blanc botrytized wines

Botrytized wines have a particular variety of aromas that range from citrus and dried fruit in young wines to orange peel in older wines, and to honey or waxy nuances in wines with oxidative ageing. The characterization of the key aroma compounds in botrytized wines derived from the two main grape varieties in the Sauternes region, Semillon and Sauvignon Blanc, has been studied. Three odoriferous compounds are present in wines made from several varieties, such as Sauvignon Blanc (Darriet *et al.*, 1995; Tominaga, *et al.*, 1996), and Semillon (Tominaga *et al.*, 2000); these are: 4-mercapto-4-methylpentan-2-one (4MMP), 4-mercapto-4-methylpentan-2-ol (4MMPOH), and 3-mercaptohexan-1-ol (3MH). The predominant role of botrytization in the development of the grape aroma potential has also been shown. In 2009, Thibon *et al.* determined the level of S-3-(hexan-1-ol)cysteine (P-3SH), the cysteinylated precursor of the most abundant volatile thiol in wine (i.e. 3-sulfanylhexanol, 3SH), during grape botrytization, and several grape over-ripening processes. Production was considerably amplified when *B. cinerea* had developed on the grapes in both on- and off-vine experiments. The P-3SH levels were increased about 100-fold between the healthy and botrytized stages within one week. It was demonstrated that the grape metabolism produced aroma precursor under post-harvest conditions. When the grapes were left to over-ripen, the vine also produced P-3SH, but the levels increased more moderately (about 10-fold). Moreover, a determination of P-3SH distribution demonstrated that *B. cinerea* was not directly responsible for the precursor formation but probably stimulated the grape metabolic pathway involved in this formation. Sweet wines made from botrytized grapes contain much higher concentrations of volatile thiols, especially 3SH (Tominaga *et al.*, 2000), than dry white wines. Three new specific volatile thiols (3-sulfanylpentan-1-ol (3SP), 3-sulfanylheptan-1-ol (3SHp), and 2-methyl-3-sulfanylbutan-1-ol (2M3SB)) were recently identified in Sauternes wines (Thibon *et al.*, 2010). Like most volatile thiols, these compounds were almost totally absent from must, mainly being formed during alcoholic fermentation. The compounds 3SP and 3SHp had a considerable impact on the overall aroma in the presence of other volatile thiols, such as 3SH, although they had low individual OAVs. Three new cysteine-S-conjugate precursors in must made from *Botrytis*-infected grapes were also reported (Thibon *et al.*, 2010); these were: S-3-(pentan-1-ol)-l-cysteine (P-3SP), S-3-(heptan-1-ol)-l-cysteine (P-3SHp), and S-3-(2-methylbutan-1-ol)-l-cysteine (P-2M3SB). Concentrations were considerably higher when *B. cinerea* had developed on the grapes. In botrytized must, the mean levels of P-3SP, P-3SHp and P-2M3SB were in the vicinity of 700, 50, and 500 nM, respectively, whereas concentrations in healthy must ranged from 0 to 50 nM. This indicated that these three new sulfanyl alcohols, responsible for the characteristic aroma of botrytized wines, were formed by the yeast metabolism during alcoholic fermentation from the corresponding non-volatile cysteine-S-conjugate precursors. The enantiomeric distribution (R and S enantiomer ratios) of 3-mercaptohexan-1-ol (3MH) and 3-mercaptohexyl acetate (3MHA) in Sauvignon Blanc and Semillon wines was determined by Tominaga *et al.* (2006). The R and S enantiomer ratios of these two thiols in dry white Sauvignon Blanc and Semillon wines are approximately 30 : 70 for A3MH, and 50 : 50 for 3MH. In sweet white wines made from grapes affected by noble rot due to the development of *B. cinerea* on ripe grapes, the proportion of the R and S forms of 3MH is in the vicinity of 30 : 70. During alcoholic fermentation, a change in the ratio of the two enantiomers of 3MH in dry white wines was observed. At the beginning of the fermentation (around a density of 1.08), the S form represented over 60%; then, at a lower density, as fermentation proceeded, the enantiomeric ratio approached 50 : 50. The ratio of the two 3MHA enantiomers remained constant throughout fermentation. The distribution of the two 3MH enantiomers changed very little during fermentation of the botrytized sweet wines. The perception thresholds for the R and S forms of 3MH in hydroalcoholic model solution are similar (50 and 60 ng L^{-1}), but these two enantiomers have quite different aromas: the R form is fruitier, with a zesty aroma reminiscent of grapefruit, while the S form smells more of passion fruit. The perception thresholds of the R and S enantiomers of 3MHA are slightly different (9 and 2.5 ng L^{-1}). The less odoriferous R form is reminiscent of passion fruit, while the S form has a more herbaceous odour of boxwood. However, the volatile compounds involved in botrytized wine typical flavours were studied by Sarrazin *et al.* (2007); two odorous zones, with grapefruit or curry nuances, were found to be apparently specific to botrytized wine. The well-known furanone, sotolon [3-hydroxy-4,5-dimethylfuran-2(5H)-one], was found in botrytized wine extracts. This compound, reminiscent of curry, has already been identified as a key odorant in botrytized wine. Among the fruity odoriferous zones, three were reminiscent of peaches and coconut. They were identified as g- and d-lactones, already known to be characteristic components of botrytized wine aromas. Moreover, two other fruity odoriferous zones were identified; these were ß-damascenone and raspberry ketone. Other zones, detected in both types of wine, were reminiscent of honey and tobacco. An identified odoriferous zone was phenylacetaldehyde, a key odorant in honey (Bicchi *et al.*, 1983) and four odoriferous zones with empyreumatic and sulphur nuances were also detected in botrytized wine, with 2-methyl-3-furanthiol. The quantification of homofuraneol®, furaneol®, norfuraneol®, phenylacetaldehyde and methional in botrytized wine, mostly at levels above their perception thresholds, confirmed their contribution to the aroma, and more particularly to caramel. Increased concentrations of some of these odorants were shown to be associated with grape botrytization, partially through the desiccation process. The three 3(2H)-furanones, especially homofuraneol® were found at concentrations far higher than their perception thresholds in young botrytized wine (up to 185 μg L^{-1} for furaneol, 324 μg L^{-1} for homofuraneol, and 3260 μg L^{-1} for norfuraneol). Phenylacetaldehyde was assayed in concentrations 2–4 times higher than its olfactory threshold in young botrytized wine (up to 136 μg L^{-1}). The same was true for methional, which was present in higher concentrations in botrytized wine (up to 15.9 μg L^{-1}), despite its odour of cooked potato; this compound contributed to the overall aroma of the wine. In older wines, the 3(2H)-furanone and phenylacetaldehyde levels are lower, indicating that they were probably degraded during wine ageing and may not make a great contribution to the overall aroma of aged botrytized wine. On the contrary, older wines had a higher methional contents (up to 49.8 μg L^{-1}), suggesting that methional levels increase with ageing and that it contributes more to the aroma of older wines (Table 12.1). Bailly *et al.* (2009) showed that most polyfunctional thiols (3-sulfanylpropyl acetate, 2-sulfanylethyl acetate, 3-methyl-3-sulfanylbutanal, etc.) proved to be completely degraded after two years of bottle ageing in a cellar. Only 3SH was still found in aged samples at concentrations above its threshold value. Most other key odorants found in the young noble rot wine were still detected 5–6 years after harvest; these were: varietal aroma (a-terpineol), sotolon, fermentation alcohols (3-methylbutan-1-ol and 2-phenylethanol) and esters (ethyl butyrate, isobutyrate, hexanoate, and isovalerate), together with oak maturation-related compounds (guaiacol, vanillin, eugenol, β-damascenone,

Table 12.1 Levels of aroma in different Sauternes botrytized wine vintages and olfactory perception thresholds (Sarrazin *et al.*, 2007).

Vintage	3SH (ng L)	3SPOH (ng L)	3SHpOH (ng L)	Furaneol® (μg L)	HomoFuraneol® (μg L)	Norfuraneol® (μg L)	Phenylacetaldehyde (μg L)	Methional (μg L)
2001–2003	2450–7033	91–299	26–72	121–185	125–324	771–3260	67–136	7.3–15.9
1975–1999	/	/	/	60–149	8–118	237–1242	23–115	6.8–49.8
Olfactory perception threshold	60*	900*	35*	60**	40**	2000**	30**	2.4**
Type of aroma	Passion fruit Grapefruit	Grapefruit/ citrus zest	Grapefruit/ citrus zest	Caramel	Caramel	Caramel	Honey	Baked potatoes

Olfactory perception threshold:
Comparison with their olfactory perception thresholds: *(ng L^{-1}) and **(μg L^{-1}) in model solution (5 g L^{-1} tartaric acid, 12% v/v ethanol, pH 3.5)
3SH, 3-sulfanylhexanol
3SPOH, 3-sulfanylpentan-1-ol
3SHpOH, 3-sulfanylheptan-1-ol

trans-non-2-enal, β-methyl-γ-octalactone, γ-nonalactone, and furaneol), and aromas exhibiting interesting cake, honey-like, and dried apricot odours, such as homofuraneol, theaspirane and γ-decalactone. Abhexon (honey/spicy compound), never previously mentioned in sweet wines, was found to be synthesized during bottle ageing with levels close to its threshold value (up to 7 $\mu g\ L^{-1}$ after 5–6 years), thus suggesting a key role of this strong odorant in aged Sauternes wines.

12.4.2 Riesling botrytized wines

A study elucidating the odour potency of aroma compounds in Riesling was carried out by Bowen *et al.* (2012). Five odour-active volatiles, 2-phenylethyl acetate, ethyl cinnamate and β-ionone, were identified as well as decanal and β-damascenone. Decanal had the highest odour potency value (OSV = 100) in Riesling wine. It was described as having a petroleum, vinyl/plastic, citrus, green aroma. β-Damascenone had the highest odour potency in Riesling wine, and it was described as a distinct pear aroma (Bowen and Reynolds, 2012). Ethyl 3-methylbutyrate, 1-hexanol, 1-octen-3-ol, ethyl hexanoate and 1-octanol were the five most odour-potent compounds in Riesling wines. 1-Hexanol was found to be the second most odour-potent aroma. Other odour-potent compounds found in most wines were: cis-rose oxide, phenethyl alcohol, nerol oxide, ethyl phenylacetate and 4-vinylguaiacol. The clove-smelling 4-vinylguaiacol and floral-smelling geranyl acetone were more odour-potent in Riesling and acetophenone had a stronger odour. 1,1,6-Trimethyl-1,2-dihydronaphthalene (TDN) is known to convey 'petrol' aromas to aged Riesling wines, but its prevalence and contribution to young Riesling was not yet well understood. Recently, Sacks *et al.*, (2012) measured TDN concentrations in 1–3-year-old varietal wines produced from Riesling (28 wines) and Sauvignon Blanc (6 wines). TDN concentrations in the Riesling wines, 6.4 ± 3.8 $\mu g\ L^{-1}$, were significantly higher than in all the other varietals, 1.3 ± 0.8 $\mu g\ L^{-1}$. The odour detection threshold for TDN was determined to be 2 $\mu g\ L^{-1}$ in both model wine and a neutral white wine. This TDN sensory threshold was a factor of 10 below the previously reported odour threshold. The OAVs odour activity values for TDN was higher than for the monoterpenes linalool and geraniol in 25 of 28 Riesling wines. The competitive formation and degradation of TDN with a kerosene, bottle-age aroma and 2,2,6,8- tetramethyl-7,11-dioxatricyclo[6.2.1.0 1,6]undec-4-ene (Riesling acetal) under wine storage conditions have been investigated by Daniel *et al.* (2009). At 45°C in model wine, approximately 40% of Riesling acetal was converted to TDN after 60 days at pH 3.2, whereas 80% was converted after 60 days at pH 3.0. Under more vigorous conditions, this conversion was completed and the TDN thus formed was stable. Riesling acetal is a precursor to TDN at wine pH, rather than a competing end-product formed by the acid-catalyzed hydrolysis of precursors common to both compounds. TDN is chemically stable to wine acid. In past analysis of Riesling wine, it was reported that the most potent odorant in both polar and non-polar extracts was a C13 norisoprenoid, β-damascenone, having a 'cooked apple' aroma (Chisholm *et al.*, 1994). However, this compound is nearly ubiquitous in wines, and its impact in the presence of other odorants is greatly diminished (Sefton *et al.*, 2011). Other odorants identified in this earlier study, (Chisholm *et al.*,1994), and a more recent study of Riesling wine, (Komes *et al.*, 2006), include linalool as well as several fermentation-derived odorants, such as ethyl 2-methylbutanoate, ethyl hexanoate, 2-phenylethanol and isoamyl acetate, which are commonly observed in wine analyses (Ferreira and Cacho, 2009). Fresh grapes frozen using refrigeration and wine produced from grapes partially dried using microwave vacuum dehydration have been tested for late harvest sweet dessert wines from Riesling grapes (Clary *et al.*, 2006). Significant differences were detected: the wine made from the dehydrated grapes exhibited lower fresh fruit aroma, higher fusel oils and oxidation, and the flavour notes, including citrus/grapefruit, and acidity were lower. Recently, Cui *et al.* (2012) investigated the effects of four different technologies to arrest fermentation on the microbial stability and quality of low-alcohol sweet white wine, with Italian Riesling grapes as raw materials; single SO_2 treatment, combined high voltage electrostatic field (HVEF)/SO_2 treatment, ultrasound/SO_2 treatment and pasteurization/SO_2 treatment were compared for the lethality of yeasts, enological parameters, and sensory evaluation of the low-alcohol sweet white wines. The results showed that the most successful technology to arrest fermentation was the one combined with ultrasound (40 Hz per 20 min)/SO_2 (40 mg L^{-1}) treatment, which had a higher total lethal rate of *S. cerevisiae*, and the wines treated in this way were rich in typical Riesling grape variety flavour and attractive aromas, with a pleasant fruity taste and microbial stability.

REFERENCES

Bailly, S., V. Jerkovic, A. Meurée, *et al.* 2009. Fate of key odorants in Sauternes wines through aging. *Journal of Agricultural and Food Chemistry*, 57(18):8557–8563.

Bely, M., P. Stoeckle, I. Masneuf-Pomarède, *et al.* 2008. Impact of mixed Torulaspora delbrueckii-Saccharomyces cerevisiae culture on high-sugar fermentation. *International Journal of Food Microbiology*, 122(3):312–320.

Benda, I. 1982. Wine and brandy. In *Prescott and Dunn's Industrial Microbiology*, 4th edn (ed. G. Reed), AVI Publishing Co. Westport, CT, pp. 310–311.

Bicchi, C., F. Belliardo, and C. Frattini. 1983. Identification of the volatile components of some Piedmontese honeys. *Journal of Apicultural Research*, 22:130–136.

Blomberg, A. and L. Alder. 1992. Physiology of osmotolerance in fungi. *Advanced in Microbiology and Physiology*, 33:145–212.

Bock, G., I. Benda, and P. Schreier. 1985. Biotransformation of linalool by *Botrytis cinerea*. *Journal of Food Science*, 51:659–662.

Bock, G., I. Benda, and P. Schreier. 1986. Metabolism of linalool by *Botrytis cinerea*. In *Biogeneration of Aromas* (eds T.H. Parliament and R. Crouteau), *ACS Symposium*, Ser. No. 317:243–253. American Chemical Society, Washington, DC.

Bock, G., I. Benda, and P. Schreier. 1988. Microbial transformation of geraniol and nerol by *Botrytis cinerea*. *Applied Microbiology and Biotechnology*, 27:351–357.

Boidron, J.N. 1978. Relation entre les substances terpéniques et la qualité du raisin (rôle du *Botrytis cinérea*). *Annales de Technologie Agricole*, 27:141–145

Bornet, A. and P.L. Teissedre. 2005. Intérêt de l'utilisation de chitine, chitosane et de leurs dérivés en œnologie. *Journal International des Sciences de la Vigne et du Vin*, 39(4):199–207.

Bornet, A. and P.L. Teissedre. 2007. Reduction of toxins and contaminants with biological tools. *Bulletin de l'OIV*, 80(917-919):471–481.

Bornet, A. and P.L. Teissedre. 2008. Chitosan, chitin-glucan and chitin effects on minerals (iron, lead, cadmium) and organic (ochratoxin A) contaminants in wines. *European Food Research and Technology*, 226(4):681–689.

Bornet, A., S. Gautier, V. Maquet, *et al.* 2011. *Patent : 'Chitosan powder'*, 2011. N° WO2011FR51368 20110616, WO2011157955 (A2), also published as FR2961512 (A1) -2011-12-23.

Bowen, A.J. and A.G. Reynolds. 2012. Odor potency of aroma compounds in Riesling and Vidal blanc table wines and icewines by gas chromatography-olfactometry-mass spectrometry. *Journal of Agricultural and Food Chemistry*, 60(11):2874–2883.

Bowyer, P.K. 2009. Technical update: MANNOSTAB – for potassium bitartrate stabilisation. *Australian & New Zealand Grapegrower & Winemaker*, 550:105–106.

Bowyer, P.K. and V. Moine-Ledoux. 2007. MANNOSTAB: the award-winning new potassium bitartrate stabilisation product. *Australian & New Zealand Grapegrower & Winemaker*, June:57–62.

Chatonnet, P., D. Dubourdieu, J. Boidron, *et al.* 1992. The origin of ethylphenols in wines. *Journal of the Science of Food and Agriculture*, 60:165–175.

Chatonnet, P., D. Dubourdieu, J. Boidron, *et al.* 1993. Synthesis of volatile phenols by Saccharomyces cerivisae in Wines. *Journal of the Science of Food and Agriculture*, 62:191–202.

Chatonnet, P., D. Dubourdieu, and J. Boidron. 1995. The influence of *Brettanomyces/Dekkera* sp. yeasts and lactic acid bacteria on the ethylphenol content of red wines. *American Journal of Enology and Viticulture*, 46:463–468.

Chauvet, S., P. Sudraud, and T. Jouan. 1986. La cryoextraction selective, premières observations et perspectives. *Bulletin de l'OIV*, 667-668:1021–1043.

Chisholm, M.G., L.A. Guiher, T.M. Vonah, *et al.* 1994. Comparison of some French–American hybrid wines with white Riesling using gas-chromatography olfactometry. *American Journal of Enology and Viticulture*, 45(2):201–212.

Ciani, M., L. Ferraro, and F. Fatichenti. 2000. Influence of glycerol production on the azrobic and anazrobic growth of the wine yeast Candida stellata. *Enzyme and Microbial Technology*, 27:698–703.

Clary, C., A. Gamache, M. Cliff, *et al.* 2006. Flavor and aroma attributes of riesling wines produced by freeze concentration and microwave vacuum dehydration, *Journal of Food Processing and Preservation*, 30(4):393–406.

Cobbold, D. 2001. *Sauternes et Barsac*, Coll. Autour d'un vin, Ed. Flammarion.

Crowell, E.A. and M.F. Guymon. 1975. Wine constituents arising from sorbic acid addition and identification of 2-ethoxyhexa-3,5-diene as a source of geranium-like off-odor. *American Journal of Enology and Viticulture*, 26:97–102.

Cui, Y., W. Lv, J. Liu, *et al.* 2012. Effect of different ending fermentation technologies on microbial-stability of Italian Riesling low alcohol sweet white wine, *Advanced Materials Research*, 393-395:1165–1168.

Daniel, M.A., D.L. Capone, M.A. Sefton, *et al.* 2009. Riesling acetal is a precursor to 1,1,6-trimethyl-1,2-dihydronaphthalene (TDN) in wine. *Australian Journal of Grape and Wine Research*, 15(1):93–96.

Darriet, P., T. Tominaga, V. Lavigne, *et al.* 1995. Identification of a powerful aromatic component of *Vitis Vinifera* var. Sauvignon wines: 4-methyl-4-mercaptopentan-2-one. *Flavour and Fragrance Journal*, 10:385–392.

Donèche, B.J. 1992. Botrytized wines. In *Wine Microbiology and Biotechnology* (ed. G.H. Fleet), Harwood Academic Publishers, Chur.

Dubourdieu, D., K.H. Koh, A. Bertrand, *et al.* 1983. Mise en évidence d'une estérase chez *Botrytis cinerea*. Incidence technologique. *Comptes Rendus de l'Académie des sciences Paris, Ser. C*, 296:1025–1028.

Ferraro, L., F. Fatichenti, and M. Ciani. 2000. Pilot scale vinification process using immobilized *Candida stellata* cells and *Saccharomyces cerevisiae*. *Process Biochemistry*, 35:1125–1129.

Ferreira, V. and J. Cacho. 2009. Identification of impact odorants of wines. In *Wine Chemistry and Biochemistry* (eds Moreno-Arribas, M. V., Polo, M. C.), Springer, New York, pp. 735.

Fleet, G.H. 2003. Yeast interactions and wine flavour. *International Journal of Food Microbiology*, 86:87–99.

Fleet, G.H., S. Lafon-Lafourcade, and P. Ribéreau-Gayon. 1984. Evolution of yeasts and lactic acid bacteria during fermentation and storage of Bordeaux wines. *Applied and Environmental Microbiology*, 48:1034–1038.

Frezier, V. and D. Dubourdieu. 1992. Ecology of yeast strain Saccharomyces cerevisiae during spontaneous fermentation in a Bordeauw winery. *American Journal of Enology and Viticulture*, 43:375–380.

Genovese, A., M. Ugliano and L. Moio, L. 2002. Powerful odorants of wines obtained from late harvested and botrytized non-aromatic grapes. In *Proceedings of the 10th Weurman Flavour Research Symposium*, 25–28 June, Beaune, France.

Hesford, F. and K. Schneider. 2004. Discovery of a third ethylphenol contributing to Brettanomyces taint. *Obst- und Weinbau*, 140:11–13.

Hesford, F., K. Schneider, N. Porret, *et al.* 2004. Identification and analysis of 4-ethyl catechol in wines tainted by *Brettanomyces* off-flavor. *American Journal of Enology and Viticulture*, 55:304A.

Holloway, P., R.A. van Twest, R.E. Subden, *et al.* 1992. A strain of *Candida stellata* of special interest to oenologists. *Food Research International*, 25:147–149.

Jolly, N.P., O.P.H. Augustyn, and I.S. Pretorius. 2003. The effect of non-Saccharomyces yeasts on fermentation and wine quality. *South African Journal of Enology Viticulture*, 24:55–62.

Komes, D., D. Ulrich, and T. Lovric. 2006. Characterization of odor-active compounds in Croatian Rhine Riesling wine, subregion Zagorje. *European Food Research and Technology*, 222(1–2):1–7.

Lafon-Lafourcade, S., V. Lucmaret, A. Joyeux, *et al.* 1981. Utilisation de levains mixtes dans l'élaboration des vins de pourriture noble en vue de réduire l'acidité volatile. *Comptes Rendus de l'Académie d'agriculture*, 67:616–622.

Landrault, N., P. Poucheret, P. Ravel, *et al.* 2001. Antioxidant capacities and phenolics levels of French wines from different varieties and vintages. *Journal of Agricultural and Food Chemistry*, 49:3341–3348.
Larcher, R., G. Nicolini, D. Bertoldi, *et al.* 2008. Determination of 4-ethylcatechol in wine by high-performance liquid chromatography-coulometric electrochemical array detection. *Analytica Chimica Acta*, 609:235–240.
Licker, J.L, T.E. Acree, and T. Henick-Kling. 1999. What is 'Brett' (*Brettanomyces*) flavour? A preliminary investigation. In *Chemistry of Wine Flavour. ACS Symposium Series* (eds A.L. Waterhouse and S.E. Ebeler), American Chemical Society, Washington, DC, pp. 96–115.
Loureiro, V. and M. Malfeito-Ferreira. 2003. Spoilage yeasts in the wine industry. *International Journal of Food Microbiology*, 86:23–50.
Mills, D.A., E.A. Johannsen, and L. Cocolin. 2002. Yeast diversity and persistence in Botrytis-affected wine fermentations. *Applied and Environmental Microbiology*, 68:4884–4893.
Moore, K.J., M.G. Johnson, and J.R. Morris. 1988. Indigenous yeast microflora on Arkansas White Riesling (*Vitis vinifera*) grapes and in model systems. *Journal of Food Science*, 53:1725–1728.
Naumov, G.I., H. Naumova, Z. Antunovics, *et al.* 2002. Saccharomyces bayanus var. uvarum in Tokaj wine-making of Slovakia and Hungary. *Applied Microbiology and Biotechnology*, 59:727–730.
Nikolantonaki, M., I. Chichuc, P.L. Teissedre, *et al.* 2010. Reactivity of volatile thiols with polyphenols in a wine-model medium: Impact of oxygen, iron, and sulfur dioxide. *Analytica Chimica Acta*, 660(1–2):102–109.
Nikolantonaki, M., M. Jourdes, K. Shinoda, *et al.* 2012. Identification of adducts between an odoriferous volatile thiol and oxidized grape phenolic compounds: Kinetic study of adduct formation under chemical and enzymatic oxidation conditions. *Journal of Agricultural and Food Chemistry*, 60:2647–2656.
Norris, L. 2004. Unraveling the mystery of *Brettanomyces* flavor [abstract]. *American Journal of Enology and Viticulture*, 55:304A.
Peynaud, E. and S. Domercq. 1959. A review of microbiological problems in wine-making in France. *American Journal of Enology and Viticulture*, 10:69–77.
Peynaud, E. and S. Lafourcade. 1952. Les conditions d'emploi de l'anhydride sulfureux dans les vins liquoreux. *Bulletin de l'OIV*, 252:110–120.
Rementeria A., J.A. Rodriguez, A. Cadaval, *et al.* 2003. Yeast associated with spontaneous fermentations of white wines from the 'Twakoli de Bizkaria' region (Basque country, North Spain). *International Journal of Food Microbiology*, 86:201–217.
Renou, A., A. Bornet, L. Pic-Blateyron, *et al.* 2010. Prévention du risque *Brettanomyces* par l'utilisation d'un biopolymère d'origine fongique: le chitosane (OIV, P.II.04 N°34P). *33rd World Congress of Vine and Wine and 8th General Assembly of the OIV*, 20–27 June 2010, Tbilisi, Georgia.
Renouf, V., A. Lonvaud-Funel, and J. Coulon. 2007. The origin of *Brettanomyces bruxellensis* in wines: A review. *Journal International de Science de Vigne et Vin*, 41:161–173.
Ribéreau-Gayon, J. and E. Peynaud. 1960. *Traité d'oenologie tome I*, pp. 217–223.
Ribéreau-Gayon, P. 1985. New developments in wine microbiology. *American Journal of Enology and Viticulture* 36:1–10.
Ribéreau-Gayon, P., D. Dubourdieu, B. Donèche, *et al.* 1998. *Traité d'oenologie. Tome I: Microbiologie du vin – Vinifications*, Ed. Dunod, Paris.
Robertson G.L., R. Eschenbruch, and K.L. Cresswell. 1980. Seasonal changes in the pectic substances of grapes and their implication in juice extraction. *American Journal of Enology and Viticulture*, 31: 162–166.
Romano, A., M.C. Perello, A. Lonvaud-Funel, *et al.* 2009. Sensory and analytical re-evaluation of 'Brett character'. *Food Chemistry*, 114:15–19.
Sacks, G.L., M.J. Gates, F.X. Ferry, *et al.* 2012. Sensory threshold of 1,1,6-trimethyl-1,2-dihydronaphthalene (TDN) and concentrations in young Riesling and non-Riesling wines. *Journal of Agricultural and Food Chemistry*, 60(12):2998–3004.
Sarrazin, E., D. Dubourdieu, and P. Darriet. 2007. Characterization of key-aroma compounds of botrytized wines, influence of grape botrytization. *Food Chemistry*, 103:536–545.
Saurez, R., J.A. Saurez-Lepe, A. Morata, *et al.* 2007. The production of ethylphenols in wine by yeasts of the genera Brettanomyces and Dekkera: A review. *Food Chemistry*, 102:10–21.
Sefton, M.A., G.K. Skouroumounis, G.M. Elsey, *et al.* 2011. Occurrence, sensory impact, formation, and fate of damascenone in grapes, wines, and other foods and beverages. *Journal of Agricultural and Food Chemistry*, 59(18):9717–9746.
Sponholz, W.R. and T. Huehn. 1994. 4,5-Dimethyl-3-hydroxy-2,5-furanon, sotolon, an indicator of Botrytis infection. *Wein-Wissenschaft, Wiesbaden*, 49:37–39.

Thibon, C., D. Dubourdieu, P. Darriet, *et al.* 2009. Impact of noble rot on the aroma precursor of 3-sulfanylhexanol content in *Vitis vinifera* L. cv. Sauvignon blanc and Semillon grape juice. *Food Chemistry*, 114(4):1359–1364.

Thibon, C., S. Shinkaruk, M. Jourdes, *et al.* 2010. Aromatic potential of botrytized white wine grapes: Identification and quantification of new cysteine-S-conjugate flavor precursors, *Analytica Chimica Acta*, 660:190–196.

Tominaga, T., R. Baltenweck-Guyot, C. Peyrot des Gachons, *et al.* 2000. Contribution of volatile thiols to the aromas of white wines made from several Vitis vinifera grape varieties. *American Journal of Enology and Viticulture*, 51(2):178–181.

Tominaga, T., P. Darriet, and D. Dubourdieu. 1996. Identification de l'acétate de 3-mercaptohexanol, composé à forte odeur de buis, intervenant dans l'arôme des vins de Sauvignon. *Vitis*, 35(4):207–210.

Tominaga, T., Y. Niclass, E. Frérot, *et al.* 2006. Stereoisomeric distribution of 3-mercaptohexan-1-ol and 3-mercaptohexyl acetate in dry and sweet white wines made from *Vitis vinifera* (var. Sauvignon blanc and Semillon). *Journal of Agricultural and Food Chemistry*, 54(19):7251–7255.

Wirz, D.O., H. Heymann, and L.F. Bisson. 2004. Descriptive analysis of *Brettanomyces*-infected Cabernet Sauvignon wines. *American Journal of Enology and Viticulture*, 55:303A.

Part 3 The Wines

13 Amarone

Daniele Accordini
Winemaker, Cantina Valpolicella Negrar, Negrar Verona, Italy

13.1 THE TERRITORY

Amarone wine is produced in Valpolicella, a charming region that stretches along the province of Verona's foothills and has a total surface area of 30 000 km^2. This area includes the so-called 'Classico' area (Sant'Ambrogio, S. Pietro in Cariano, and the valleys of Fumane, Marano and Negrar), the Valpantena area and the eastern part of the province of Verona (the valleys of Squaranto, Tramigna and Illasi). It is from this highly diverse and distinctive *terroir* that the wines of Valpolicella derive their unique character and their typicity. Meadows, pastures, as well as beech and fir forests, cover the limestone mountain area known as the Lessin Mountains; a hilly area extends from the southern spurs of the mountains up to 600 m above sea level, with slightly elongated summits that separate the different valleys. The vineyards are situated on these hills, often on artificial terraces built with dry stone walls. The floodplains are made of alluvial soils from the Adige river, while streams (locally known as 'progni') descend from the valleys. Each valley has a slightly different geology, but all are characterized by a good concentration of limestone and clay in a rocky soil. Overall, the climate of the Valpolicella area is never really cold as it is protected to the north by the Lessina Prealps, and benefits from the favourable exposure of the valleys in the south and the warmer temperatures rising from Lake Garda, which ensure mild winters. In these valleys, vine-growing has remarkably similar traits, because of their geographical proximity and shared cultural influences.

13.2 HISTORY

In the Valpolicella area, the dehydration technique was widespread in ancient times. The Romans used this technique for making different types of wines and possibly also for grapes used in the production of Rhaetian wine (*vinum Rhaeticum*). As proof thereof, a Roman patrician villa dating back to the early centuries AD was found in Ambrosan (S. Pietro in Cariano). Some of the rooms had under-floor heating (hypocaust); two of them were used for drying agricultural products, particularly grapes. The first written evidence explicitly mentioning the production of wines using partially dehydrated Valpolicella grapes in Verona dates back to the fifth century AD. It is in a letter written by Cassiodorus, King Theodoric's minister in Ravenna, to the senators of the Venetian canonry, to obtain the procurement of a

Sweet, Reinforced and Fortified Wines: Grape Biochemistry, Technology and Vinification, First Edition.
Edited by Fabio Mencarelli and Pietro Tonutti.

wine called 'Acinaticum, whose name derives from the word berry (*Acinaticum, cui nomen exacino est. . .*)'. Thus, Acinaticum is the archetype of Recioto or, better still, Reciotos, since the same name was used to identify the white wines made with dehydrated grapes that were also common in the Valpolicella region and the hilly areas of Verona in general (Recioto di Soave). So, the practice of dehydrating grapes is ancient and various techniques were used in the past. The bunches were hung from wires or from ceiling beams, with metal hooks. They were usually hung upside down, which kept the berries separated, in order to improve aeration.

The name Recioto appears in Verona towards the end of the nineteenth century. Its etymological origins are unclear. It may come from the Latin word *recis* (detached clusters), or *racemus* (chosen cluster). Nevertheless, according to the most likely and fascinating theory, Recioto derives from the dialect word *recia* (ear). Apparently, Recioto was given this name because traditionally, it was only made with the side clusters (the 'ears'), which are considered to be the most mature and sugar-rich of the bunch. For centuries, Recioto was the most popular, sought-after and valuable wine of the Valpolicella area. Nevertheless, towards the end of the nineteenth century, sweet wines became less fashionable, and people started to appreciate dry wines, such as Amarone.

The official birth of Amarone wine dates back to 1936, in the wine cellars of Villa Mosconi in Novare di Arbizzano (Negrar). The oldest existing bottle with the Amarone label is dated 1939 and is kept at the Cantina Valpolicella in Negrar (then established in Novare). The Denominazione di Origine (DOC) 'Valpolicella' was created in 1968, both for Valpolicella Recioto (a sweet wine) and Valpolicella Recioto Amarone (a dry wine), because Amarone was considered to be the 'child' of Recioto. Back in those years, only small quantities of Amarone were produced, since the large majority of dehydrated grapes were still dedicated to the production of Recioto, while today Amarone has the highest production of the two.

13.3 VINEYARD TRAINING

Traditional vine-growing in Valpolicella is based on typical Verona-style pergolas (approximately 80%). However, double and simple Trentino-style pergolas are also to be found, though to a lesser extent. Among the espalier training systems, the most common is the Guyot training system. The double and simple Trentino-style pergola (the latter being only justified in the presence of steep slopes) is particularly useful since the indigenous cultivars have a low basal fertility that requires long pruning. The perpendicular spatial arrangement of the clusters makes the selection of grapes for the dehydration process easier. In hilly areas, this training system ensures a good-quality production, since it allows for good light exposure and easy management. With pergola or pergoletta training systems, the vine spacing is between 3.3×0.7 and 3.8×0.7 m. For the Guyot training system, the average vine spacing is 2.5×0.8 m.

13.4 THE AMARONE GRAPE VARIETIES

Corvina: it is the most important variety of the grape blend for the production of Valpolicella wines, so much so that the new DOCG (Denominazione di Origine Controllata e Garantita) rules accept the use of a variable percentage between 40 and 80% (which also includes the cv. Corvinone). It is a vigorous late-flowering variety, fairly cold hardy. There are many biotypes of Corvina (small, medium or large Corvina), the main difference being the weight of the cluster, which fluctuates between 200 and 350 g. This variety is often subject to *Botrytis cinerea* attacks during the dehydration process due to the compactness of the clusters and the thickness of the skins. Monovarietal wines made from Corvina grapes are deep-ruby red, full bodied and tasty, with a fresh and delicate bouquet.

Rondinella: this variety is cold hardy and drought-resistant. It has a low sensitivity to diseases, especially fungal diseases, which makes it suitable for the dehydration process. The cluster is medium sized, cylindrical and more or less compact, with one or two side clusters; the berry has a spheroidal shape with a black-purplish bloom-covered skin. Bud-break is between the first and second decades of April, and ripening towards the end of September. The wine produced from the Rondinella variety is ruby red, pleasantly aromatic with a subtle floral and fresh red-fruit nose (reminiscent of cherry and marasca cherry). While full bodied, it is not particularly tannic. According to the new DOCG rules, Rondinella can be used to between 5 and 30%.

Corvinone: while mistakenly considered a biotype of Corvina Veronese, it is a different variety, so much so that the new DOC rules expressly refer to it as an autonomous variety, which can form up to 50% of the grape blend. Its thick skin and juicy berries make it an ideal variety for dehydration. It is affected by the yields per hectare; in high yields, it produces a poor-quality wine with diluted flavours, while in smaller yields, it produces a rich, mellow and intensely fruity wine. Therefore, this variety is extremely interesting, especially when cultivated in hilly areas, on poor soils and on well-exposed slopes.

Optional varieties: Other red varieties, among those that are authorized and recommended for the province of Verona, may be added to the blend with an overall maximum of 15%, and 10% per single variety. This list includes some native varieties such as Molinara (under the previous DOC rules, it was compulsory to add between 5 and 25% of this variety), Croatina, Barbera and Sangiovese, and international varieties such as Merlot, Cabernet Sauvignon and Syrah. With the introduction of the DOCG rules in 2010, 10% of indigenous Valpolicella varieties were added. Those varieties, such as Dindarella (or Pelara), Oseleta, Negrara trentina (or Terodola), Cabrusina and Forsellina, were rediscovered and brought to new light because of their distinctive characteristics.

In Valpolicella, the grape's ability to shrivel is the most decisive factor in the choice of the variety. This characteristic mainly depends on genetics and production techniques, which may tend to reduce yields while encouraging ripening. The grape's physiological suitability to dehydration is mainly based on the structure of the berry, which must be highly resistant to the separation of the pedicel, and should have a thin and elastic skin that is not subject to cracking or to common rot. At the same time, the berries must have a low resistance to dehydration, that is, the evaporation of water through the lenticels whose number and size are of fundamental importance in the dehydration process, as is the thickness of the cuticle, the epidermis and the hypodermis, which 'harbour' the aromatic compounds, the anthocyanins and the fine tannins.

13.5 THE NEW DOCG RULES FOR AMARONE

After more than 15 years, the producer's association (Consorzio tutela vini Valpolicella) obtained DOCG (Denominazione di Origine Controllata e Garantita) status for Valpolicella Recioto and Amarone wines in April 2010. Production rules were significantly changed. The main changes are summarized as follows:

- Corvina Veronese + Corvinone varieties: between 40 and 80% (with a maximum of 50% for Corvinone); Rondinella: between 5 and 30%; other red berry varieties: max. 15%

of authorized varieties in the province of Verona + 10% of indigenous varieties (with a maximum of 10% per single variety).
- Bud load 60 000 max. (for the pergola Veronese training system). A minimum of 3300 vines per hectare (excluding vineyards already belonging to the appellation and terraced vineyards with dry stone retention walls).
- Max. grape yield 12 t ha^{-1} + 20 %; grapes for Amarone: max. 65% of the total allowed production, with a grape-to-wine ratio of 40%.
- Minimum alcohol content of the grapes: 11% at harvesting before dehydration.
- Minimum alcohol content of the grapes: 14% at the end of the dehydration process.
- Vinification of the grapes is not allowed before 1st December.
- Grapes are dehydrated without heating.
- Minimum ageing time: 25 months (50 months for Amarone Riserva).
- Minimum alcohol content of the wine is 14% with a maximum allowable residual sugar amount of 12 g L^{-1}; minimum total acidity 5 g L^{-1}; net dry extract 26 g L^{-1}, and 35 g L^{-1} in the Riserva version.
- Obligation to indicate the vintage on the label.
- Bottling must mandatorily take place in the area of production except for the holders of a temporary or permanent authorization.

13.6 THE DEHYDRATION TECHNOLOGY IN VALPOLICELLA

13.6.1 The main factors in the dehydration process

In Valpolicella, before dehydrating Amarone grapes, enological objectives must be set. A proper working protocol and the correct sizing not only of the dehydration loft, but also of the necessary equipment, are to be defined according to these objectives. For the sake of brevity, all viticultural considerations will be left aside. Grape production is very important, but would require a whole chapter in itself. Our starting point will be the arrival of the grapes at the winery. Undoubtedly, the first decision to be made is of an enological nature, and has to do with deciding on the type of product to be obtained in terms of the desired sugar levels (sweet or dry). The available equipment has also to be taken into account (one or more cycles). We will then have to choose the technology (natural, integrated or artificial) and the materials to be used (wooden slats, wooden or plastic trays, or super plateaux); the quantity of grapes to be treated; the desired length of stay in the dehydration loft (from 90–150 days); the arrangement in the dehydration loft, which may vary from 25 kg m^{-3} to a maximum of 75 kg m^{-3}; the electric power available, which in some cases becomes a limiting factor; the number of desired air changes (from a minimum of 5 changes h^{-1} to the optimal 15 changes h^{-1}); the external air speed (from a minimum of 1 m s^{-1} to a perfect 4 m s^{-1}); the working temperature and humidity, and naturally, the grape variety.

13.6.2 The dehydration techniques adopted in Valpolicella

In the Amarone production chain, dehydration is a key factor in developing the typical organoleptic characteristics of this wine. In the last 20 years, the dehydration technique in Valpolicella has undergone a substantial change. Mats made from reeds have given way to wooden trays and then to the exclusive use of plastic trays. The trays are small (60 cm × 40 cm), light (1.1 kg), easy to handle, sturdy and, most importantly, hygienic, since they can be washed. These stackable trays are also easily transportable. However, the revival of the dehydration technique was mainly linked to the ability to actively manage the shrivelling by controlling the dehydration process, and thus reducing the risk of *Botrytis* attacks which can considerably limit the amount of grapes being laid to rest during unfavourable years, and can also endanger the product during exceptional vintages. Today in Valpolicella, almost all producers use the assisted dehydration technique, as the economic stakes are exceptionally high.

The first attempts at controlled dehydration were accomplished in Valpolicella during the early 1970s by Usseglio-Tomasset and Paronetto and were subsequently pursued by Ferrarini. The new controlled dehydration systems used today are essentially made up of dehumidifiers working at ambient temperatures with important air changes (10–15 changes h^{-1}). More specifically, they comprise two ventilators with a cold plate for the condensation of the water vapour coming from the dehydration loft. There is also a compressor cooling plate that brings the dehumidified air temperature to ambient values. The dehumidifiers have a humid air inlet turned towards the dehydration loft and an opening at the top that conveys the air above the trays in which the grapes are laid to dehydrate. There are another two openings on the sides, one of which takes in the air at ambient temperature, while the other releases the hot air resulting from the cooling of the hot plate inside the dehumidifier. This hot air is conveyed externally so that the temperature of the loft does not rise. The air in the dehydration loft is then circulated by large fans (120 cm × 120 cm), which are positioned in order to distribute this dry air evenly over all the grapes. When this system is properly managed, external air is only circulated in the dehydration loft during cool and sunny days, while on wet and rainy days, when humidity is close to saturation, the humidity levels are kept between 60 and 70% within the dehydration loft (Figure 13.1).

Figure 13.1 Big dehydration room to store 500 tons of grape in Valpolicella area.

All rooms are equipped with probes that measure the humidity outside the dehydration loft. The dehumidifier is automatically enabled according to the measured values, and windows that let the air in and out are automatically opened and closed.

Temperature and relative humidity (RH) play a crucial role, especially in the early stages of the grape dehydration process. Once low levels of inner water in the grapes are reached, the ambient conditions (temperature and humidity) can no longer affect the success of the dehydration process, except in the sense of increasing or decreasing its duration. There are basically three processing and air distribution techniques in use.

1. *Use of a pre-dehydration chamber for 48–72 hours and subsequent moving of the grapes into the dehydration room.* With this method, there is a weight loss of 10–12% in 72 hours with a temperature of approximately 27–30°C. A massive ventilation and dehumidification power is applied to a limited quantity of grapes. The aim of this process is to act immediately by removing the water from rachids (vegetative water) and superficially drying the grapes in case of recent rainfalls. This system, while quickly eliminating humidity and preventing the risk of *Botrytis* attacks during the dehydration process, is exceedingly costly in economic terms, because of the double handling of the grapes and the excessive use of heat, which can also alter the varietal and flavour notes of the wine (Figure 13.2).
2. *Only one room available*. This system is the most commonly used in Valpolicella. Grapes are stored in the dehydration loft. The arrangement of the trays allows dehumidified air to be evenly distributed with the help of the ventilators which convey it in all directions. This system needs greater cooling capacity in order to deal with the first postharvest period, when high levels of atmospheric humidity together with noteworthy levels of vegetative water could possibly create situations conducive to the development of *Botrytis*. When

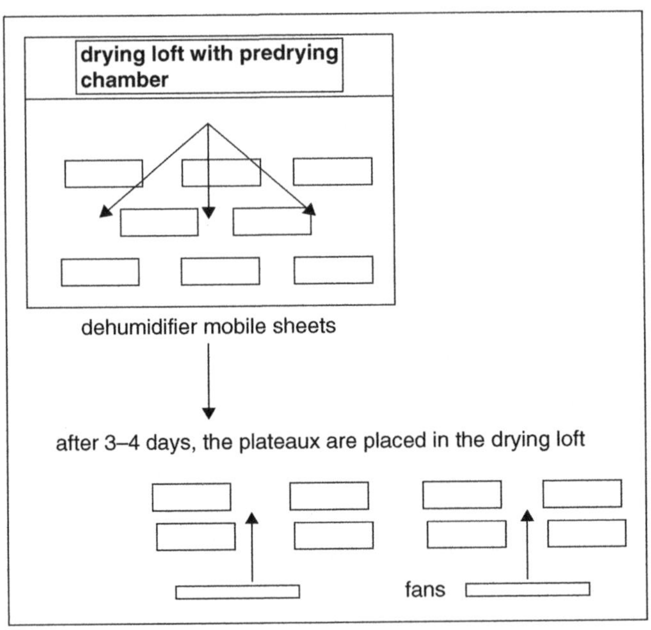

Figure 13.2 Dehydration loft with pre-drying chamber.

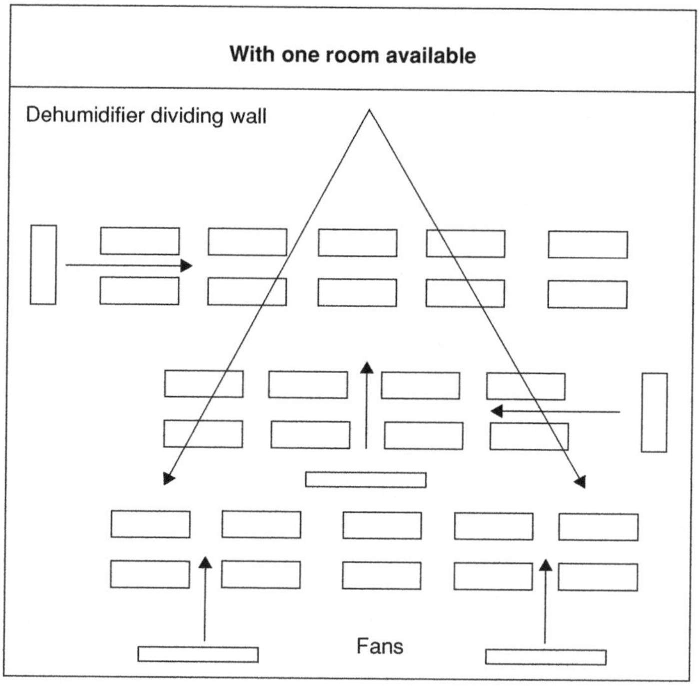

Figure 13.3 One drying room.

the grapes have reached a sufficient weight loss (10–15%), the dehumidification systems are turned off, and only the ventilation is used. Then, the dehumidifiers are turned on only when it is raining, foggy, or when there are high levels of atmospheric humidity (Figure 13.3).

3. *Vacuum treatment chamber*. The third system is inspired by the treatment used to preserve fruit. The grapes are stacked in a room without corridors. Air is first pushed perpendicularly through the bunches, then it is vacuum aspired. So, the air sent to the back end of the room returns and is evenly distributed over all the grapes. The advantage of this system is that grapes are moved only once. Because of the absence of corridors, this stacking system is also more space-efficient (only 25% of the space is used for manoeuvring instead of the 40% used in the other systems). Once again, after the first month of using dehumidifiers, it is possible to resort to the system as an aid, only with high humidity conditions. Few vacuum systems are used in Valpolicella. Their advantage is in the reduced use of property investments, since the available surface is used to a maximum. On the other hand, they have a high energy consumption.

13.7 EVOLUTION OF AMARONE GRAPES DURING DEHYDRATION

In Amarone wine, besides the dehydration process, the metabolic effects of the berry's cellular chemistry (over-ripeness) have also to be taken into account. These phenomena characterize wines made from over-ripe grapes and make the entire process very different from mere dehydration. In fact, the time factor (90–120 days) plays a pivotal role in the evolution of the grapes. The producers are well aware of this, so much so that in Valpolicella, people do not say that the grape is being dehydrated, they say, rather, that the grape has been put to 'rest'. In general, during the first 60 days of dehydration, the physical phenomena of dehydration (bunch weight loss and sugar concentration) are predominant although not exclusive, and then in the 60 subsequent days, the berry biochemical phenomena are induced even by *Botrytis* presence. Some of these biochemical phenomena are also favoured by the change in temperature from the initial 25–30°C to the final 0–5°C.

13.7.1 Biochemical aspects

The analysis of the changes within the berry shows that there is a general decrease in all the constituents, except for the products deriving from the metabolism of the moulds and in particular, glycerine and gluconic acid. Due to the concentration that has taken place, the final juice has a very high sugar content. If the sugars increase by 27–30%, there is no corresponding increase in acids, which are oxidized, both because of respiratory phenomena and the intervention of the moulds, mostly *Botrytis cinerea*. After approximately 100 days of dehydration, a sweeter and less acidic grape must is obtained (Usseglio-Tomasset *et al.*, 1981). During dehydration, glucose undergoes a more pronounced oxidation than fructose; the percentage increase of glucose is around 12%, against an increase of fructose of about 43%, while the glucose-to-fructose ratio goes from a value of 1 in the juice of fresh grapes to 0.80 when the dehydration of the grapes is complete. This is of paramount importance to explain the sweet perception typical of this wine, since the fructose is twice as sweet as the glucose, all amounts being equal. It is also probably worth highlighting that sugar enrichment through the dehydration process causes a loss of approximately 30% of the initially available sugars (−38% of glucose and −22% of fructose), and therefore, also considering that 5–10% of the sugars are transformed into glycerine (through the moulds), the preservation of the grapes in the dehydration loft significantly affects production costs. Concerning the acidity, the total acidity decreases by 36%, tartaric acid by 33% and malic acid by 70%, the latter degraded during the dehydration of the grapes (Usseglio-Tomasset *et al.*, 1980). During dehydration, cell wall degradation provokes an intense migration into the cell sap of calcium, magnesium and above all, potassium. The dehydration process also has an impact on phenolic compounds because, with the degradation of cell compartmentalization, these are readily attacked by tyrosinase and other oxidative enzymes (Ferrarini *et al.*, 2007). Like polyphenols, the varietal flavours are also preserved in the early dehydration stages, while they undergo degradation afterwards.

13.7.2 Amarone and *Botrytis*

The composition and the specific and distinctive enological profile of Amarone are mainly determined by *Botrytis cinerea*, which plays a key role in the evolution of some important compounds. The *Botryitis* can derive both from the vineyard and the dehydration process. At the end of ripening, if the year's weather patterns encourage the alternation of wet and dry periods, at a relatively low air temperature, the so-called noble rot can make an appearance, thus encouraging a certain dehydration of the bunches or parts of the bunches. If the dehydration process is carried out in humid places or in case of heavy rains, the grapes may develop grey rot. In drier conditions, the development of *Botrytis* will be more moderate, mostly in the noble form (Tosi *et al.*, 2005).

The various grape varieties, such as Corvina (35–45%), Molinara (30–41%) and Rondinella (27–40%), have different dehydration speeds and sensitivity levels to fungal attack because of the differing thickness of the berry's skin (Usseglio-Tomasset *et al.*, 1980). Rondinella is rather indifferent towards the attack of *Botrytis*, while Corvina is more sensitive (Corvina has an average skin thickness of 145 μm while Rondinella reaches 170 μm). A moderate and controlled noble *Botrytis* infection is a good thing, but a large-scale *Botrytis* attack, even in the noble form, may lead to an unpleasant enological profile. *Botrytis cinerea* is known to spread at a temperature ranging between 12 and 18°C, with a RH as close as possible to 100%. The Amarone musts coming from botrytized grapes have a concentration of D-gluconic acid between 1 and 5 g L^{-1}, while musts coming from grapes affected with grey mould have levels that largely exceed 5 g L^{-1}. The glycerol-to-D-gluconic acid ratio is a value used to determine the type of development induced by *Botrytis cinerea*. Thus, the *Botrytis* determines a certain level of glucose consumption; in fact, the dehydration of the berry causes an increase in the sugars, as all the berry's components are concentrated. Regarding the main organic acids, approximately 70–90 % of the tartaric acid and 50–70% of malic acid are consumed. Through an enzymatic degradation of the proteins, the fungus can release the nitrogen contained in the amino acids, in order to metabolize it during its own development. This nitrogen is absorbed by *Botrytis cinerea* which then synthesizes the metabolic and structural proteins needed for its growth. The dehydration process is also positively correlated with resveratrol, which increases in musts and wines produced from grapes affected by *Botrytis*. The content of resveratrol in grapes seems to be the result of the balance between the vine synthesis in response to the pathogen attack and the degradation carried out by the pathogen. The contents of this stilbene in the wine also depend on the metabolism of bacteria and yeasts, in addition to technological factors (Tosi *et al.*, 2009). Flavour changes are also important. In particular, an increase in botrityzed berries also implies an increase in the following compounds:

- Isoamyl acetate and beta-phenylethyl acetate (ripe white fruit notes)
- Ethyl phenylacetate (honey notes)
- Ethyl hydroxyacids, except ethyl 3-hydroxybutyrate (winey-floral/fruity aroma)
- Diethyl 2-hydroxyglutarate and 4-carbethoxy-gamma-butyrolactone
- 1 octane-3-ol (mushroom notes)
- N-(3 methylbutyl)-acetamide (pungent almost vinegary note)
- Sherry lactone isomer 1
- 3-methyltiopropanol or methionol (cooked potato)
- 4-terpineol (balsamic earthy note); among the monoterpenes, beta-damascenone (hay-floral note); and among the norisoprenoids, 3-keto-alpha-ionol (hints of tobacco)
- Ethylvanilate, acetovanillone and homovanillic alcohol among benzenoids
- Phenylacetaldehyde (honey-hyacinth floral aroma) and benzaldehyde (notes of bitter almond).

A reduction of the ethyl esters (C10–C6) and their relative fatty acids in relation to the percentage of *Botrytis* present was observed. If the *Botrytis* attack is light, floral and fruity notes are dominant, but if the attack is heavier, more complex aromatic notes will be noticed, with hints of bitter almond, resinous and balsamic flavours, and a stronger persistence (Tosi *et al.*, 2009).

13.8 TECHNOLOGICAL INNOVATIONS IN VALPOLICELLA: PREDICTIVE DEHYDRATION

The advances in the technical management of the vineyard, together with the climate changes that have occurred in the last two decades, have led to an earlier ripening, and consequentially an earlier dehydration in dehydration lofts. The length of the dehydration process and the thermal conditions within the dehydration lofts are also different from the past. In the past, the traditional harvesting period in Valpolicella was October, and the dehydration process lasted 90–120 days. Today, grapes are harvested about a month before and are made into wine in December; the vinification process rarely exceeds 100 days (Ferrarini *et al.*, 2007). The extreme variability of the vintage in terms of maturity, the obligation to carry on the dehydration process until 1st December, according to the appellation rules and, above all, the extreme difficulty in controlling the dehydration speed with the use of dehumidifiers, often cause excessive sugar concentrations in the grapes (300–350 g L^{-1}) with consequent residual sugar in wines due to stuck fermentation. The high osmotic pressures also cause high levels of volatile acidity, with an alcohol content over 18%, thereby reducing the balance and the pleasantness of Amarone. In order to positively control these climatic and technological factors for the production of Amarone della Valpolicella, a study carried out by the University of Verona measured the distinct effects of the compositional, biological, morphological and physical variables, that is, the effects of temperature, RH and ventilation on Corvina, Corvinone and Rondinella varieties. The study formulated specific mathematical models that could measure, control and predict with reasonable accuracy the evolution of the dehydration process. A dehydration coefficient ('W') was calculated for each variety by measuring the average weight loss rate at each condition (temperature and humidity), thus defining the average weight loss rate for each of them.

A value exceeding 1 indicates a faster dehydration compared to the average rate, while a value lower than 1 indicates lower dehydration rates (Ferrarini *et al.*, 2007). In the Amarone blend, Rondinella is the variety with the highest W coefficient (Table 13.1), while Corvinone is the slowest, because it has larger berries. The vintage also influences the dehydration rate of each single variety. In rainy years, the dehydration rate increases, because grapes contain more vegetative water and have thinner skins. The study of the ventilation also represents an important element in dehydration management, as the dehydration rate increases only if the temperature exceeds 20°C. Lower temperatures, even with high air speeds, do not make a significant difference. In order to evaluate the efficiency of air distribution in industrial dehydration lofts, the speed of the air at various points was measured (among the pallets and in the stacks).

Table 13.1 Dehydration coefficient (W) of different grape varieties.

Variety	W
Cabernet F.	1.02
Corvina	0.82
Corvinone	0.73
Rondinella	0.84
Garganega	0.93
Muscat	0.89
Raboso	0.99

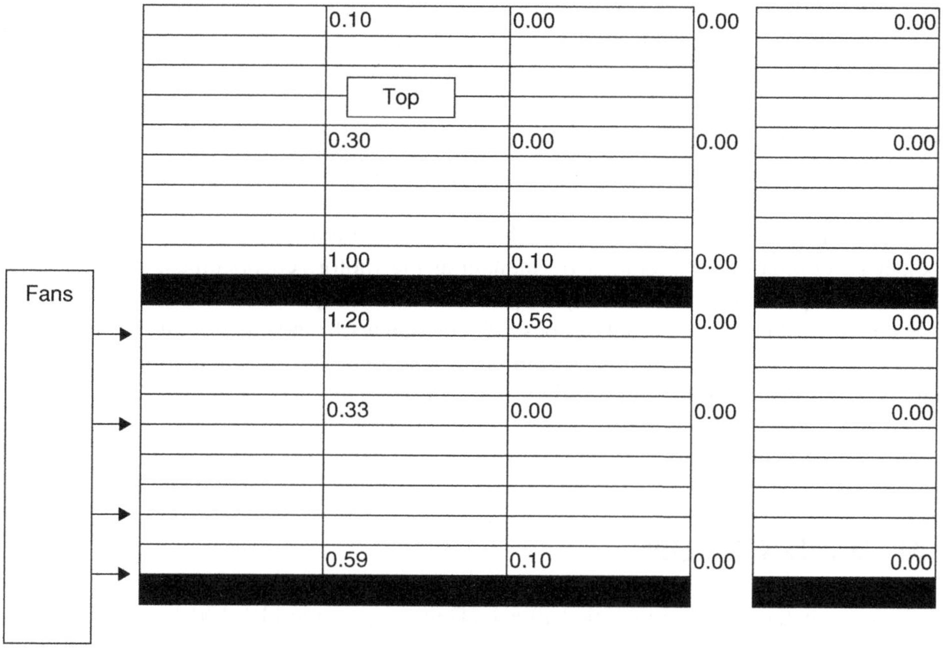

Figure 13.4 Analysis of the air speed in the stacks.

The analysis of ventilation has revealed several inefficiencies in the distribution of air in the dehydration lofts (Figure 13.4). If air distribution has been considered of little influence until now, it has been shown that the correct positioning and sizing of ventilators, the direction of the outlet and intake flow, the maximum stacking height, the different flow of penetration depending on the tray characteristics and the analysis of the moisture pockets in the dehydration loft, substantially improve the dehydration management, due to a better knowledge of the process. This knowledge of the incidence of the individual factors for each variety has led to the development of mathematical models and industrial applications that are able to predict the probable evaporation and sugar concentration by entering the relative humidity % (RH) and temperature °C (T), the dehydration coefficient (W) and the speed of ambient air. When programming the dehydration, after entering the sugar levels of the incoming grapes, the desired sugar level at the end of the period and the duration of dehydration in days, the software suggests the dehydration conditions (T° - RH) by considering the critical humidity levels that may generate the development of *Botrytis*. A warning is issued if the operator wishes to work in harsh conditions. The operator can either accept the proposal made by the software or decide to change the temperature and humidity conditions. This change will consequentially generate a new dehydration curve in order to reach the desired conditions at the end. Considering the interannual variability of the dehydration rate and the difficulty in standardizing the conditions in large industrial premises, the software has a system that electronically checks the weight on some of the pallets, thus giving the operator the possibility to measure and compare in real time the correspondence of the value that has been entered into the software with the actual dehydration rate. As a consequence, the software either increases or decreases ventilation rate, temperature and humidity in the dehydration loft, in order to bring it close to the requested conditions. The sugar level is also monitored during the dehydration process by taking periodic samples (every 7 days) from the dehydration loft. The values are entered in the calibration curve to ensure that the set rate and probable sugars in the period are those expected. A very interesting aspect in this new management of the dehydration process in Valpolicella is the attention given to the reduction of energy consumption, which is particularly burdensome in the dehydration lofts. In fact, the software intelligently uses the air by favouring outside air whenever it has the necessary requirements (T° - RH). When the air is too dry or too humid (because of the time, the conditions and the speed determined by the grapes dehydration curve), it is automatically excluded by managing the motorized openings. Of course, all these forecasting methods can be very helpful, but do not replace human know-how, since we are dealing with a raw material, grapes, which changes from year to year and during the dehydration process. A careful observation of the interplay and the interactions of all these variables is always required. Nevertheless, these instruments are particularly useful for an accurate monitoring of this delicate process, to ensure that grape quality will not be affected by the uncertainty and oddities of autumn weather as was the case in the past.

13.9 VINIFICATION

As we have seen, the grapes are vinified after 1st December, some 110–120 days from the start of the dehydration process. In the past, the pressing usually started in January or February. Nowadays, because of the climate change over the last 20 years, the vinification process has to be started in December, in order to avoid an excessive increase in sugar concentrations. At the end of the dehydration process, it is quite normal to have sugar values of 280–300 g L^{-1} and an alcohol potential close to 17%. The grapes are manually emptied from the trays on the grape-sorting tables in order to eliminate clusters attacked by vulgar *Botrytis*. The grapes are then crushed and destemmed. Some parts of the dried stems (20–30%) may enter the fermentation tanks without causing any damage; on the contrary, the tannins of the stems fix the laccase and help stabilize the colour. Sulfur dioxide quantities are chosen according to the sanitary state of the grapes. A SO_2 addition of 40–50 mg L^{-1} is enough, since the low fermentation temperatures inhibit the bacteria growth. At the beginning of fermentation, the outside temperature is very low, which means that the grapes may enter the tanks when temperatures are as low as 0°C. Low fermentation temperatures together with high sugar levels make the choice of yeast extremely important in the making of Amarone. In the past, the majority of wineries successfully fermented with native yeasts. Today, the conditions have changed, and the choice of the strain and its preparation with an adequate starter are of paramount importance for success. Good yeast for Amarone production must be able to start fermentation at low temperatures, with a high production of glycerol, a low production of acetic acid, a high alcohol production (18%) and a low degradation of malic acid (which is normally present in small quantities). At the same time, yeast must possess the ability to enhance the aromatic traits with floral and spicy flavours. Because of the high quantity of pomace in relation to must, all smaller sized fermentation systems are to be preferred to the profusion of technologies proposed by the market. With this particular vinification (which takes place during the cold season, and thus at low temperatures), there are two types of maceration: a cold maceration of crushed grapes, before fermentation, and a second maceration during the fermentation process. The fermentation usually starts 7–10 days after crushing and can last up to 30 days before sugar completion. The induction period is quite slow (with a lag time of 7–10 days) unless aided by an exogenous increase in temperature up to 10°C. The second step is the tumultuous fermentation that goes from a 3–4% to a 13–14% alcohol content. This second phase is quite fast and can last 10–15 days. Finally, in the last phase, which is undoubtedly the most difficult and dangerous, the alcohol content goes from 14 to 17%. Heat may be applied during this final phase, which can last 5–7 days. According to the desired Amarone style, the fermentation temperature should be between 14 and 18°C for traditionalists looking for elegance and finesse, and between 22 and 26°C for the modernists who look for colour, fruitiness and concentration. The long fermentation period and the high fragility of the skins of the native varieties used for making Amarone require a lot of attention and careful handling of the maceration phase. The focus should be placed on time rather than on intensity. At racking, there are usually a few grams of residual sugar (5–12 g L^{-1}) and the fermentation is generally completed in barrels.

13.10 AGEING IN WOOD AND IN BOTTLES

Ageing is crucial for Amarone, as it is for all premium red wines. The wooden cooperage used in Valpolicella are both large barrels made of Slavonian oak, which gradually evolve the wine and highlight the fineness of the varietal notes, and French or American barriques, for modern, fruity wines. Both barrels and barriques give Amarone wines that are full of character, tannic and full bodied. Both wine styles meet commercial success even if, generally speaking, there is a more conscious and careful use of wood and of all the winemaking techniques, focusing on recognizability and territorial originality rather than the fads of wine critics. After its ageing period in wood, either in large or small barrels, the wine is bottled and bottle-aged for a few more months, in order to achieve a greater organoleptic balance. Amarone cannot be traded until 25 months after harvest for the 'standard' version, and until 50 months for the riserva version.

13.11 THE 'RIPASSO' TECHNIQUE

According to a locally established practice, the Valpolicella wine is 'repassed' on the Amarone pomace. This technique consists of leaving the Amarone pomace intact (without pressing) in the fermentation tank and 'repassing' an equal amount of Valpolicella wine for 7–10 days. This pomace, which still contains 10–15% of Amarone wine, is rich in extractive substances. It also contains a few grams of residual sugar, which start a short fermentation. This technique allows for an increase in alcohol content (1–2%) according to the residual quantity of Amarone in the pomace, an increase in the dry extract (28–30 g L^{-1}), an increase in the colour intensity, a decrease of the total acidity and a more fruity and intense aroma (Usseglio-Tomasset *et al.*, 1980). Overall, the wine that undergoes this second fermentation becomes suitable for ageing. Nowadays, the Valpolicella Ripasso is a great success for the entire territory of Valpolicella.

13.12 AMARONE'S SENSORIAL CHARACTERISTICS

In recent years, one of the most distinctive traits of Valpolicella's Amarone has been the plurality of sensorial characters which may be found in the wine, both horizontally (i.e. between wines from several wineries) and vertically (i.e. in the same wine), in the dynamic sequence of sensations detected. Visually, purple and orange hues are equally present in different Amarone, all belonging to the most-recent vintage on the market from different wineries. The nose is traditionally reminiscent of jam, cherries in alcohol and fresh fruit (marasca cherry, cherry and wild berries), with notes of herbs and tobacco and more spicy and vanilla notes (Ugolini, 2003). The analysis of Amarone's aromatic compounds has shown a significant presence of 4-terpineol (up to 0.5 mg L^{-1}), both in young and aged wines (the latter with their typical balsamic, spicy, earthy and resinous notes). Some hydrocarbons are also to be found, such as monoterpenes (p-cymene, alpha-terpinene, alpha-terpinolene and gamma-terpinene, but not limonene). Other monoterpenes are also present in small and homogeneous amounts, thus highlighting that Corvina is a neutral variety. Another particularity is the benzaldehyde concentration which can reach up to 200 mg L^{-1}; this probably expresses the varietal character, since it is present in considerable amounts even in grapes that are not dried (Versini *et al.*, 1999). Its characteristic note of cherry is typical of Valpolicella wines.

However, the factor that characterizes Amarone and makes it more modern and sought after by the consumer is its ability to express strong olfactory and tactile emotions in a disarming drinkability. The sweet, balsamic and dried fruit olfactory notes harmonize well with an alcohol content that frequently exceeds 16%. On the palate, the highly polymerized tannins from the native varieties together with the soft and glycerine notes combine power and elegance (Ugolini, 2003). The territorial diversity translates into various styles of Amarone; for example, the marly limestone in the Sant'Ambrogio valley, deriving from the red limestone known as *scaglia rossa*, gives very elegant Amarone wines, while the calcareous sandstone of Fumane gives fuller bodied and long-lived Amarone wines. The basalts (eruptive rocks) of the valley of Marano, known as *Toar*, give silky Amarone wines with good alcohol content. The alluvial soils and detritic soils of San Pietro in Cariano give spicy Amarone wines and, finally, the nummulitic limestone of the Negrar valley gives very elegant Amarone wines with a rare finesse. Hence, Valpolicella and Amarone are multifaceted and highly emotional, all the more so because the reading of the landscape and its wines can hardly be contained and constrained by precise formulas and schemes.

13.13 THE AMARONE PHENOMENON

In the last few years, Valpolicella has gone through a period of great success and important changes. For almost 30 years, Valpolicella has dehydrated some 3000–3500 tons of grapes per year. Since 1995, this amount has continuously increased, until reaching 12 500 tons in 2001 and approximately 30 000 tons in 2010. In contrast with the rest of Italian viticulture, where the surface area is receding, Valpolicella is growing at a rate of 200 ha $year^{-1}$, up to a total of 6500 hectares and more than 410 dehydration lofts in 2010. This increase in grape quantities for Amarone production has been accompanied by a parallel increase in prices. In the last 15 years, the price has increased sixfold, reaching €2.5 kg^{-1} during the 2010 grape harvest. Production is 10 times higher than it used to be, while conceptually remaining craftsmanlike. The quality has not suffered from this increase; on the contrary, the improvement in grape management has brought about an improvement in consistency, which used to be very critical in the past. The sales of this commercially successful wine have increased from 1.5 million bottles in 1997 to 14 million in 2011. Amarone is greatly appreciated on foreign markets, which now absorb 80% of the produced quantities, and is mainly present in North America, northern Europe and in the historical markets of Switzerland and Germany. This success is the result of the skills and perseverance of the wine entrepreneurs of Valpolicella, who have been able to innovate, while combining craftsmanship and industrial factors in the wine production. At the same time, producers have brought about a reduction of production costs that has allowed them to compete in a more modern and effective way on international markets. In Valpolicella, the analysis of the production structure shows a good balance of large production entities (both cooperative and private wineries) and small and medium wineries. This mix is very characteristic. Fragmentation is indeed a major limiting factor for the Italian wine sector, whose competitiveness is hindered. However, in the Valpolicella area, this problem is somewhat reduced in scale. There is a correct balance between small, medium and large wineries that have been able to develop an interesting competitive model in different markets (Pomarici *et al.*, 2006). In fact, the smaller companies focus on the craftsmanship of their products and on their important relational skills. The medium-sized companies not only produce the wines of the area, but also distribute national products, their activity being more aimed at foreign markets. Finally, the large companies, whether they are private companies or cooperative wineries, are very robust and innovative. They manage a large range of products through various distribution channels, both at the international and national levels. Hence, the character of Valpolicella is a synthesis of all these human, historical, social and economic components. All are the expression of a common denominator, made of originality and regional recognition.

REFERENCES

Ferrarini, R., A. Padovani, F. Peretti, *et al.* 2007. Problematiche dovute a modificazioni climatiche nel caso dell'appassimento delle uve nella produzione di vino Amarone della Valpolicella. In *Proceedings of the 30th Congress of Vine and Wine*, OIV, Budapest, Hungary, 10–16 June 2007.

Pomarici E., S. Raia, and L. Rocco. 2006. *Analisi della filiera vitivinicola del vino Valpolicella Doc*, Consorzio di Tutela dei Vini Valpolicella, Edizioni L'Informatore Agrario.

Tosi E., M. Azzolini, G. Zapparoli, *et al.* 2009. Incidenze della Muffa nobile sulle caratteristiche chimiche e sensoriali dell'Amarone. *L'Enologo*, 10:81–89.

Tosi E., P. Malacrinò, D. Accordini, *et al.* 2005. I principali effetti della *Botrytis* sulla composizione dell'Amarone. *L'Informatore Agrario*, 44:65–69.

Ugolini, A. 2003. Il Volto dell'Amarone. *L'Assaggio*, 4:22–29.

Usseglio-Tomasset, L., P.D. Bosia, C. Delfini, *et al.* 1980. I vini Recioto e Amarone della Valpolicella. *Vini d'Italia* 125:85–97.

Usseglio-Tomasset, L., G. Paronetto, and G. Morsiani. 1981. I vini Recioto e Amarone della Valpolicella. *Cassa di Risparmio di Verona Vicenza e Belluno*, 5:46.

Versini, G., R. Schneider, S. Carlin, *et al.* 1999. Characterisation of some Northern Italian passiti wines through aroma and stable isotope analyses. In *Proceedings of the 12th International Enology Symposium*, Montreal, Canada, 31 May–2 June 1999, International Association for Enology, Management and Wine Marketing, Breisach, pp. 544–571.

14 Moscato Passito

Daniele Eberle

Terredavino SpA, Barolo, Italy

14.1 INTRODUCTION

Piedmont wines made from Moscato Passito, from various denominations of origin, are obtained from 100% Moscato bianco (White Muscat) grapes. White Muscat is cultivated throughout Italy and has the synonyms: 'Moscato d'Asti' or 'Canelli' in Piedmont, 'Chambave' in the Aosta Valley, 'Trani' in Puglia, 'Noto' or 'Siracusa' in Sicily, 'Tempio Pausania' in Sardinia, and 'Moscatello di Montalcino' in Tuscany.

Outside Italy, it is called 'Moscatel menudo blanco' in Spain, 'Muscat à petits grains', 'Frontignan', 'Lunel' or 'd'Alsace' in France, 'Gelber muskateller' in Germany, and 'Tamioasa' in Romania. The reference names are 'Piemonte Moscato Passito DOC', which has the broadest regional character, 'Loazzolo DOC' for the province of Asti, and 'Strevi DOC' for the province of Alessandria. It is the second variety grown in Piedmont after Barbera, and is produced primarily in the south of the region, in the provinces of Cuneo, Asti and Alessandria. The oldest documents testifying to the cultivation of White Muscat date back to the early thirteenth century, but it is not until the sixteenth century, when Duke Emmanuel Philibert of Savoy was forced to impose limits on imports in order to protect local production, that we appreciate how widespread cultivation had become. However, it was during the nineteenth century that White Muscat became a prominent grape variety, not just in Piedmont but also throughout Italy and abroad. At that time, Canelli was the main area of cultivation, followed by Santo Stefano Belbo, Calosso, Strevi, Castiglione Tinella, Acqui Terme and Ricaldone. The wine types produced in those years were 'Moscato d'Asti', a sweet and slightly sparkling wine, and 'Asti Spumante', a sweet sparkling wine. Asti's success led to the gradual downsizing in production of other types of wine made from White Muscat. The growth of production in Piedmont was characterized by small family businesses (average 2.5 ha). Within these businesses in the traditional areas of cultivation of Muscat grapes, a part of the grapes had always been destined for dehydration and use in the production of Passito wine, which over the years that 'Asti Spumante' developed, had fallen into decline. However, during the 1990s, some producers decided to revive the ancient traditions and reintroduce 'Passito' wine. To this end, one of the territory's most important cooperative wineries initiated a project that would look at the choice of a territory, research to deepen their knowledge of the aromatic part of the White Muscat grape, and the application of new technologies for dehydrating grapes, and consequently, winemaking techniques appropriate to the type of product. The cooperative began to establish vineyards, provided

Sweet, Reinforced and Fortified Wines: Grape Biochemistry, Technology and Vinification, First Edition.
Edited by Fabio Mencarelli and Pietro Tonutti.

by associate wineries, within the extended area, and to undertake experiments in dehydration and winemaking. In the late 1990s, the current zone for grape production was defined, and a series of investigations into the photosynthetic efficiency of plants began to evidence the limit of their foliar activity period. Further studies followed on the synthesis of the aromatic properties of the grapes, and on the evolution of the wine during production and refinement in the bottle. Cultivation techniques were totally reviewed, and new agricultural practices were established to reduce yields, guarantee perfect grape health and maturity of the fruit. Until the late 1990s, areas of cultivation had remained modest, a maximum of three hectares depending on the vintages, because the method of production was by slow dehydration on the vine and a late harvest. In 2000, at the end of a period of favourable harvests, the changes in climatic conditions and their unpredictability forced producers to rethink the philosophy behind production. In order to ensure that the aromatic characteristics of the initial product (i.e. the grapes) were maintained without modification due to uncontrolled dehydration, they introduced the practice of dehydration in controlled rooms.

14.2 THE VINE

White Muscat is a white grape variety with a medium-sized berry (2.2 g), spheroidal or slightly flattened in shape; it has a thin skin, not very waxy, yellow-greenish in colour, particularly sensitive to sunburn, which becomes golden yellow or amber when exposed to the sun (Figure 14.1). It has an intensely aromatic-flavoured berry characterized by low olfactory threshold terpenes (volatile linalool content 300 $\mu g\ L^{-1}$); the aroma of the wines obtained depends on the richness of the grapes. The bunch is medium-sized (c.350 g), cylindrical, slightly elongated, winged, compact or compact on average; the stalk is short-medium, green.

The plant has average vigour with internodes of medium length and an erect growth habit; it has long tendrils that facilitate a foothold to trellises and wires. Fertility (1.3 to 1.6 clusters per shoot) and yield is generally good and constant; fertile in basal buds and lateral shoots. It is usually trained to a vertical shoot positioned canopy; mixed pruning using the Guyot system with a single cane results in 8–12 buds. However, it is well suited to short pruning (cordon spur), but is less suitable, due to its medium vigour, to high and expanded training systems (cordon cane). Typical plant density is 4000 vines per hectare.

Bud break takes place at the average time (second decade of April), as well as flowering (first decade of June). Veraison is medium-early (first decade of August) and the ripening of grapes is early or mid-early (first half of September) with a collection of wine that varies, depending on its production fate, from sparkling to Passito.

14.3 AREA OF ORIGIN

Muscat vineyards for the production of Passito wine are grown in the Alto Monferrato area, at 45° N latitude on a hilly ridge between the Bormida and Belbo rivers, at an altitude of between 200 and 350 metres above sea level, in particular in the villages of Cassine, Strevi and Ricaldone (Alessandria), Alice Bel Colle and Castel Rocchero (Asti). The soil is classified as Entisol (according to the US Department of Agriculture Soil Taxonomy): a very young, fine-weaving hill soil that does not show horizons. Characterized by a limited expression of pedogenetic processes and generally, by a surface horizon poor in organic

Figure 14.1 White Muscat grape cluster before the harvest for Passito wine. (For a colour version of this figure, please see the colour plate section.)

matter, clear and thin, above compact lithoid substrates or recent alluvial deposits. The absence of horizons is due to the lack of a sufficiently long time for their formation and the type of source rock. In Piedmont, the entisols of hill and mountain areas are among the poorest soils from the point of view of potential agricultural productivity but are suited to quality wine production. Soil analysis highlights a medium-clay-silty type, pH 7.7, total and active limestone to medium values of 14% and 7%, respectively, medium-low organic matter equal to 1.4%, a low available phosphorus (19 ppm), and exchangeable magnesium and potassium in good quantities, 230 ppm and 300 ppm, respectively. Cationic exchange capacity is considered to be medium at 20 meq, with a ratio of Mg : K equal to 2.5, also considered to be medium.

The climate of the area falls within the Po valley type with limited thermal excursions and rainfall concentrated in spring and autumn. It is also characterized by the absence of prevailing winds and a high value of relative humidity that is always above 60%. The annual mean temperature is 13.0°C with peaks to 30° in the months of July and August; the Huglin Index (HI, growing degree-days in °C) ranges between 1800 and 1900. The annual mean rainfall is about 900 mm, half of which falls during the period 1st April to 31st October. In this environment, the water reserves (available water capacity (AWC)) are almost exhausted by the end of July and the outcome of the harvest is negatively correlated to the intensity of the rainfall before harvesting.

14.4 VINEYARD MANAGEMENT

Vineyard management in the production of Muscat grapes for dehydration aims to achieve maximum maturity of the grapes, in terms of sugar accumulation and synthesis of aromatic substances, in harmony with the climate of the vintage. Consequently all techniques should focus on making the crop less sensitive to attack from pests, fungi and insects, which might cause product loss, decrease the quality of the grapes and endanger subsequent dehydration.

14.4.1 Crop load

The crop load depends on the number of buds left at pruning, the number of shoots left after the bud break and, possibly, on the number of bunches left after thinning. In our areas, the best-quality White Muscat grapes for dehydration are obtained from a yield in the range of 1.8 to 2.4 kg per plant, that is, from 6 to 8 clusters per plant, with bunches of 300 g. In vineyards with 4000 plants per hectare, yield is about 6.0 ton ha^{-1}. This level of production allows good sugar accumulation, maintenance of a good level of malic acid, and the synthesis of high amounts of terpenes in glycosylated form. In the case of superior wine production, it is necessary to thin the grapes.

14.4.2 Canopy management

Vines must be nourished in order to develop enough leaf surface for the grape to ripen but not to excess so as to avoid too much vine vigour and consequent dilution of the grapes. Values of pruning weight between 0.7 and 1.0 kg per plant are optimal, which corresponds to a maximum of 8 shoots per plant. Using the Guyot pruning system, this would be 6 buds on the cane and 2 on the spur. As well as the influence of bud load on production levels, and therefore the quality of the grapes, the timing of vegetation management is crucial for an aromatic-based variety such as White Muscat. Excessive shading of clusters due to excessive vigour, large leaves, the emission of side shoots, badly positioned shoots in lush vegetation, all determine the appearance of a herbaceous character and poor health in the grapes, and excessive exposure to sunlight causes dangerous sunburn and loss of flavour. Our experimental work reveals that the best time to remove the leaves in the area of the clusters is between flowering and cluster closure (i.e. between late May and late June). If it is performed earlier and if too many leaves are removed, defoliation can reduce the number of berries per cluster, even though White Muscat is not a variety that is sensitive to flower drop. It is possible to remove the leaves early, during flowering or later at fruit set, with machines that operate with a flow of compressed air, or before cluster-closure using machines that operate by suction. The timing of leaf removal is critical to reduce the injury that can be caused to the berries (Skinkis *et al.*, 2010). Removing 4–6 leaves from the shoot base is mainly to improve the cluster's defence against powdery mildew (a pathogen to which White Muscat is particularly sensitive). Consequently, sensitivity to the development of *Botrytis* is also decreased. A well-exposed position of clusters is also hostile to grapevine moths. *Botrytis* and moths are a particularly important risk factor for the grapes assigned to dehydration after harvest.

14.4.3 Timing of cluster thinning

In particularly favourable years with successful flowering and high fertility rates, the thinning of the grapes is worthwhile. Damaged clusters or those that are distal on the shoot should be removed leaving just the basal cluster. To obtain the best results from thinning, that is, complete ripening of the grapes (particularly in the production of grapes intended for dehydration), it is fundamental to complete the operation before veraison. The phenology of White Muscat in the Alto Monferrato area provides 60 days between flowering and veraison, with flowering beginning in early May and veraison at the end of July. Removing clusters after veraison is not recommended since some of the flavour characteristics that have already accumulated in the berries will be lost. The operation should start in the third or fourth week of July depending on the year. Monitoring grapes for berry softening, colour change, sugar content, gives an indication of when veraison will begin.

14.4.4 Harvest

Grapes from vineyards managed as described earlier, are hand-picked and placed in boxes for dehydration; clusters that show obvious health problems are discarded. Timing of the harvest is crucial to achieve a high-quality product. Even though dehydration is carried out under controlled conditions, it is essential that grapes are collected at an optimum stage of ripening and health. The operations carried out previously in the field should provide a grape of excellent quality, but climatic conditions before harvest time can negatively affect the outcome of earlier work. Monitoring the level of maturity and consulting weather forecasts enables a risk assessment of whether to harvest or wait further.

Given that White Muscat is a typical terpene aromatic variety, and that terpenes of importance in the production of grapes for dehydration are in glycoside form and provide longevity to the product, they must be carefully monitored during maturation to allow them to accumulate in high quantities (Di Stefano *et al.*, 1983). For example, at the time of harvest of White Muscat grapes intended for the production of sparkling wine (Asti DOCG), the glycoside and free linalool content are similar, between 300 and 500 $\mu g\ L^{-1}$, while at the time of harvesting of grapes intended for dehydration, the ratio between glycoside and free linalool varies between 5 : 1 and 10 : 1 with a glycoside linalool content above 1000 $\mu g\ L^{-1}$ (Figure 14.2). The accumulation of sugars varies accordingly. A risk assessment is carried out on the basis of weather forecasts for the Alto Monferrato area; given the typical Po valley climate with rains in autumn, rain intensity and duration have to be taken into consideration. In case of forecasts of prolonged rain, even if not intense, harvesting must be done because typically in autumn even light rain is not followed by windy days and low relative humidity. The quality of the grape and wine is very much influenced by the final phase of maturity.

14.5 POSTHARVEST DEHYDRATION TECHNIQUE

The technique of dehydration involves harvesting grape clusters in high-density polyethylene (HDPE) boxes. The material is suitable for food contact, and is reusable, washable and

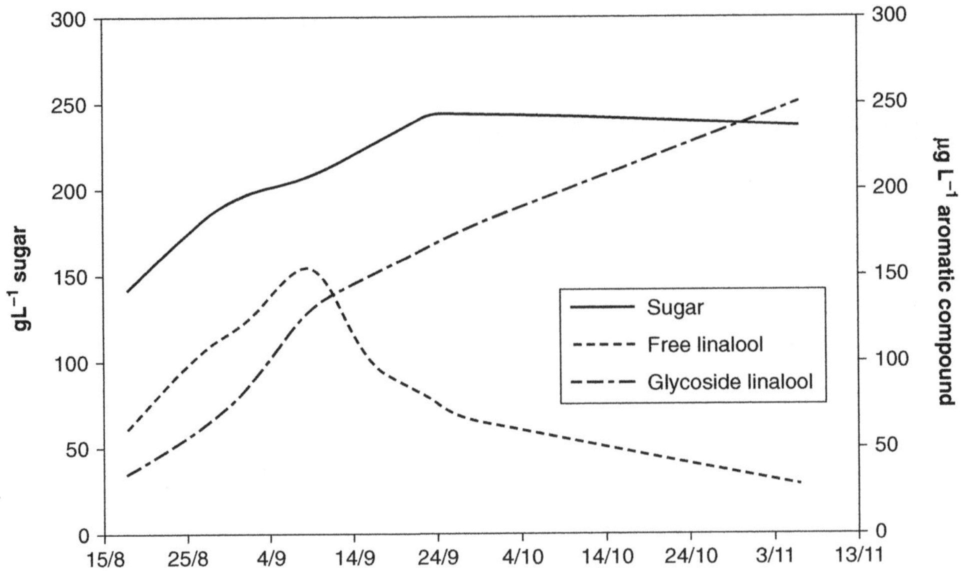

Figure 14.2 White Muscat berry aroma compound evolution during ripening. Compared with sugar, the aroma glycoside compound continues to increase due to the berry activity during the maturation.

load-resistant. The box dimensions are 400 × 600 × 150 mm (L × W × H), weight 1.2 kg, perforated at the bottom and on the sides. If possible, the grape clusters must be harvested in the absence of dew and with dry soil to avoid dirtying the bottom of the box. The boxes must be filled in a single layer (5 to 6 kg of bunches) to allow the air to flow among the grape clusters during the dehydration process. Loading the boxes is done in the field; they are then stacked and transported to the dehydration facility. Before being placed in the dehydration room, pallets of 40 boxes are formed, 4 per layer, which are secured with straps and numbered with a unique code that identifies the vineyard and winegrowers. The pallets are stored in the dehydration room, one beside the other, three pallets high, with a total capacity of 192 pallets, equivalent to 40–45 tons of fresh grapes. The dehydration room is closed and thermally isolated to achieve the best outcome in grape dehydration (Bellincontro *et al.*, 2004). Loading the room takes place over 2–3 days, as the growers proceed with the harvest, and the dehydration room is turned on at night to cool down the grapes. The dehydration temperature varies between 10 and 15°C with relative humidity lower than 60%. The measured air velocity is equally distributed in the room and through the boxes at 1.0 m s^{-1}. The dehydration process is controlled by verifying the decrease in weight of 20 boxes placed on a scale with readings every 24 hours, and by recovering and measuring the amount of water in the dehumidification circuit (Figure 14.3). Every 7 days berry samples are taken for analysis from the same pallets but at different heights. Considering that the initial sugar content of grapes varies between 280 g L^{-1} and 300 g L^{-1}, and the target is 400 g L^{-1}, a variable weight loss of 25–30% is required. Dehumidification systems are sized to determine an average 1% drop in weight per day for a period of about 30 days. Daily weight loss varies from 1.3% in the first week to up to 0.8% in the fourth week. Weight loss in the first week, especially in the first three days, is determined by the loss of water from the stalk, and in the slowing down of the final stage, by the concentration of sugars.

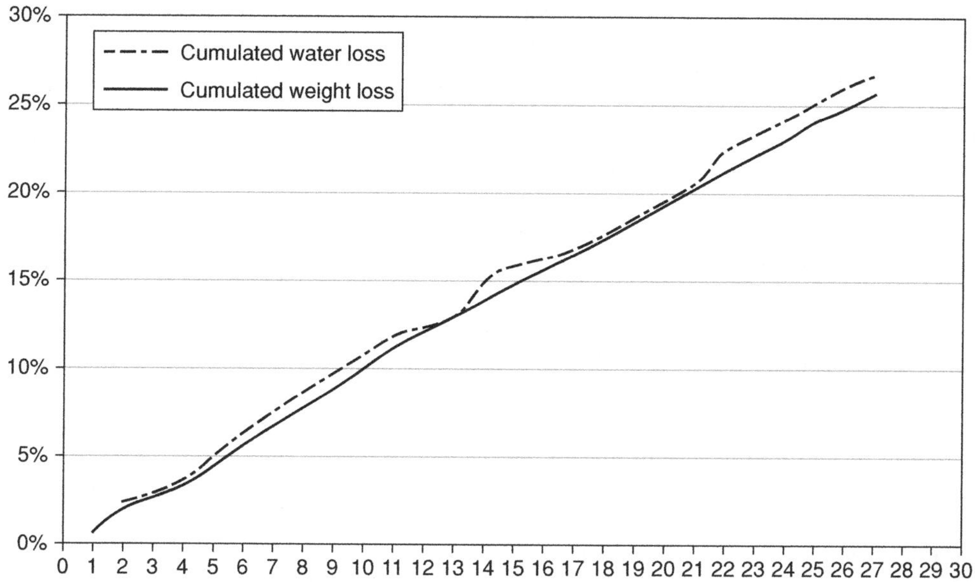

Figure 14.3 White Muscat grape weight and water loss during dehydration. Cumulated water loss of the grape contained in the dehydration chamber compared to the cumulated weight loss of the grape on the load cell.

14.6 VINIFICATION

Once the right degree of water loss has been obtained, before opening the dehydration chamber, the weather forecast is checked. It is important to move the grapes from the dehydration room to the crusher in the absence of rain or high humidity. The day before, the grapes are cooled to a temperature of between 0°C and 10°C. Pressing takes place without removal of stalks. Since the berries are soft and with thin skins, all handling operations should be performed with the utmost delicacy. Once crushed, pectolytic enzymes are added and the must is macerated for 12 hours at a low temperature (10–15°C). The pneumatic presses are filled to half of their capacity to improve work efficiency. At the end of the maceration period the pressing cycle starts, which involves limited cracking and a maximum pressure of 2.0 bar. Pectolytic enzymes and SO_2 are again added but at lower doses (5 ppm), and the must is moved into a storage tank for 24 hours. It is then sent to a rotary filter containing a perlite panel to remove suspended particles. After grape dehydration, must yield compared to fresh grapes should not exceed 50%.

The particularity of the vinification of musts originating from dried grapes is the high sugar concentration (400 g L^{-1}), which places osmotic stress on the yeasts. The choice of the yeast strain is fundamental; in case of a tendency towards active dry yeasts (*Saccharomyces cerevisiae*), it is best to choose yeasts with high alcohol tolerance, low production of volatile acidity, low acetaldehyde production so as not to combine with SO_2, medium-low nitrogen needs, and able to withstand the combined stress of alcohol and sugar. The fermentation process should be steady.

Preparation of the starter must respect the fundamental rules without neglecting the addition of nutrients. Dosage recommendation: one dose of yeast to 20 g hL^{-1} of grape must, diluted in 10 L of water per kg of yeast to which are added 15–50 g of sugar per litre of water and complex activators (ammonium salts, thiamine, yeast residues); temperature 35–38°C with continuous shaking and insufflation of air for 30 min to enhance growth and multiplication of the yeast. The ideal fermentation temperature is 18–20°C. The alcoholic fermentation takes place in stainless steel tanks, with temperature conditioning and an agitator. The control of fermentation is carried out daily by the production of alcohol, which proceeds at the rate of 1% per day, and the development of volatile acidity. When 6% alcohol is reached, the air of the entire volume of the tank should be replaced to create oxygen and add complex nutrients to prevent an arrest of fermentation. Generally, after reaching 9% alcohol, yeasts start to produce acetic acid, and at 12% alcohol acetic acid will not exceed 1 g L^{-1}. At this point, with a sugar content residue equal to 150–180 g L^{-1}, the alcoholic fermentation is interrupted by cooling the wine to 5°C, the shaking is stopped to allow the sediments to settle, and after 24–48 hours the wine is racked. Sulfur dioxide is then added, and the wine kept in contact with the fine lees at a low temperature in temperature-controlled stainless steel tanks. It is stirred constantly for the first month, and then once a week until the end of the fining period. During the fining process, which lasts one year or more depending on the grape harvest, the level of SO_2 is maintained. Bottling is done after slight clarification, addition of SO_2, and cartridge filtration.

Before being released to the market for consumption, the wine must stay in the bottle for at least 6 months, at cellar temperature, to allow its features to develop fully (Figure 14.4). The result is an intense yellow-coloured wine, aromatic with a strong varietal stamp, complex aromas, notes of citrus fruits such as orange blossom and candied lime peel, fruit such as melon and pear and honey, a sweet taste but not cloying, with a balanced acidity, persistent and long, suitable for ageing up to eight years.

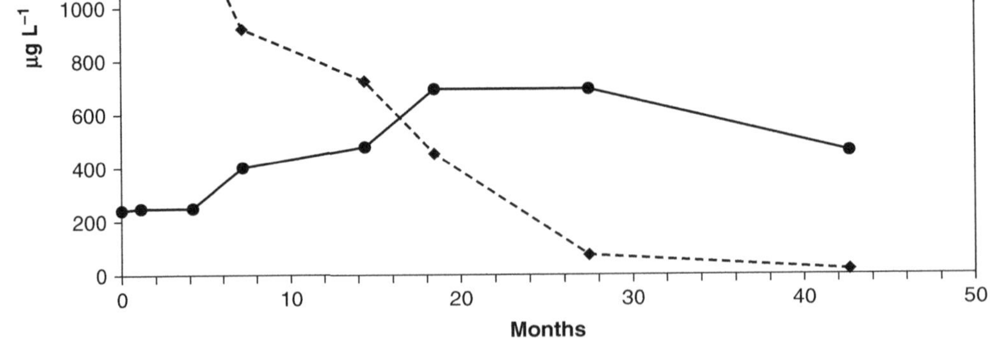

Figure 14.4 Muscat wine aroma compound evolution during ageing. The content in volatile linalool continues to be over the threshold until glycoside linalool starts to decline.

ACKNOWLEDGEMENT

The author would like to thank Roberto Eberle for help in translating this chapter.

REFERENCES

Bellincontro, A., D. De Santis, R. Botondi, *et al.* 2004. Different postharvest dehydration rates affect quality characteristics and volatile compounds of Malvasia, Trebbiano and Sangiovese grapes for wine production. *Journal of the Science of Food and Agriculture*, 84:1791–1800.

Di Stefano, R. and G. Maggiorotto. 1994. Evoluzione dei composti terpenici durante il processo di appassimento dell'uva Moscato bianco. *Rivista di Viticoltura e di Enologia*, 6:12–18.

Di Stefano, R., L. Corino, and P.D. Bosia. 1983. Evoluzione dei composti terpenici del moscato bianco durante le maturazione in relazione alla carica gemmaria. *Rivista di Viticoltura e di Enologia*, 4:22–28.

Skinkis, P.A., B.P. Bordelon, and E.M. Butz. 2010. Effects of sunlight exposure on berry and wine monoterpenes and sensory characteristics of Traminette. *American Journal of Enology and Viticulture*, 61:2–10.

15 Italian Passito Wines

Attilio Scienza

Department of Agricultural and Environmental Sciences, University of Milan, Milan, Italy

This chapter is a collection of technical notes describing the features of the best-known Italian wines made from dehydrated grapes. The majority of these wines are sweet, but they can also be dry. The wines can be produced from late harvest grapes, grapes dehydrated in the sun (a practice defined as 'drying'), or grapes dehydrated in closed facilities, well ventilated, with or without automatic control of the environmental parameters (a practice called 'dehydration', as described in earlier chapters of this book). Drying and dehydration are the techniques used in the production of Italian Passito wines.

15.1 ALBANA DI ROMAGNA DOCG

15.1.1 History

The first written evidence of Albana di Romagna wine dates from the beginning of 1300 in the *Trattato di Agricoltura* by Pier de Crescenzi. In 1782, there is a second reference from the Abbot Giovanni Battara, who describes a procedure for drying grapes and making wine. In 1967, Albana di Romagna was awarded DOC status, but it was not until 1987 that production of Albana di Romagna Passito became regulated under DOCG protocols.

15.1.2 Production area

The production area of Albana di Romagna stretches along the hills in the provinces of Forlì-Cesena, Ravenna and Bologna. The soil there is comprised of limestone and marl-sand deposits of marine origin, due to the particular climate influenced by the Apennine hills, the wind and its proximity to the sea; such *terroir* contributes to the wine's unique features.

15.1.3 Grape variety

The grape variety Albana was known among the ancient Romans. The name possibly comes both from Colli Albani, the hills surrounding Rome, or more likely from the quality of the wine, appraised among the best from white (Albane) grapes.

Sweet, Reinforced and Fortified Wines: Grape Biochemistry, Technology and Vinification, First Edition.
Edited by Fabio Mencarelli and Pietro Tonutti.

15.1.4 Dehydration and vinification

Generally the grapes are harvested between September and October. The law imposes a minimum sugar level in the grapes of 23%. In order to reach such a concentration on the plant, harvest takes place about 30 days later than usual. Grapes can also be dried using traditional methods, for example by storing them in a warehouse equipped with temperature and humidity control. Either way, the law states that grapes can be dried until 30th March of the year after the vintage, so that sugar levels are not less than 284 g L^{-1}. Vinification cannot start before 15th October of the vintage year and wooden casks can be used for maturing the wine. The wine is then stabilized, bottled and aged for six months. In most cases, however, wines will be aged for 2–4 years.

15.1.5 Organoleptic features

COLOUR: gold, with amber shades
NOSE: intense, with clean velvety fruit (and noble rot) notes
PALATE: pleasantly medium sweet or sweet

15.1.6 Food matching

Pastries, shortbread, dried fruit, honey, candied orange and medium seasoned cheese.

15.2 ALEATICO DELL'ELBA DOC

15.2.1 History

The origins of viticulture on the island of Elba date back to the Graeco-Roman period. In 800 AD, there was a large demographic development on the island and a consequent extension of the vineyard area; it soon became the Tuscan island with the highest wine production.

15.2.2 Production area

The territory of the island of Elba is characterized by the presence of Monte Capanne (1016 m) which guarantees good climatic variability according to its different aspects. The soil is characterized by many geological substrates made of eruptive, sedimentary and metamorphic rocks.

15.2.3 Grape variety

Aleatico: this grape variety was probably first introduced by the Greeks, but it was widely diffused in many Italian regions during the Middle Ages, in order to imitate the wines of Greek origin. It results from a spontaneous cross with White Muscat (Moscato bianco).

15.2.4 Drying and vinification

Harvesting time varies according to ripening in the different *terroir*, but it generally takes place between September and October. Bunches are dried in 4–5 kg crates placed on drying racks in the sun for 15–20 days. During the night or in case of rain, they are covered with large plastic sheets. Maceration on the skins lasts no longer than 5–6 days and fermentation lasts 7–10 days. After racking, fermentation continues slowly in stainless steel tanks or in small wooden casks of 2–5 hL. Maturation can take place in stainless steel or in wood. Some wineries bottle the wine the year after the vintage, while others prefer to mature it up to 24 months.

15.2.5 Organoleptic features

COLOUR: ruby to deep red
NOSE: pronounced, typically dried rose and citrus notes
PALATE: medium sweet to sweet, full bodied, balanced

15.2.6 Food matching

Cream desserts, such as panna cotta (a traditional Italian cream pudding) with strawberry sauce, Bavarian cream with fruits of the forest, chocolate cream waffle with black cherry sauce.

15.3 CHAMBAVE MOSCATO PASSITO DOC

15.3.1 History

The origin of this wine dates back to the introduction of several traditional grape varieties in the Aosta Valley under the kingdoms of Burgundy and the Francs (from the sixth to the tenth centuries AD), followed by the Savoia (during the eleventh century).

15.3.2 Production area

The production area of Chambave Muscat is within the boundaries of the DOC 'Valle d'Aosta', along the mountainside of the Dora Baltea Valley from Pont-Saint-Martin to Morgex, especially on the orographic (left) side. Cultivation is at 600 m above sea level on morenic soils with very steep south-facing aspects.

15.3.3 Grape variety

White Muscat on light, deep soils with a dry and windy climate, characterized by summer drought. The best training system is vertical shoot positioning, but the canopy must not be spread too wide.

15.3.4 Dehydration and vinification

The best Muscat bunches are dried in particular aerated environments away from the sun, or in conditioned environments with artificial ventilation, until the sugar levels are no less than 26%. The minimum natural alcohol by volume (ABV) level in the grapes before drying must be at least 10.5%. The grapes are processed as for white winemaking, and the wine is left to rest until the December after the harvest. In order to claim the appellation 'Valle d'Aosta', the wine must mature for at least five months from 1st December of the vintage date. When it is sold, Chambave Moscato Passito must have 16.5% potential alcohol and the actual alcohol content must be at least 13%.

15.3.5 Organoleptic features

COLOUR: gold, tends to become amber
NOSE: intense, aromatic (Muscat), acacia flower, thyme, sage, withered violet, yellow peach
PALATE: sweet, aromatic, typical of Muscat

15.3.6 Food matching

Seasoned and blue cheese, pastries, foie gras.

15.4 ERBALUCE DI CALUSO PASSITO DOC

15.4.1 History

The first reports of wines from the Canavese area date back to ancient Rome and to the medieval period, as evidenced by the Archbishops' archives of Turin and Ivrea. In 1530, Sante Lancerio, Pope Paul Farnese's cellarman, relates that the Pope was a great wine enthusiast and loved the fortified wine produced in the Ivrea area. He called it 'very perfect'. We also know that the wine was appreciated by Napoleon III and Vittorio Emanuele II. The origins of Passito di Caluso should, however, be researched through the tradition of the local rural families, who used to produce a sweet wine for important festive occasions.

15.4.2 Production area

Erbaluce di Caluso is produced in Piedmont, in the area called Canavese, within the provinces of Turin, Biella and Vercelli. The soil is predominantly hilly with glacial reliefs (moraines) that have an elongated aspect, mainly made of clay, sand and pebbles.

15.4.3 Grape variety

Erbaluce: the grape variety is native of the Canavese area, in the subalpine region where there are zones that are neither too dry nor too sunny. The name Erbaluce comes from the Latin '*Alba lux*' as the grapes 'shine like dawn'. Canes need to be pruned quite long, as the plant is very vigorous. Normally, it is pergola trained.

15.4.4 Dehydration and vinification

Grapes are usually harvested between the end of September and the first days of October, according to the weather conditions of the vintage. The healthiest grapes are carefully selected and put in low, wide boxes, in order to avoid pressure on the berries. After that, they are transported to closed and aerated warehouses called *sulè*. The grapes dry from the beginning of October until the end of February or the beginning of March. Fermentation is never tumultuous either for the high sugar levels or for the environmental conditions. It ends in the summer, starts again in autumn and stops in the winter. On average, re-fermentations continue up to the second year. The slow fermentations, together with siphoning (twice a year) can also continue for up to five years in small casks or demijohns. Eventually, the wine is fined, filtered and bottled.

15.4.5 Organoleptic features

COLOUR: from gold to dark amber
NOSE: delicate perfume, typical
PALATE: sweet, balanced, full, velvety

15.4.6 Food matching

Biscuits, zabaglione, amaretti, mature gorgonzola cheese.

15.5 FRASCATI CANNELLINO OR FRASCATI DOLCE DOC

15.5.1 History

The first descriptions of the so-called wine of Tuscolo, the hill above the area of Frascati, are contained in *De agricoltura* by Marcus Porcius Cato. Sweet Cannellino is one of the traditional styles of the DOC 'Frascati'. There was a time when this wine was produced due to the climatic influence of Agro Pontino, a marshland which had not then been drained. Its humidity would reach the first hills of the so-called Castelli Romani, encouraging the development of noble rot on the berries with a consequent concentration of sugars: just like Sauternes. Today, most Cannellino is late harvested.

15.5.2 Production area

The production region is located south-east of Rome and covers the municipalities of Frascati, Monte Compatri, Monte Porzio Catone and parts of Rome. The soils are predominantly volcanic (basalts, tophus and pozzolanic ash) and rich in mineral elements.

15.5.3 Grape variety(ies)

White non-aromatic Malvasia of Candia (50%), Tuscan Trebbiano and Green Trebbiano (10–20%), Malvasia Puntinata (10–40%). Vineyards are wire trained, although pergola or tendone can be found.

15.5.4 Drying and vinification

Grapes are late harvested in mid October with a natural alcohol of 10.5% ABV. Grapes can be attacked by noble rot in the right atmospheric conditions. White winemaking procedures are then used, either with or without pellicular maceration.

15.5.5 Organoleptic features

COLOUR: lemon with shades of gold
NOSE: typical, reminds of ripe fruit, honey, dried fruit
PALATE: typical, sapid, sweet with fruity notes

15.5.6 Food matching

Creamy desserts, ice-cream, cake.

15.6 GRECO DI BIANCO DOC

15.6.1 History

The first news of Greco di Bianco is found in the book by Father Giovanni Fiore published in 1734, titled *Calabria Illustrata*. The author reports a quotation by Barro 'here there is a wine which is extraordinarily nice, white, strong, always improving and it is produced in great quantity'. A work by Doctor Raso in 1843, titled *Quadro statistico de' Distretti di Palmi e Gerace nella prima Calabria*, describes the wine Greco as follows: 'This is one of the most desired wines, the best in the whole province and maybe in the whole kingdom. It is produced exclusively in the Locri area, but planting of Greche vines is now spread, although the ancestors considered it an exclusive privilege of their soil'.

15.6.2 Production area

The area within the appellation comprises the municipality of Bianco and a part of Casignana, in the province of Reggio Calabria, in hot and sunny locations. Soils are mainly made of clay, sometimes mixed with sand.

15.6.3 Grape variety(ies)

Greco di Bianco could have been introduced during the Magna Greece age, when the first settlers from the central Greek region of Locride arrived at the beginning of the seventh century BC near the Zefirio mountain, today called Capo Bruzzano. It is called with various names in the different Mediterranean winegrowing regions: Malvasia delle Lipari, Malvasia di Sitges, Malvasia di Bosa. Production is abundant and consistent. It benefits from bush training or wire training with a short pruned cane.

15.6.4 Drying and vinification

Greco di Bianco grapes are picked when sugar concentration is around 22%. Grapes are placed on drying racks in the sun for 10–15 days until their sugar level reaches 26–30%. At night or when it rains, they are protected with plastic sheets. Grapes are softly whole bunch pressed, then destemmed by hand and pressed a second time. Must is fermented in 15–30 hL casks or stainless steel containers. The wine is racked when it still has a good level of residual sugar and is siphoned a couple of times towards the end of winter. It is bottled 2–4 years after the harvest.

15.6.5 Organoleptic features

COLOUR: gold with amber nuances
NOSE: dried grapes, citrus notes and a moderate maderized hint
PALATE: medium sweet or sweet, very light tannic feel, warm, dried grapes, a vaguely bitter finish

15.6.6 Food matching

Biscuits, marzipan cake, seasoned hard cheese.

15.7 LACRIMA DI MORRO D'ALBA DOC

15.7.1 History

The area has very ancient origins: first, the Piceni and later, the Senones cultivated grapevines. The wine was almost certainly known during the Middle Ages, as the tradition relates that Frederick Barbarossa, who had his headquarters near the Castle of Morro d'Alba during the siege of Ancona in 1167, had the opportunity to taste and appreciate the wine produced from such grapes.

15.7.2 Production area

Lacrima di Morro d'Alba is produced in a delimited area in the province of Ancona, to the north of the river Esino. The grapes are cultivated on the best hillsides, with clay and/or sandy soils rich in minerals. Grapes cannot be grown on the plains and hills facing the sea in the municipality of Senigallia.

15.7.3 Grape variety

Lacrima nera: it has ancient origins and it was cultivated only in Morro d'Alba and a few nearby places. It is a very vigorous grape, not very long-lived, with an inconsistent production. In spite of this, it is very suitable for the temperate climate of the Marche coast.

15.7.4 Dehydration and vinification

Grapes are harvested towards the middle of October, when they reach a sugar concentration of 20–21° Babo. Ripe grapes are placed on drying racks or wooden cases until January–February when they reach a sugar level of 27–28°, in closed, well-ventilated facilities. Vinification cannot start before 1st December of the vintage year, with a maceration on the skins that lasts about 10 days. Maturation takes place in stainless steel and in bottle for 6–7 months. Some producers prefer to mature the wine in small casks for 3–4 years.

15.7.5 Organoleptic features

COLOUR: light to deep ruby, sometimes almost garnet
NOSE: typical, medium to high intensity
PALATE: balanced, velvety, ripe cherry and spices

15.7.6 Food matching

Main courses with tomato sauce or meat sauce, regional dishes, bluefish starters with the same wine marinade, white meat.

15.8 MALVASIA DELLE LIPARI DOC

15.8.1 History

Malvasia delle Lipari is one of Sicily's oldest wines. Diodorus Siculus, in the first century BC, says that the wine was introduced by the Greeks, even if the origins of viticulture in the Aeolian islands are even more remote.

Towards the second half of the nineteenth century, production of dried grapes increased but the situation declined because of powdery mildew and Phylloxera, which reached Sicily, including Salina, in 1880. Unfortunately, viticulture on the island was drastically reduced. Malvasia delle Lipari was rediscovered in the mid 1970s; in 1973, it was granted DOC status.

15.8.2 Production area

Malvasia delle Lipari is produced only in Salina, an island in the middle of the Aeolian archipelago in the centre of the southern Tyrrhenian Sea to the north-east of Sicily. The vineyards thrive mainly on very loose volcanic soils (pumice, ash, lapillus) very rich in minerals and potassium.

15.8.3 Grape variety

Malvasia delle Lipari is genetically identical to Malvasia di Bianco or di Bosa, a very old variety introduced into various zones of the Mediterranean by the first Greek settlers between 588 and 577 BC. The most suitable pruning systems are those with a medium canopy expansion, mixed and long cane pruning, very low pergola to cope with the sea wind, bush trained or fan trained.

15.8.4 Drying and vinification

The grapes are harvested when they are slightly over-ripe. Sometimes, they are placed in the sunshine for 10–15 days on drying racks made of canes or metal, and they are covered during the night. The sugar level in the grapes must be of 25–28%, sometimes even of 30–32%. Fermentation takes place in 8–10 hL vessels and goes on until January–February when the must is racked in the open air. The actual alcohol content at this stage is 13–15°, residual sugar is about 8–13% and the glycerine content is quite high (12–14 g L^{-1}.). The wine will also have a high content of ashes, sulphates, potassium, calcium and magnesium. The wine must be matured for a minimum of 6 months, but it will age up to 10 years.

15.8.5 Organoleptic features

COLOUR: deep gold or amber, sometimes with pink nuances
NOSE: delicate, exceptionally complex, notes of eucalyptus, honey, wild flowers, aromatic herbs and privet
PALATE: sweet and elegant, velvety, a light taste of apricot, very long aromatic finish

15.8.6 Food matching

Panna cotta pudding, zabaglione, sbrisolona cake (Italian crumb cake).

15.9 MOSCADELLO DI MONTALCINO PASSITO DOCG

15.9.1 History

The production of Moscadello has very antique origins. There is evidence dating back to 1540 and 1591 and in some documents in the secret Vatican archives. Other quotations about this precious wine are found between the seventeenth and nineteenth centuries thanks to many Tuscan poets, such as Francesco Redi and Ugo Foscolo. The DOC status was acknowledged in 1984.

15.9.2 Production area

Montalcino is situated in the heart of southern Tuscany, 40 km south of Siena. The growing region is in a large area comprising one large hillock at 567 m above sea level. Soils are made of clay, rich in limestone with ample tophus areas and mainly thin. The presence of hills with different aspects, the consistent modulation of the hills and the difference in the altitude between the valleys and the highest hill (Poggio Civitella) determines a variety of very different climates.

15.9.3 Grape variety

White Muscat or Moscatello selvatico: it is probably a variety of Greek origin, spread throughout the Mediterranean. It prefers marl-limestone soils, not too humid, deep and without much clay, dry and windy climates; it is summer drought resistant. The most suitable pruning system is cordon trained, spur pruned with a medium dense canopy.

15.9.4 Dehydration and vinification

Grape harvest begins in October, 20 days after normal ripening. Grapes stay on the vines for about 20 days longer and sometimes they undergo even longer dehydration in suitable warehouses. Must is fermented in small wooden vessels for at least one year in conditioned environments in order to control fermentation temperature. Some producers age the wine for two years in small wooden casks, a further year in bottle is necessary to complete the evolution of the wine.

15.9.5 Organoleptic features

COLOUR: lemon to yellow
NOSE: typical, delicate, persistent, clary sage with persistent citrusy notes
PALATE: aromatic, sweet and balanced, very soft

15.9.6 Food matching

Biscuits, puddings, jam pastries.

15.10 MOSCATO DI SCANZO DOCG

15.10.1 History

The first historical document referring to this wine is dated 8 June 1347: it is the will of Alberico da Rosciate, a notorious jurist. A trace can be found also in 1372, in the correspondence of the feudatory Bishop of Tribulina di Scanzo. In 1700, the architect Giacomo Quarenghi from Bergamo introduced the Tsarina Catherine of Russia and her English allies in the war against Napoleon to the quality of Moscato di Scanzo. The first news about Moscato di Scanzo Passito comes from the London Stock Exchange in 1850 because it was the first wine to be quoted.

15.10.2 Production area

The production area of Moscato di Scanzo falls within the province of Bergamo, comprising the municipality of Scanzorosciate in Valcalepio. The soils to the east are made of limestone and clay, while to the west there is more silica, clayey-schist.

15.10.3 Grape variety

Moscato di Scanzo: the grape variety might be of Hellenic origin, like the White and Black Muscat which still extends over many Balkan and Mediterranean regions. Nowadays, Moscato di Scanzo is cultivated only in the area of Scanzorosciate. The vines grow better on dry, hilly soils; the most ancient ones are in shielded areas with thin soils rich in skeletal matter.

15.10.4 Dehydration and vinification

The wine is produced only in the best vintages with suitable weather conditions. Generally speaking, harvest takes place in September–October. Grapes are carefully selected and dried for at least 40 days in suitable warehouses with temperature and moisture conditioning and ventilation by fans. The must macerates on the skins for 3–5 days; it is then racked and siphoned. Wine matures in bottle for at least 3 months.

15.10.5 Organoleptic features

COLOUR: medium to deep ruby; it can have pink or garnet nuances
NOSE: delicate, intense, persistent, typically of dried rose and Muscat
PALATE: sweet, pleasant, balanced, a slight almond aftertaste

15.10.6 Food matching

Tarte tatin (apple tart), jam cookies, chocolate dessert.

15.11 MOSCATO DI SIRACUSA PASSITO DOC

15.11.1 History

According to the famous historian and winemaker Saverio Landolina Nava (1743–1814), Moscato di Siracusa is the ancient Pollio Siracusano, obtained by *biblia* grapes (from the Biblin mountains in Thrace). It was introduced to Syracuse by Pollis, the tyrant of the city. If this is correct, its origins go back to between the eighth and seventh centuries BC.

15.11.2 Production area

The production area is in the eastern part of Sicily, in the province of Syracuse. Plyocenic soils with silica and limestone, medium clay content, a dry and moderate climate are the best conditions for cultivating Muscat.

15.11.3 Grape variety

White Muscat.

15.11.4 Drying and vinification

White Muscat grapes are dried for a short while both on the plant and on drying mats in the sun. In either case, the drying process is not taken too far in order to enhance the elegance and finesse of the floral descriptors. Berries are carefully separated from the clusters and softly pressed. Must is fermented until it reaches an alcohol level of between 14 and 16%. The wine is stabilized in stainless steel containers, matured and bottled.

15.11.5 Organoleptic features

COLOUR: old gold with amber nuances
NOSE: delicate, typically of flowers and Muscat, honey and dried fruit
PALATE: sweet, velvet, pleasant

15.11.6 Food matching

Small cakes, white fruit pastries, blue cheese.

15.12 MOSCATO DI STREVI PASSITO DOC

15.12.1 History

The first evidence dates back to the beginning of 1800, when several growers, like Buzzone and Marenco, started to produce and trade such wine. In 1990, thanks to the Marenco Company, researchers tried to select a biotype of Muscat with the most suitable shape for drying.

15.12.2 Production area

The area of production of Moscato di Strevi DOC is situated in northern Monferrato, in the province of Alexandria in Piedmont. The area is situated to the north-west of Strevi, on the hills between the river Bormida and Tanaro-Po. The hills are not very high, ranging between 160 and 337 m above sea level. The Bagnario valley is the ideal environment for the production of Moscato Passito, given that such hills have a soil made of white marl mixed with sand, which confers to the wine its quite unique features.

15.12.3 Grape variety

White Muscat.

15.12.4 Drying and vinification

Grapes for Passito are picked a few days earlier than the regular harvest: the healthiest and loosest bunches are chosen. Sugar levels in the must are usually 28–30° Babo. Grapes are put into baskets or chestnut cases called ‘gerbe’ and are gradually dried in the sun or in the open air. Drying the grapes in a closed environment is not permitted. After a period of drying, which can vary according to the environmental conditions, dried grapes are pressed with a 40% yield. Fermentation lasts for 6–8 months and stops at around 13.5–14% alcohol. The wine is matured in stainless steel or in small barrels for six months, followed by six months in bottle. It can be traded after two years of ageing.

15.12.5 Organoleptic features

COLOUR: medium to deep gold, amber
NOSE: very complex, typical of Muscat and orange blossom
PALATE: sweet, balanced, typical, sometimes pronounced ripe fruit

15.12.6 Food matching

Biscuits or local seasoned cheese.

15.13 MOSCATO DI TRANI

15.13.1 History

Moscato di Trani is cited by Pliny, Cato and Columella but it is only around the year 1000, thanks to the Venetians, that it acquires commercial relevance.

15.13.2 Production area

It is cultivated on the slopes that reach to sea level from an altitude of between 200 and 300 m. The ideal soils are marl and tophus, limestone and clay or limestone and silica. They can be deep but have to be rather dry; it is not by chance that the abandoned tophus pits (*tufare*) have become the most suitable location to grow Muscat.

15.13.3 Grape variety

White Muscat is locally called Moscato Reale in order to distinguish it from the other varieties with less intense aromatic notes. Grapevines are trained along wires, mainly facing south-east and north-west. There are also examples of bush- or pergola-trained vines.

15.13.4 Drying and vinification

After a short period drying on the vine, grapes are softly pressed and destemmed. The yield in juice is about 60% and it is immediately chilled to block fermentation. After static decanting and siphoning, the must is fermented and fermentation is blocked through the help of filtration, which maintains adequate levels of residual sugars. After this, wine is transferred into small oak barrels for at least three months, and when residual sugar reaches 4–5 % it is filtered.

This wine has a considerable sugar level (160–180 g L^{-1}) and a minimum alcohol content of 12.5%; the law imposes a maturation period of four months. The Moscato di Trani Liquoroso (fortified) style must age for 12 months and have 18% total alcohol, of which at least 16% is actual alcohol.

15.13.5 Organoleptic features

COLOUR: lemon, becomes gold as it ages
NOSE: aromas of sage, rose, honey, apple, spices and vegetal notes
PALATE: sweet flavour, slightly bitter with fruit notes

15.13.6 Food matching

Almond biscuits, honey biscuits, local fruit tarts.

15.14 MOSCATO ROSA PASSITO DOC

15.14.1 History

During the Middle Ages, viticulture survived in the region due largely to the important role played by convents and monasteries, especially the ancient Benedictine monasteries. Trading with Austria and Germany was a great incentive for improving the quality of sweet wines as well.

15.14.2 Production area

The DOC 'Trentino' comprises the whole of the Trent province. The cultivated areas are mainly on the hills with various soil types like porphyry, glacial moraines, dolomitic and alluvial soils; with a good exposure and an altitude of between 200 and 450 m.

15.14.3 Grape variety

Moscato Rosa (Rosenmuskateller): the origins are quite obscure; it certainly belongs to the set of oriental Muscats. It has female flowers that reduce the yield but enhance the drying process on the plant. The name comes from the perfume of rose, conferred to the wine by the high level of nerol.

15.14.4 Drying and vinification

Grapes are naturally dried on vine with a slight over-ripening, then undergo the red vinification process: maceration on the skins lasting for only a few days. The must ferments under a controlled temperature of 14–16°C until the desired residual sugar level is reached. It matures for a minimum of 10 months in bottle at low temperature; it is then filtrated and fined in bottle. Moscato rosa must have a minimum total alcohol of 15%, of which at least 11% must be actual.

15.14.5 Organoleptic features

COLOUR: pale garnet rosé
NOSE: delicate, pleasantly aromatic, typical note of tea rose
PALATE: sweet, full, pleasant, rather alcoholic

15.14.6 Food matching

Red fruit desserts, biscuits, ricotta cheesecake, pastries with fruits of the forest jam, doughnuts, poppy seed strudel, very strong blue cheese.

15.15 NASCO DI CAGLIARI PASSITO DOC

15.15.1 History

Avicenna, an Arab philosopher and doctor who lived around the year 1000, talks about Nasco as a vigorous variety, but at the same time as delicate and fine as a woman. This wine represented the island of Nasco at the Universal Exposition of Vienna in 1873, at the time of its maximum viticultural expansion (75 000 ha at the end of the nineteenth century). In 1972, Nasco was awarded DOC status, which recognizes the ‘sweet natural’ and the ‘fortified sweet natural’ styles.

15.15.2 Production area

The area of production of Nasco is situated in the Campidano zone. There are alluvial soils of the Quaternary age, with deep layers and good permeability. Their composition is similar to that of the rocks which constitute the alluvial deposits. There can be a relevant percentage of limestone, but there are also soils with subacidic reactions. The climate is typically Mediterranean, with warm, dry summers.

15.15.3 Grape variety

Nasco: it is considered one of the eldest varieties in Sardinia and it was already known during the Roman age, when it was named (*nuscus*, moss), due to the slight moss aroma that the wine shows when it ages. It is mainly bush trained or high wire trained.

15.15.4 Drying and vinification

The grapes are harvested 15 days later that their natural ripening time, so that they dry naturally on vine. The grapes are pressed and destemmed, and are fermented after 2–3 days of maceration. After racking, fermentation goes on until the wine reaches a minimum total alcohol of 14.5%, of which 12% has to be actual. Natural sweet Nasco (non-fortified) has an average alcohol volume of 13%, 5% residual sugar and a total acidity of 5.3 g L^{-1}.

15.15.5 Organoleptic features

COLOUR: lemon to gold
NOSE: delicate, with a light aroma of grapes and fresh fruit
PALATE: pleasant, a bitter touch, typical

15.15.6 Food matching

Dried biscuits, cakes with cream and jam, almond biscuits.

15.16 ORVIETO DOLCE DOC

15.16.1 History

The grape-growing and wine-making traditions of the Orvieto area go back to the Etruscans. There are many cellars dug into the tophus substrate, where the low temperatures regulate wine fermentation and maturation. During the Medieval and Renaissance periods, the sweet wine Orvieto was the favourite of Pope Paul Farnese III.

15.16.2 Production area

The production area of 'Orvieto' DOC is situated between Umbria and Lazio. The vineyards are on the hills and the soils are of marine sedimentary origin, with clay, rich in limestone, sometimes volcanic with tophus. The climate is warm continental, quite humid and in autumn it allows *Botrytis* to develop on the berries.

15.16.3 Grape variety

Grechetto (minimum 40%); Tuscan Trebbiano (20–40%). Vineyards are wire trained.

15.16.4 Drying and vinification

Harvesting is delayed about 15–30 days compared to natural maturation. In some cases, it goes on until December in some vineyards that have particular mesoclimatic conditions which favour noble rot: one of the main causes of aromatic complexity in these wines. After crushing and fermentation on the skins, the wine is softly pressed. Maturation and conservation are performed in small oak containers.

15.16.5 Organoleptic features

COLOUR: lemon to gold
NOSE: floral, delicate peach perfumes, apple, tropical fruit, apricot and citrus fruit
PALATE: sweet, balanced, persistent, ripe fruit with almond notes

15.16.6 Food matching

Seasoned Pecorino cheese and honey.

15.17 PASSITO DI PANTELLERIA DOC

15.17.1 History

Pantelleria can be considered the heart of the Mediterranean, as it is at the crossroads of different cultures. Evidence of the Arab domination is still preserved by the names of the most famous locations on the island. Passito, the wine of the local tradition, was recently rediscovered, and owes its success to the growing care that producers have taken in the last few years.

15.17.2 Production area

On the island of Pantelleria, in the province of Trapani, the vineyards are situated in two very different environments from a climatic point of view: the early harvest grapes on the western side and the late harvest on the eastern side. Soils are volcanic, loose, with basalt and rich in potassium. Vines are bush trained: so low that they are almost creeping. They are situated in deep caves, in order to retain rainwater, night humidity and shelter the plants from the salty wind, which blows in from the sea.

15.17.3 Grape variety

Zibibbo or Muscat of Alexandria was already cultivated in ancient times: probably, the Arabs from Cape Zibibbo brought it in. According to a different etymology, the name comes from the Arab 'zabib', a word meaning dried grapes. Zibibbo, this very generous grapevine, ripens on the island in a very uneven way: there can be up to a month of difference between the first and the last harvest. Vines are bush trained, with or without a support, with 3 or 4 canes up to one metre long. At the extremities there is a spur with 1–2 gems. Generally, vines are kept within the traditional 'caves'. Plant density varies in the range between 3460 and 2500 plants per hectare.

15.17.4 Drying and vinification

The harvest of bunches that are selected for drying takes place between August and September, according to the different areas of the island. Grapes are left to dry for about 4 weeks on drying racks or on a floor made of rock. In the later harvest areas, the bunches – called 'sganguna' – are harvested one month later, they are pressed and fermented together with a small quantity of dried grapes. After this stage, the wine is mixed with berries from bunches that have been destemmed, dried but not pressed. They are left in contact with the wine for 4 months, during which sugar and flavours are released. Wine reaches 14.5% alcohol and almost 200–300 g L^{-1} of sugar, then it is stabilized and filtered.

15.17.5 Organoleptic features

COLOUR: gold with amber shades
NOSE: very intense, aromatic (typical of Muscat), dried fruit notes, honey, ginger
PALATE: sweet, soft, pleasant, very persistent

15.17.6 Food matching

It can be enjoyed on its own and it will match the traditional Sicilian desserts.

15.18 COLLI ORIENTALI DEL FRIULI PICOLIT DOC

15.18.1 History

The first news of Picolit dates back to a document of 1682, but it becomes more precise towards the end of the eighteenth century, thanks to the Count Fabio Asquini, called 'the father of Picolit', who started producing this wine for trade. According to Giorgio Gallesio (an Italian botanist), the name comes from 'the small berries produced by the grapes', but it might refer to the term *pecol*; in the Friuli region, this means the top of the hill, where grapes ripen better.

15.18.2 Production area

The area of production of Picolit is located in the Friuli–Venezia Giulia region, within the DOC 'Colli Orientali del Friuli', along the eastern hillside of the province of Udine, and the DOC 'Collio', along the northern side of the province of Gorizia on the border with the Republic of Slovenia. The soil is constituted by alternating marl (limestone clay) and sand (limestone sand), and flysch of marine origin, locally called 'ponca'.

15.18.3 Grape variety

Picolit: it has female flowers and the pollens have a very low germinability; it should therefore be planted together with other pollinating varieties, such as *Verduzzo Friulano*. Given the vigour of the grape variety, the most popular training system is 'doppio capovolto' or 'alla cappuccina'.

15.18.4 Dehydration and vinification

Picolit is a late ripening variety and drying time is very variable from zone to zone, ranging from 2 weeks to 2 months. The over-ripe grapes are picked and put in naturally aerated warehouses. Fan-conditioned places can be used and some producers also use conditioned cells. Artificial systems must ensure circulation of dry air at a temperature ranging from 25 to 35°C. In vinification of dried grapes, it is hard to extract juice and sugar because of the thickness of the skin. In order to deal with such problems, producers perform a mixed vinification technique: usually, 50% of the grapes undergo the white vinification process and the fermenting must is used to macerate the rest of the dried grapes. Fermentation usually continues for the whole of the winter season. The use of barriques is quite widespread, both during fermentation and maturation. Ageing is according to the producer's choice, from 2 to 4 years.

15.18.5 Organoleptic features

COLOUR: medium to deep lemon
NOSE: delicate, fine, pleasant, floral, citrusy
PALATE: medium sweet to sweet, warm, balanced, honey, good acidity

15.18.6 Food matching

Blue cheese, foie gras, shellfish.

15.19 PRIMITIVO DI MANDURIA DOC

15.19.1 History

It is commonly acknowledged that Primitivo arrived in Apulia from Dalmatia at the end of the seventeenth century and it underwent a growing expansion during the nineteenth century, especially in the area of Gioia del Colle, where pedoclimatic conditions were particularly favourable for this grape variety. The denomination Primitivo comes from the fact that its grapes ripen particularly early.

15.19.2 Production area

The cultivation area of Primitivo di Manduria is situated to the south-west of the Apulia region, along the Ionic Coast of Salento. Environmental conditions are very varied and are influenced by the sea and the soil, which can be more or less sandy.

15.19.3 Grape variety

Primitivo is a spontaneous cross of Croatian varieties like Plavac Mali and Plavina, and it corresponds to the Californian Zinfandel. Especially in the area of Taranto, the grapes are bush trained: this, together with other environmental factors, allows the production of high-quality wines.

15.19.4 Drying and vinification

Harvesting takes place at different times, according to the pedoclimatic features of the growing area. On the flat lands, such as Manduria, grapes are harvested during the second decade of September; in nearby Sava, it is 10–15 days later; while near Bari, it may start in late September or early October. Through natural maturation, grapes reach 20–23% sugar level, but they are left on the plant for several weeks, until sugar is above 30%.

Grapes are crushed, destemmed, and fermentation, with maceration, lasts for about 4–5 days at a temperature of 28–30°C, with frequent pumping over. Actual alcohol varies between 16 and 18°C, residual sugar ranges between 1.5 and 8%.

If the grapes produce a wine with a natural 15% alcohol level, Primitivo di Manduria can be produced in the following styles: natural sweet, natural fortified sweet, fortified dry. Wine is matured in cement tanks and stainless steel from 2 to 5 years, and then bottled.

15.19.5 Organoleptic features

COLOUR: garnet, with orange hue

NOSE: dried fruit, plum, black cherry, kirsch

PALATE: medium sweet, balanced acidity, a hint of carob, full bodied, high but balanced tannins, generous, austere

15.19.6 Food matching

Rich desserts with unleavened dough, such as dried fig stuffed with almonds, pastries with fruits of the forest jam, mostaccioli (almond biscuits), fried cheese with honey and spices, ciambella (lemon-flavoured) cake, amaretti. The natural fortified style matches cakes with rich, alcoholic creams, while the dry fortified wine is best with oven-baked cakes and tozzetti.

15.20 RAMANDOLO DOCG

15.20.1 History

Viticulture in this area greatly expanded around the first to second centuries AD, as a result of the Romans making Aquileia an important centre for collecting wine for trade towards northern Europe. The first news about Verduzzo di Ramandolo dates back to a praising testimonial by Giovanni Colonnelli in 1893 and, in 1825, it is quoted by Acerbi.

15.20.2 Production area

The growing area of Verduzzo di Ramandolo is situated within the DOC 'Colli Orientali del Friuli', in the province of Udine. Within this area, a particular zone in the municipality of Nimis was awarded DOCG status in 2001, with the denomination 'Colli Orientali Ramandolo'. The soil shows alternating marl layers (limestone clays) and calcified sand that look very typical. Vineyards grow up to 380 m above sea level: the highest growing altitude for this denomination.

15.20.3 Grape variety

Verduzzo friulano yellow biotype: it is a native variety from Friuli. The typical training system is 'doppio capovolto' or 'alla cappuccina' with middle expansion and a medium long pruning. It performs better on well-exposed hills, with unfertile soils and a dry climate.

15.20.4 Dehydration and vinification

For the Passito style, harvesting takes place between 15th and 30th October, according to the course of the growing season. Grapes can dry on the plant and be late harvested or in suitable warehouses, with air conditioning or fans. Grapes are placed on a drying rack; they are selected and left to dry for 8–10 days.

Maceration is quite long: fermentation is slow both because winters are very cold and for the type of indigenous yeasts. Barrels are permitted both for vinification and maturation of Ramandolo.

15.20.5 Organoleptic features

COLOUR: pale to deep gold
NOSE: intense, acacia flowers and green apple
PALATE: pleasant, velvety, low to medium tannin, full bodied

15.20.6 Food matching

Blue cheese, foie gras, biscuits.

15.21 RECIOTO DI GAMBELLARA DOCG

15.21.1 History

There are various theories about the origin of the name Gambellara. One early hypothesis comes from 'land of grapevine', from the dialect 'ambeli', hence 'ambelaria' from the Greek 'ampelos' meaning grapevine. Another source links the name to the Gens Camillaria people, who during the Roman age lived near Ravenna and moved to Veneto in the Byzantine period.

The origins of Recioto go back to the Roman period; we have written evidence of a '*Vinum Suave, nobile, pretiosum*'. Actually, the Romans were experts in the preparation of *vinum passum*, as Pliny states in his *Historia Naturalis*.

The first analytical data about Recioto are from 1873.

15.21.2 Production area

The area of production of Recioto di Gambellara comprises the municipalities (or parts of them) of Gambellara, Montebello Vicentino and Montorso, in the province of Vicenza, and Zermeghedo in the province of Verona, in the Veneto region. The soil is made of volcanic rocks of basaltic origin and the vineyards are on hills.

15.21.3 Grape variety

Garganega (minimum 70%). The grape variety is ancient, and is related to the Sicilian Grecanico. De Crescenzi describes it for the first time during the fourteenth century as a common variety of the Po valley. Production is abundant and consistent; long-pruned pergola is the most popular training system.

15.21.4 Dehydration and vinification

Grapes for dehydration are picked before the beginning of the main harvest, sugar levels must not be lower that 16° Babo and the most suitable grapes are carefully selected. Grapes are placed in dry and perfectly aerated warehouses, either on a plateau or hung on the walls (the traditional *picai*) for several months, usually until the beginning of December. Towards the end of November or in early December, grapes are softly processed with a horizontal or vertical press. Must is placed in a cold place for decanting, but a part of it is left in a warmer environment to encourage the beginning of fermentation. Usually, fermentation takes place in small barrels; it must be long, in order to allow the development of all the typical perfumes of Recioto. It generally lasts until March or April.

When residual sugar is about 8–10%, the juice is siphoned to block fermentation and fine the wine. After drying, grapes must have a natural potential alcohol of 12%.

15.21.5 Organoleptic features

COLOUR: lemon to gold
NOSE: intense, raisins and often candied fruit
PALATE: typical, balanced, a slightly passito taste, medium sweet or sweet, more or less lively with a delicate bitter finish

15.21.6 Food matching

Recioto di Gambellara is very good with desserts (except chocolate desserts), ice cream, zabaglione and all sorts of biscuits. The traditional food match is with Brasadelo, a typical dry cake in the shape of a doughnut.

15.22 RECIOTO DI SOAVE DOCG

15.22.1 History

The origins of Recioto go back to the Roman presence in the Retia region, within the area of Verona, as stated in the descriptions of the grapes' drying techniques. The name comes from *recie*, parts of the bunch that ripen better than others, which are selected for dehydration. Between the first and third centuries, sources report for the first time about *acinatico*, a sweet, dense and robust wine produced from grapes that had been stored on drying racks during winter. Cassiodorus, the minister of Theodoric, King of the Goths, describes in detail such wine. Around 1300, in his *Ruralium Commodorum*, Pier de Crescenzi writes about a 'garganica' grapevine, yielding 'round and white, amazingly sweet grapes'. Around 1400, people from this region began to call sweet wines 'vini natalini' or, as in Tuscany, 'vini santi'. Such a denomination spread throughout the Italian territory and in the area of Verona the ancient *acinatico* also started to be called by these names.

15.22.2 Production area

The area of production of *Recioto di Soave* is situated within the Veneto region, in the province of Verona. The area comprises all the hilly sites within the boundaries of the 'Soave' DOC denomination. There are two distinguished zones: Soave Classico DOCG and Recioto di Soave DOCG.

15.22.3 Grape variety(ies)

The grape varieties allowed for Recioto are at least 70% Garganega with Trebbiano di Soave, Pinot Blanc and Chardonnay. A long-pruned pergola is the most suitable training system.

15.22.4 Dehydration and vinification

After a careful bunch selection, grapes are picked slightly earlier compared to normal harvest, but with a sugar level no lower that 16° Babo. Grapes are places on drying racks and stored in a dry and aerated warehouse for 4–6 months. Grapes are constantly controlled to eliminate any rot or mould. Towards the end of February, they are softly pressed.

Nowadays, warehouses are equipped with conditioning plants with warm air that will accelerate the drying process, provided that temperatures reflect the natural conditions required to dry grapes.

A part of the must is kept in a cool place for cold stabilization, while another part will stay in a warm place to encourage fermentation and will be used to inoculate the stabilized must.

After dehydration, sugar levels must not be lower than 24–25° Babo. Fermentation often occurs in small barrels and it must be long in order to develop all the traditional aromas of Recioto. It generally lasts until the following spring. When residual sugars reach 8–10%, fermentation is blocked by siphoning. After drying, the grapes for the production of Recioto di Soave must have a minimum natural potential alcohol of 14%.

15.22.5 Organoleptic features

COLOUR: light to deep gold
NOSE: pleasant, intense, fruit notes, citrusy and white flower notes
PALATE: medium sweet or sweet, velvety, balanced, sometimes it has a pleasant balsamic note imparted by the oak

15.22.6 Food matching

It matches dry biscuits, blue cheese (salted or spicy), foie gras and all foods that will enhance its soft and fragrant features.

15.23 SAGRANTINO DI MONTEFALCO PASSITO DOCG

15.23.1 History

The first news of Sagrantino di Montefalco is dated around 1800, although it is likely that the Roman Itriola, the Greek renaissance wine that was offered to the Princes, and the Muscat wines of the eighteenth century can be identified with the modern Sagrantino di Montefalco. The name of the wine, from *sagrestia* (sacristy), reveals its ecclesiastic use for celebrating Mass. Such wines were usually sweet (like vini santi), both in accordance with the Classical World tradition of offering the gods the wines from dried grapes, and because this guaranteed their preservation.

15.23.2 Production area

The hilly area for the production of Sagrantino di Montefalco comprises, besides the municipality of Montefalco, parts of Bevagna, Castel Ritaldi, Giano dell'Umbria and Gualdo Cattaneo, in the Umbria region. Soils are of marine and lake origin, clayey with a high sand content. Climate is subcontinental with cold winters and very warm summers.

15.23.3 Grape variety

Sagrantino: it is probably related to the variety called Itriola, whose Greek origin recalls sugar. It is quoted by Pliny the Eldest in Book XIV of his *Naturalis Historia*. According to other sources, the variety was introduced during the Early Middle Ages by the Byzantine monks coming from Greece because they were attracted by the fame of Saint Francis. The most suitable training systems are those with medium expansion and medium long pruning. It is trained with various systems like Palmetta Modificata, double spurred cordon, high spurred cordon and classic spurred cordon.

15.23.4 Dehydration and vinification

Sagrantino grapes for making Passito are picked when they are over-ripe, towards the beginning of October, with a sugar concentration of 25–30%. Grapes are dried on racks or piled in wooden boxes for a period of almost two months in closed, well-ventilated, facilities.

Towards December–January, the grapes are crushed and destemmed; maceration lasts for about 15 days and it is interrupted when the wine reaches an alcohol volume of 14–16 % and a minimum of 30 g L^{-1} residual sugar (which generally reaches 180 g L^{-1}). It is fined in stainless steel for 12 months and each producer decides about oak maturation. Ageing must be of at least 30 months.

15.23.5 Organoleptic features

COLOUR: intense garnet

NOSE: red and black fruits, raisins, wild blackberry and redcurrant, floral notes and a spiced background of vanilla, cocoa and sweet tobacco

PALATE: medium sweet and full bodied, slightly bitter finish

15.23.6 Food matching

Oven-baked sweet biscuits, very mature spicy cheese.

15.24 SCIACCHETRÀ CINQUE TERRE DOC

15.24.1 History

The first source on viticulture in the Cinque Terre area is Pliny, who praised the quality of these wines. Sciacchetrà, however, has lesser known origins; in the local vernacular language '*sciacca*' means to squeeze, and '*trà*' means to save up, to age, so the name would suggest a wine that can be preserved for a long time.

15.24.2 Production area

The denomination 'Cinque Terre' includes the oriental coast of Liguria, about 15 km long. It goes from the municipality of Levanto to La Spezia. Soils are rich in shells and are subacidic, with clayey schist and sand. Vines are grown on small terraces on the sea with an east to west exposition, up to 600 m altitude.

15.24.3 Grape variety

Bosco (minimum 40%), Vermentino (maximum 40%), Albarola or Bianchetta Genovese (maximum 40%). Vines are low pergola trained (to cope with the strong, salty sea winds), with a very thick canopy, which does not allow any mechanization.

15.24.4 Dehydration and vinification

At harvesting, only the best grapes with a sugar content of 18–22% are chosen. They are left on the vine for 15–20 days after natural ripening and they are dried on suitable wooden frames (*teà*) made of canes and wire, or on modern plastic webs or even in wooden or plastic boxes. Grapes are left on the teà in aerated warehouses for 40–60 days, then they are destemmed, crushed and macerated in 2–3 hL barrels for several days. The marc is pressed and the resulting liquid is added to the free-run juice; the wine is then siphoned with the aim of fining the mass and stopping fermentation. Wine is fined for a minimum of one year; it is bottled and traded only after 1st November of the year after the vintage.

15.24.5 Organoleptic features

COLOUR: antique gold with definite amber highlights
NOSE: very intense and persistent, very complex with elegant notes of apricot, acacia honey and dried fruit
PALATE: sweet, intense, but never too sweet: a slight salinity makes it harmonious and balanced

15.24.6 Food matching

Spicy blue cheese, creamy cakes, canestrelli biscuits, pandolce Genovese (a sweet bread).

15.25 SFORZATO DELLA VALTELLINA DOCG

15.25.1 History

The name Sforzato appears for the first time towards the end of the eighteenth century, even if we know that in the past there was a wine called 'aromatico', obtained by local dried grapes, which can be considered the ancestor of Sforzato. Sforzato is produced with dried black grapes pressed at the beginning of winter. The must rests for a couple of days, and then is placed in the highest room of the house to ferment. As it ages, it gains its distinctive aroma. The name 'sforzato' comes from the verb to force, because wines are not dried on the plant but in wooden boxes (*mantovale*) placed in warehouses or farmhouses.

15.25.2 Production area

The production area of Sforzato is Valtellina, in the province of Sondrio, to the very north of the Lombardy region. Valtellina is a long and wide valley stretching from east to west in the centre of the Alps. It borders with the Orobic Alps to the south and with the Retic Alps to the north-east. The vineyards face south and are at an altitude of between 300 and 700 m above sea level. Soils are of morenic origin deriving from schist and gneiss, subacidic, rich in calcium and mineral elements.

15.25.3 Grape variety

Nebbiolo, which in Valtellina is called *Chiavennasca*, from 'ciù vinasca', which means 'suitable for transforming wine'. The name might derive from the abundant presence of bloom on the berry of the ripe grapes, to the extent that they look as if they were covered in 'fog', or from the late maturation of wine, which occurs during the autumn fogs.

Vineyards are located on small artificial terraces and are trained with a sort of modified Guyot, called 'archetto' (little arch) in Valtellinese. Plant density is of about 5880 plants per hectare and rows run from north to south following the steepest aspect, so that the plant benefits from the highest light exposure.

15.25.4 Dehydration and vinification

Generally grapes are harvested during the second decade of October selecting the loose and healthy bunches, which are then placed in wooden boxes and stored in dry and well-aerated rooms. Because of the vineyards' features, harvesting is long and very hard in Valtellina. The grapes that are selected for dehydration must be harvested with a minimum potential alcohol of 11%. Grapes are subsequently stored for dehydration in ventilated warehouses with an east–west opening called 'fruttai'.

The dehydration process lasts 110 days and, during this period, grapes are controlled periodically in order to eliminate any rotten bunches. At the end of January, grapes will have lost 40% of their weight and will have concentrated their sugars reaching a level of 26–27%. Crushing cannot take place before 10th December of the vintage year, and grapes are crushed and destemmed when they are really cold (0–3°C) in order to obtain a soft pressing. Fermentation with maceration lasts between 10 and 30 days, with a yield of 40–50%.

The ageing and maturation period starts from 1st April of the year after the vintage; it takes place in barriques or in 30 hL barrels and lasts for 15–18 months. Before being traded, the wine usually matures in bottle for one year. According to the law, Sforzato must have matured for at least 20 months, at least 12 of which should be in wooden barrels.

15.25.5 Organoleptic features

COLOUR: ruby to garnet
NOSE: intense with ripe fruit, very complex with notes of spices, vanilla, cloves, dried fruits and flowers; in the most aged wines there are also hints of carob and liquorice
PALATE: very soft, dry, with structure and character, might have some wood influence

15.25.6 Food matching

Sforzato should be served with main meat dishes like stews, game and braised meat; and also with local cheeses, such as Bitto and mature Casera.

15.26 TORCHIATO DI FREGONA COLLI DI CONEGLIANO DOC

15.26.1 History

Torchiato di Fregona is an ancient wine probably born by chance in 1600, thanks to a farmer who, in order to enhance the ripening of the grapes during a poor vintage, hung them in his granary. He obtained a very sweet must, and fermented it in small barrels. After some time, this practice spread in the territory situated on the mountainside of Cansiglio, in the municipalities of Fregona, Cappella Maggiore and Montaner, to produce a style of Vinsanto.

15.26.2 Production area

The area of cultivation of Torchiato di Fregona is situated in the DOC 'Colli di Conegliano', between the Prealps and the Venetian plane, in the province of Treviso. The specific area of Torchiato di Fregona is in the morenic zone of Vittorio Veneto, San Giacomo and Colle Umberto, with red deep soils (rich in iron), alternating with more superficial, steeper soils with more erosion.

15.26.3 Grape variety(ies)

Torchiato di Fregona is obtained from Prosecco (30% minimum), Verdiso (30% minimum), Boschera (25%) and other white grapes authorized within the province of Treviso. It was once trained on trees or in the so-called *scarasse*, in promiscuous agricultural forms. Nowadays, it is trellis trained, vertical shoot positioned.

15.26.4 Drying and vinification

Grapes are usually picked with a sugar level of about 15–17 g L^{-1}; they can either be hung or placed on drying racks in the sunshine. The length of the drying process varies according to the level of concentration that the producer wants to reach, but it generally lasts until the spring of the year after the vintage. During the week before Easter, the grapes are crushed in a cask and squeezed with the 'becanéa', a short wooden cylinder with two handles. The must is subsequently pressed for several times, until the marc is dry. Must is fermented in 1–2 hL small oak or chestnut barrels, which are not full in order to deliberately oxidize the wine. Fermentation goes on until August; the wine is then siphoned into other barrels until March–April. Fining occurs during this period. In the barrels, wine can age up to 2 or 3 years. According to the regulations, Torchiato di Fregona cannot be consumed before 1st December of the year after the harvest, with a maturation period of at least 3 months in bottle.

15.26.5 Organoleptic features

COLOUR: deep gold
NOSE: intense, typical, honey and linden
PALATE: from dry to sweet, round, full, persistent and delicately maderized

15.26.6 Food matching

Ideal with desserts such as pastries, dry homemade biscuits, chestnut desserts with honey, raisins or hazelnut. It is also suitable with mature blue cheese.

15.27 TORCOLATO DI BREGANZE DOC

15.27.1 History

The history of Torcolato is linked to the history of viticulture in the area of Vicenza dating back to the Roman age, as it is witnessed by the remains of a Roman villa in Montebello Vicentino, where the remnants of prehistoric pips were found. In 1610, Antera Scoto writes about Breganze describing the 'sweet and flavourful wines'. More recent news is found in 1754, in the wine guide of the province of Vicenza entitled *Il Roccolo Ditirambo*, by Aureliano Acanti, about many grape varieties such as Vespaiolo Breganzino and a 'delicious vinsanto' that 'is spontaneously called Torcolato di Breganze'.

15.27.2 Production area

The wine is obtained in the DOC 'Breganze' in the province of Vicenza. The vines are grown on the hills; soils are morenic, volcanic and from tophus. On the plains, there are areas with alluvial white pebbles, which are a great substrate for vines.

15.27.3 Grape variety

Vespaiola or Bresparola: the variety is cultivated mainly in the area of Vicenza, the skin is thick and it is quite acid. The origins are still unknown.

15.27.4 Dehydration and vinification

Grapes are picked early, with a sugar content of 16–20%. Bunches are placed in low boxes and subsequently selected and hung on the ceilings with strings. This is where the name originates: in order to be hung up, bunches are twisted on a couple of strings. The verb that indicated this operation was *torcolare*, hence *torcolato*. It is important that the room is well ventilated; the grapes stay there for 5 months. This is when *Botrytis* develops, concentrating sugars and forming the typical aromas. Usually, the grapes dry until January.

Nowadays, only a small part of the grapes is hung; most are left to dry in suitable warehouses on dehydration racks. Modern producers make the wine searching for a maderized style: grapes are crushed with horizontal presses and SO_2 is immediately added for antioxidasic purposes in 30–35 hL tanks where a white vinification process is performed. Fermentation is quite rapid, and it stops when the wine has an alcohol content of 15% and a residual sugar of 7–10%. Torcolato is aged from a minimum of one and a half years up to four years in barrels or in barriques before bottling. Its ageing potential is great: it can age for over 20 years.

15.27.5 Organoleptic features

COLOUR: gold to deep amber
NOSE: intense, typical, honey and raisins
PALATE: medium sweet to sweet, balanced, velvety, definite, with or without a balsamic wooden influence

15.27.6 Food matching

Dry cakes but also foie gras appetizers, mature blue cheese.

15.28 VERDUZZO PASSITO COLLI ORIENTALI DEL FRIULI DOC

15.28.1 History

Verduzzo Friulano is an ancient variety from Friuli and it is named after the light green colour of the ripe grapes. It is described by Acerbi at the beginning of the nineteenth century with the name Verduz. We have very few historical sources about the wine, as it has always been produced for family consumption.

15.28.2 Production area

The area of production of Verduzzo is situated in the Friuli–Venezia Giulia region, within the DOC 'Colli Orientali del Friuli', which stretches along the oriental hillside of the province of Udine. Soils are characterized by marl (limestone marl) and calcified sand. Together with the aspect of the vineyards, these are the factors that make this area a great *terroir* for quality viticulture.

15.28.3 Grape variety

Verduzzo Friulano: there are several biotypes of this variety which can be distinguished by the colour of the berries (yellow biotype, suitable for over-ripening; green biotype, suitable for still wines). Within the yellow biotype there are three types. One with a small, compact bunch, one with a semi-loose bunch called *ras'ciè* and the ordinary one called *vert*. The typical training system is the 'doppio capovolto' or 'alla cappuccina', with a medium dense canopy and a medium long pruning. It performs better on well-exposed hills, unfertile soils and dry climate.

15.28.4 Dehydration and vinification

Grapes are dried partly on the plant and/or in suitable warehouses either air conditioned or with fans. Harvesting takes place between 15th and 30th October, according to the course of the growing season, with a natural sugar concentration of 25–30° Babo.

Grapes are placed on drying racks for 8–10 days; they are carefully destemmed in order to avoid stems absorbing any alcohol and giving the wine any off-flavours, and then crushed. Maceration is quite long as fermentation is slow, due both to the temperatures of the winter period and also to the features of the indigenous yeasts.

The regulations allow the use of wooden barrels, even new, both for vinification and maturation.

15.28.5 Organoleptic features

COLOUR: deep gold, clear and bright

NOSE: intense, with delicate honey and vanilla notes from the barrique, notes recalling apricots and/or acacia flowers

PALATE: sweet, full bodied and robust with a slight tannic feeling, typical of the grape variety

15.28.6 Food matching

Dry cakes, pandoro (sweet yeast bread), panettone, custard cakes with fruit, blue cheese.

15.29 VERNACCIA DI ORISTANO DOC

15.29.1 History

The first reports of Vernaccia of Sardinia go back to 1327. A legend says that the variety originated from the tears of Saint Giustina, the protector of Oristano. At that time, the island was afflicted by malaria, and there were many victims. The saint came down from heaven to the earth, and in front of such a miserable situation started to cry and pray to God for the unlucky people. Her tears fell on the ground, and vines grew quickly out of them, with many leaves and fruits. The people feared that the vines would disappear just as they had appeared out of the blue, and therefore they immediately picked the grapes and pressed the juice in big jars. As the sick people, burning with fever, drank the strange liquid, they immediately recovered. Since then, the Sardinian people drink Vernaccia to fight their health problems.

The origins of this variety are common to other Italian and Mediterranean Vernaccie: they are medieval wines par excellence, whose fame was linked to the commercial power of the maritime republic of Genoa.

15.29.2 Production area

The area of production of the DOC ‘Vernaccia di Oristano’ is situated in Sardinia, in the province of Oristano, to the north of the Campidano, and more precisely, at the mouth of the river Tirso to the south, up to Villanova Truschedu to the north, from Palmas Arborea to the east down to Capo Mannu to the west. Soils are of lake and alluvial origin, medium thick, quite coarse, deep, while the climate is Mediterranean, warm, sometimes humid.

15.29.3 Grape variety

Vernaccia di Oristano is also called Garnazza or Grenaccia (the same semantic root as Grenache). It is the same grape variety as cultivated in the Calabria region in Bianco and in Salina, in the Aeolian Islands, where it is called Malmsey. Traditionally, it is bush trained.

15.29.4 Drying and vinification

Harvest takes place a fortnight after the natural ripening time, and grapes have a sugar content ranging between 23 and 26%, guaranteeing a minimum natural alcohol content of 14%. After crushing and destemming, must is fermented at high temperatures until all sugars have disappeared. Wine has an alcohol content of about 15%, with traces of residual sugars. Before barrel maturation, the wine is siphoned during winter in order to allow the must to complete fermentation.

In February–March of the year after the vintage, Vernaccia is matured in chestnut or oak barrels of 7–18 hL, about 90% full. The presence of air in the barrels, after 3–4 months, determines the presence of a veil constituted by particular oxidative yeast strains called ‘flor’, which enhance the ageing process. Wine must remain in these barrels for at least 2 years before being bottled.

A Vernaccia Riserva is also produced, and if the alcohol content is greater than 15.5%, it is even possible to produce a sweet fortified or dry fortified version, with two years of ageing and an alcohol content of 16.5% for the sweet wine, and 18% for the dry wine.

15.29.5 Organoleptic features

COLOUR: gold or amber
NOSE: delicate, alcoholic, almond tree flower, slightly maderized
PALATE: fine, subtle, warm, very dry, slight aftertaste of bitter almonds

15.29.6 Food matching

It is very good with fish dishes like fish roe, mature or smoked cheese. Excellent as a dessert wine, particularly with Sardinian traditional almond cakes.

15.30 VERNACCIA DI SERRAPETRONA DOC

15.30.1 History

Vernaccia di Serrapetrona, or Black Vernaccia, has been cultivated since ancient times in the area near Macerata. More reliable reports of its origin are dated around 1800.

15.30.2 Production area

The area of production of Vernaccia di Serrapetrona is situated in the Marche region, in the province of Macerata. The production area comprises the whole territory of the municipality of Serrapetrona and parts of a few nearby municipalities.

15.30.3 Grape variety

Black Vernaccia for at least 85%. Genetically, the wine is very similar to Cannonau. The old vines are trained with a wide system like 'doppio capovolto' while the most recent ones are narrow. The climate is warm and temperate with a marine influence, the soils on the hills are of sedimentary, marl-sandy origin, quite fertile.

15.30.4 Dehydration and vinification

Vernaccia di Serrapetrona is obtained through three different fermentations: one with fresh grapes, one with dried grapes and a final fermentation in tank. The result is a sparkling wine from dried grapes.

The harvest of Black Vernaccia takes place towards the end of October to early November. A part of the harvested grapes, precisely no more than 60%, is normally macerated, while 40 % is naturally dried. After maceration, 60% of the grapes are crushed and destemmed and the must is left on the skins. Generally, the wine has an alcohol volume of 10–11°; it is rather acid and tannic: this is called base wine. The other grapes are dried in suitable warehouses.

The dehydration process begins towards the end of October and continues until December–January, when the grapes are crushed and destemmed. The must is added to the base wine produced in October; after two months the wine will have completed its second fermentation and will start to mature.

The wine is then inoculated with yeast and sugar and tank fermented for two months, followed by two more months of maturation and stabilization. According to the law, the wine cannot be consumed before 30th July of the year after the vintage. Bottle maturation varies from six months to a year.

15.30.5 Organoleptic features

MOUSSE: persistent, fine
COLOUR: intense ruby to garnet
NOSE: typical, must, slightly aromatic, yeast
PALATE: dry to sweet, a pleasant bitter almond finish

15.30.6 Food matching

Dry cakes, pastries or any dessert with fruits of the forest sauce, semi-hard cheese.

15.31 VINSANTO DI VIGOLENO DOC

15.31.1 History

According to historical sources, the name Vigoleno comes from the Latin '*Vico Lieo*', the village of Bacco, bearing witness to the vinification tradition of the area. In 1539, in the inventory of Vigoleno Castle, the consumption of local wine is mentioned. The certain birth date of Vinsanto can be dated back to the first years of the nineteenth century.

15.31.2 Production area

The production area of Vinsanto di Vigoleno comprises a small portion of the wide DOC 'Colli Piacentini', in the province of Piacenza. The area is situated on the high plains of Piacenza, where soils are alluvial conoids of the rivers descending from the Apennines: vast alluvial deposits of pebbles or sand and pebbles. In the area of cultivation, soils are particularly fertile and fresh. Vineyards are well exposed on the southern mountainside of Stirone, and to the west on the mountainside of the Ongina. Soils derive from pliocenic scaly clays and shale with a clayey-limestone origin, often pebbly and dry.

15.31.3 Grape variety/ies

At least 60% Marsanne and/or Beverdino and/or Sauvignon and/or Trebbiano Romagnolo and/or Ortrugo. Also the native Santa Maria and Melara are used. The most popular training systems are classic Guyot and a so-called Guyot Piacentino, which has two fruiting canes.

15.31.4 Dehydration and vinification

Harvest varies according to the maturation of the different varieties, but it generally takes place between September and October. Grapes are placed in aerated attics on drying racks or piled in metal or wooden boxes. Some producers hang them up on the ceiling on wooden poles, in order to avoid contact between bunches. The length of the dehydration process varies according to the vintage and to how healthy the grapes are. Other factors are the temperature and the aeration of the drying rooms, but generally it lasts for three months.

Must is pressed for 2–3 days and then placed in oak or cherry barrels. Fermentation lasts for 20–30 days; the wine is then racked and placed in small 10–20-year-old oak barrels, or 'caratelli', where fermentation takes place very slowly due to the cold climate and the scarce yeast content. In spring and summer, fermentation becomes more vigorous.

The maturation of Vinsanto, which often coincides with the fermentation phase, lasts for five or more years, and each year the wine is siphoned from a bigger to a smaller barrel, because the volume of the wine shrinks over time.

15.31.5 Organoleptic features

COLOUR: medium to deep gold or amber
NOSE: intense, aromatic, typically of dried fruit and white flowers
PALATE: sweet, balanced, full bodied, velvety, sometimes aromatic

15.31.6 Food matching

Dry cakes, jam tarts.

15.32 VINSANTO TOSCANO DOC

15.32.1 History

The first evidence of Vinsanto dates back to the medieval period: a legend from Siena tells the story of a Father who during pestilence distributed a wine that cured the sick people. In Florence, people claim the origin of this wine: until 1349 it was called 'pure wine' (not diluted with water). During the Ecumenical Council that took place in Florence under Eugene IV, the Greek Hierarch Bessarion drank the 'pure wine' in front of John VII and cried out that it was a Xantos wine, probably referring to the Greek sweet wine from Thrace. The people at the table thought that the Father said the wine had 'saint' virtues, and therefore they called the wine Vinsanto. Another possible origin of the name is due to the liturgical use of this wine: in the Umbria region it is called 'wine of the priests'. Around the start of the fifteenth century, such sweet wines began to be called 'vini natalini' because they were produced around Christmas time, while others called the wine 'Vino Forzato' (forced wine) because it was fermented in closed vessels for a long time. Other possibilities are that the wine was produced towards the beginning of November, during the feast of All Saints, or because it was bottled during the Holy Week.

15.32.2 Production area

Vinsanto Toscano is produced in Tuscany in three distinct denominations: Vinsanto del Chianti, del Chianti Classico and di Montepulciano. Soils are constituted by clayey schist, galestro (loose marl and limestone) and alberese (calcareous stony); alternated with several pliocenic sandy plaques and very small clay zones.

15.32.3 Grape variety

Tuscan Trebbiano and Malvasia lunga del Chianti, on their own or together, minimum 70% for Vinsanto del Chianti, Chianti Classico, and also Grechetto for Vinsanto di Montepulciano. Sangiovese, the red variety, can be used for a minimum of 50–70% for Vinsanto Occhio di Pernice. These varieties prefer a long pruning, but they are also suitable for spur pruning.

15.32.4 Dehydration and vinification

Grapes are imperatively harvested by hand before the natural ripening time in order to have healthier bunches. If maturation is completed, the best and loosest bunches are selected. The traditional dehydration method in Tuscany is to put the bunches in rooms called 'appassitoi', frequently refreshing the air and placing the selected bunches on cane racks or on dry chestnut racks or even in small wooden (or modern plastic) boxes in several layers. In former times, grapes were placed on cane racks called 'canicci'; on warm sunny days they were taken out into the courtyards, but at night they were put back inside. Canes had the advantage of leaving small spaces between them, which allowed air to circulate between bunches in order to keep them healthy. The racks had some wooden staves below, which made it possible to pile them up.

Another option was to hang bunches up on metal frames or on the ceilings using 'S'-shaped hooks. This system, called 'penzane', allowed a perfect aeration but was extremely expensive. Nowadays, grapes are placed in aerated rooms on special racks made of wood and wire. They are placed horizontally or vertically, and fans are permitted to a certain extent. The drying process can last for up to six months, depending on the producer.

At the end, sugar levels must not be less that 26–27%; grapes are selected, hand destemmed and crumbled in vats. Some producers have mechanized this operation. After two months, must is transferred into small chestnut or other wood 50-litre vessels called caratelli. They are 90% filled with two litres of 'mother' and 43% of must. The 'mother' is a dark and dense sediment found at the bottom of the caratelli; it is very different from Vinsanto and it contains specialized yeast strains that multiply and live well in very sugary environments.

The fermentation in caratelli generally starts towards the end of February and goes on for several years, given the temperatures between 10 and 12°C in the winter and 20°C in the summer, together with the high sugar content. The caratelli, placed in the 'vinsantaia', a dry and well-aerated room, are closed and sealed after being filled, and left for a period ranging from 3 to 10 years depending on the producer. According to the law, it cannot be traded before 1st November of the third year after the vintage (four years for the Riserva) and eight years for the Vinsanto di Montepulciano Occhio di Pernice.

After the long maturation, the wine is racked in the open air and siphoned three or four times. The sediment of the racking is never eliminated: it provides the 'mother' for the following vintage. At the end of the ageing period, the product must have a minimum total alcohol of 15.5–16% and a residual sugar of 2–4%.

15.32.5 Organoleptic features

COLOUR: from lemon to gold or deep amber
NOSE: ethereal, intense, typically of dried fruit and caramel
PALATE: balanced, velvety, dry, rounder for the medium dry, medium sweet and sweet styles

15.32.6 Food matching

On its own, with a dessert; it is ideal with the traditional 'cantuccini' biscuits.

15.33 VINO SANTO TRENTINO DOC

15.33.1 History

The origin of Vino Santo dates back to 1800 thanks to a producer (G. Angelini-Gianotti from Dro) who made this wine for commercial reasons; he was followed in 1822 by the Counts

Wolkenstein from Castel Toblino. Before this, the wine had been produced by many farmers in the Valle dei Laghi area for family consumption. During the Asburgic period, Vino Santo was sold and consumed mainly on the German and Austrian markets. At the end of the First World War, after the fall of Austria, such markets were no longer available, and Vino Santo started to struggle. It was rediscovered around the end of the 1980s and production was increased. Nowadays, it is a typical product of the Trent area.

15.33.2 Production area

Vino Santo Trentino DOC is produced in the province of Trent, in the Valle dei Laghi. Soils are made of marl and limestone and vineyards thrive on well-exposed hills; in these conditions, the Nosiola grape variety can express its finest features. The best vineyards are on the mountainside of the Calavino Mountains but also in Padergnone, Santa Massenza, Lasino, Sarche, Dro, up to Arco and Riva del Garda.

15.33.3 Grape variety

Nosiola: the genetic origin of this variety is similar to that of Groppello, in the Val di Non, as they both have a common parent in the Vallese grape variety Rezè. Nosiola, or Nosiola gentile, was cited for the first time by Acerbi in 1825, among the grape varieties cultivated within the area of Trent. The name comes from the colour of the shoots and from the taste of the wine, which recalls the hazelnut ('nocciola'). Trent pergola is the most popular training system; production is abundant but not consistent.

15.33.4 Dehydration and vinification

Grapes are harvested rather early with a sugar concentration of 18–20%. The loosest bunches are selected from the vines thriving on the least fertile soils. Grapes are packed in wooden boxes, taken to warehouses and placed on drying racks made of swamp canes (called *soleri*), on webs called *arele* or hung up on wires. The warehouse must be well aerated but not excessively, otherwise the grapes will dry too quickly; they must last for 5–6 months, until the Holy Week, after which the wine is named. Grapes reach a sugar concentration of 40%, and *Botrytis* will develop too, conferring further aromatic complexity to the wine.

The must is transferred to 2–4 hL oak barrels, and fermentation lasts for 5–6 months. During the first year, fermentation is never totally completed: it only reaches about 7–8%, and it goes on for the following 2 or 3 summers. After racking, the wine stays in these barrels for up to 3–4 years and it is siphoned twice a year. Fermentation is very slow and it stops when the actual alcohol level is around 12–13% and the residual sugar level is 8–13%. Usually the wine is bottled 4–6 years after the harvest.

15.33.5 Organoleptic features

COLOUR: medium to deep gold or amber, very clear
NOSE: pleasant, fine, delicate, yeast and raisin notes

PALATE: sweet, very well-balanced alcohol and acidity, raisins, a slight maderized note, very delicate

15.33.6 Food matching

Salty and intense cheese like blue cheese, jam and fruit tarts.

16 Pedro Ximénez and Malaga

Juan J. Moreno-Vigara[1] and Juan C. García-Mauricio[2]

[1] *Department of Agricultural Chemistry,* [2] *Department of Microbiology, University of Córdoba, Córdoba, Spain*

16.1 INTRODUCTION

Three winemaking zones under the protection of their respective Protected Designation of Origin (DOP) in Andalusia, southern Spain, produce high-quality dessert wines from grapes that have been previously dehydrated partially by direct exposure to sunlight (off-vine drying). These products fall into the category of 'special wines' and can be of three different types, depending on whether the proportion by volume of ethanol they contain, which comes from the partial alcoholic fermentation of their respective musts, is at least 4% (sweet liquor wines), 7% (natural sweet wines) or 13% (naturally sweet wines).

Among the three types, Pedro Ximénez and Malaga wines are especially popular. The former are monovarietal wines obtained from partially dehydrated grapes of the Pedro Ximénez variety, which is preferentially grown in the Montilla–Moriles region (Cordoba, S. Spain) and to a minor extent in Jerez de la Frontera (Cádiz, S. Spain). Malaga wines are renowned dessert wines consisting of a mixture of wines made from Muscat of Alexandria and Pedro Ximénez grapes, and fortified with wine alcohol and syrups obtained by heat reduction of the must. The regulations of the Regulatory Councils of the Montilla–Moriles, Jerez de la Frontera and Malaga DOPs were published in the 'Boletín Oficial de la Junta de Andalucía' in 2011, and they include a description of each type of wine, the specific authorized products and their production rules.

16.2 DESCRIPTION OF THE ZONE AND THE GRAPE VARIETIES CULTIVATED

The production zones of these wines lie in a triangle with the towns of Montilla (Cordoba), Jerez de la Frontera (Cadiz) and Malaga as vertices. These Andalusian zones fall in Region V of Winkler's classification and have a total sum of temperatures exceeding 10°C that exceeds 2597°C during the vegetative cycle of the vine. Temperatures in these zones rise above 40°C during the grape ripening period. These climatic features facilitate an early, rapid and thorough ripening of grapes resulting in the attainment of excellent dessert wines.

Vineyards in these zones are grown largely on calcareous soils (so-called 'albarizas' because of their white colour). These soils consist mostly of calcareous marls containing a

Sweet, Reinforced and Fortified Wines: Grape Biochemistry, Technology and Vinification, First Edition.
Edited by Fabio Mencarelli and Pietro Tonutti.

high proportion of calcium carbonate and variable amounts of clay. A few areas have red soils characterized by an anthropic horizon with a high proportion of sand and a deeper horizon with calcareous marls. These types of soil dictate the choice of rootstock to be used in order to facilitate growth of the aerial part of the vine. According to Valencia (1990) and López-Alejandre (2000), the rootstocks most commonly used in these zones are 41B Millardet, 161-49 Couder and 110 Richter.

The climate type is semi-arid, with the exception of a few dry subhumid areas, and a thermal efficiency of mesothermic type. The average annual temperatures range from 13.2 to 19.5°C, and the annual water precipitation from 415 to 813 mm.

The specific grapevine varieties most widely used to obtain sweet wines in the Montilla–Moriles, Jerez and Malaga zones are Pedro Ximénez and Muscat of Alexandria. Both are typically Mediterranean white grape varieties and can be grown at a maximum density of 2500–3000 plants per hectare. Traditionally, the most common vine-training system used in these grape-growing zones consists of a short blind pruning and shoots left with two exposed buds, but there is an increasing trend toward using the espalier system.

A comprehensive ampelographic description of the two grape varieties can be found on the web page of Rancho de la Merced (2011). The page highlights the fast ripening of both varieties and the high sensitivity of Pedro Ximénez to ambient humidity, which is leading to a decline in the planting of this variety in coastal areas. Their most salient enological features are a low titratable acidity and high sugar content in their musts, particularly in those from Pedro Ximénez grapes.

16.3 THE GRAPE-DRYING PROCESS. PHYSICAL AND CHEMICAL CHANGES IN THE BERRY

Off-vine grape drying is a postharvest treatment traditionally used in the Montilla, Jerez and Malaga grape-growing and winemaking zones. This treatment is carried out by direct exposure to sunlight and is known in Spanish as 'soleo' (sunning). The Pedro Ximénez variety is used in the three zones, but especially in Montilla–Moriles, whereas the Muscat of Alexandria variety is used preferentially in Malaga. The grape ripening and grape drying rates and as a consequence, the quality of the resulting must, are directly related to the pedoclimatic conditions in each location and the meteorological accidents that occur during these two processes. Thus, in some areas located in Montemayor (Cordoba), the Pedro Ximénez grapes ripen fully during the first half of August, which facilitates their off-vine drying under the typically high temperatures and sunlight hours of this month. Based on weather data for the area, the average temperature during the drying period in a typical year is 28.1 $\pm$ 2.2°C and ranges from 14 to 45°C, the average ambient relative humidity (RH) is 31.5 $\pm$ 9.2% (range, 8.7–82.7%) and the average solar radiation 25.4 MJ m^{-2} day^{-1} (range, 16.1–28.1 MJ m^{-2} day^{-1}) (Junta de Andalucía, 2011).

The soleo process begins with the harvesting of healthy and ripe grape bunches which are carefully placed in plastic boxes for transport to the 'pasera' (drying site). The pasera is usually established on terrains characterized by gentle slopes and sandy soils that are chosen so as to maximize the time they receive direct sunlight each day (see Figure 16.1). Formerly, grape bunches were placed on esparto mats; nowadays, they are spread over pieces of food-grade plastic mesh 10 m long $\times$ 0.75 m wide, which are laid on the ground. The drying process spans 3–5 days at the beginning of the harvest season in August and lasts about 7–10 days during the last days of September, if the meteorological conditions are

Figure 16.1 Grapes subjected to an off-vine drying process (soleo) in the so-called 'pasera' site. (a) Pasera located in Montemayor (Montilla–Moriles zone). (b) Pasera in the Axarquia (Malaga zone). (For a colour version of this figure, please see the colour plate section.)

favourable (dry and sunny). However, complete off-vine drying can easily take 10–20 days if the paseras are located in areas with a high RH and exposed to nocturnal dew. Ensuring that all grapes will be dehydrated to the same extent entails periodic turning over of bunches. Also, covering them at night helps to protect the grapes from excessive ambient humidity (especially in the coastal areas of Jerez and Malaga). Once the grapes show the appropriate degree of dehydration (viz. when the reducing sugar content of the must exceeds 350 $g\,L^{-1}$), the bunches are collected and stored in plastic food containers and transferred to the cellar.

The off-vine drying process dehydrates Muscat of Alexandria grapes in Malaga to a lesser extent than the Pedro Ximénez grapes in Montilla. This is largely the result of the thicker skin and greater berry size of Muscat grapes, in addition to the higher ambient humidity in the Malaga zone. In this respect, the main shortcoming of the soleo process is its strong dependence on the ambient conditions. Thus, too high a humidity, unexpected storms or a succession of rainy days can ruin a drying season, mainly because these conditions facilitate the development of fungi of the genus *Aspergillus* section *Nigri*. These fungi prevail during ripening, harvest and off-vine drying in a hot, wet climate (Valero *et al.*, 2005). Must from grapes attacked by *A. nigri* contains Ochratoxin A (OTA) at levels exceeding the $2\cdot10^{-6}\,g\,L^{-1}$ limit imposed on all types of wines; fortunately, however, off-vine drying under optimal conditions leads to OTA levels below $0.1\cdot10^{-6}\,g\,L^{-1}$ (Serratosa *et al.*, 2008).

Physical changes. Direct exposure to sunlight results in water loss from all parts of the grape bunch. Thus, it causes grape skin to gradually shrink, darken and become less elastic and more fragile. It also shrinks and darkens the berry pulp and dries the stalk. Dehydrated grapes contain little juice and are highly resistant to pressure action; as a result, the complete extraction of the must from these grapes requires the use of hydraulic presses. The colour of must changes from amber to reddish brown to jet black; its density, viscosity and sweetness increase, and its acidity decreases as the off-vine drying process is prolonged. Although its odour can hardly be perceived directly by the nose, the must leaves retronasal notes of cake, caramel and dry hay.

Chemical changes. In general, the concentration of the compounds dissolved in the must obtained from dehydrated grapes increases with respect to the must from ripe grapes. This is a consequence of the loss of water during off-vine grape drying. According to Franco *et al.* (2004), some compounds increase in concentration to the same extent as water decreases in the grape, whereas others increase or decrease as a function of metabolic and enzymatic processes that occur in the grapes and during pressing, when the grape has lost its cellular integrity and the enzymes have lost contact with their substrates.

The components of must from off-vine dehydrated grapes change to a variable extent in quantitative and qualitative enological terms. Thus, off-vine drying increases the concentrations of some *major components* in grapes (viz. sugars and acids), albeit to a different extent since some acids can even decrease in concentration (Ribéreau-Gayon *et al.*, 2000). These variations can be ascribed to conversion of the acids under the action of enzymes, which in turn depends on temperature and the physiological status of the grapes. López de Lerma *et al.* (2012) found that must from Pedro Ximénez grapes dried for 9 days had 43.3° Brix (i.e. twice the value of must from the initial ripe and undehydrated grapes), and also that the average dehydration rate decreased with increasing drying time, from 5.6% during the first two days to 3.5% over the next four, and 1.5% for the last three days (Table 16.1). Also, according to Franco *et al.* (2004), titratable acidity and pH rise with increasing off-vine drying time. However, acidity increases to a lesser extent than the sugars do (Flanzy, 2000; Bellincontro *et al.*, 2004), as a result of the loss of water and the metabolism of malic acid, which is present at levels below 1 $g\,L^{-1}$ in must from Pedro Ximénez ripe grapes (López *et al.*, 1988).

Table 16.1 Changes during the off-vine drying process (*soleo*) of Pedro Ximénez grapes in the Montilla–Moriles zone. The concentration of the volatile aroma compounds is expressed as μg L^{-1}. (Data from López de Lerma *et al.*, 2012, with permission from the authors.)

Fractions or compounds	Days of off-vine grape drying (soleo)				
	0	2	4	6	9
° Brix	22.2 ± 0.4a	30.8 ± 0.3b	36.2 ± 0.3c	40.6 ± 0.2d	43.3 ± 0.1e
% dehydration		11.2	18.8	25.7	30.3
Sugars (g L^{-1})	217 ± 6a	318 ± 5b	385 ± 5c	442 ± 4d	478 ± 3e
pH	3.85 ± 0.02a	3.97 ± 0.02b	4.19 ± 0.02c	4.30 ± 0.02d	4.44 ± 0.02e
Titratable acidity (g tartaric acid L^{-1})	3.56 ± 0.03a	3.81 ± 0.03b	4.40 ± 0.03c	4.14 ± 0.04d	4.93 ± 0.04e
∑ ***Higher alcohols***	***381 ± 19a***	***2238 ± 114b***	***8385 ± 429c***	***6197 ± 317d***	***6748 ± 345d***
Isobutanol	34 ± 2a	661 ± 34b	2049 ± 105c	1575 ± 80d	1650 ± 84d
2-Butanol	9.5 ± 0.5a	8.7 ± 0.4a	12.4 ± 0.6b	15.6 ± 0.8c	16.1 ± 0.8c
1-Butanol	0.33 ± 0.02a	1.48 ± 0.08b	2.5 ± 0.1c	1.71 ± 0.09b	1.50 ± 0.08b
Isoamyl alcohol	25 ± 1a	313 ± 16b	1087 ± 56c	903 ± 46d	859 ± 44d
1-Pentanol	0.29 ± 0.01a	6.8 ± 0.3b	14.3 ± 0.7c	10.7 ± 0.5d	10.1 ± 0.5d
1-Hexanol	46 ± 2a	65 ± 3b	236 ± 12c	62 ± 3b	63.7 ± 3.3b
(E)-2-hexen-1-ol	73 ± 4a	46 ± 2b	48 ± 2b	14.5 ± 0.7c	13.1 ± 0.7c
2-Phenyl-1-ethanol	176 ± 9a	1044 ± 53b	4748 ± 243c	3140 ± 161d	3453 ± 177d
Benzyl alcohol	15.7 ± 0.8a	91 ± 5b	188 ± 10c	474 ± 24d	679 ± 35e
∑ ***Carbonyl compounds***	***2186 ± 112a***	***12063 ± 617b***	***5759 ± 294c***	***597 ± 31d***	***532 ± 27d***
(E)-2-hexenal	308 ± 16a	172 ± 9b	53 ± 3c	15.6 ± 0.8d	11.8 ± 0.6d
5-Hydroxymethylfurfural	3.0 ± 0.2a	2.8 ± 0.1a	10.0 ± 0.5b	31 ± 2c	25 ± 1d
Furfural	1.23 ± 0.06a	1.36 ± 0.07a	2.5 ± 0.1b	3.9 ± 0.2c	12.1 ± 0.6d
5-Methylfurfural	0.33 ± 0.02a	0.44 ± 0.02a	1.70 ± 0.09b	1.23 ± 0.06c	1.45 ± 0.07d
Acetoin	1868 ± 95a	11,845 ± 605b	5553 ± 284c	460 ± 24d	395 ± 20d
2-Octanone	1.66 ± 0.08a	4.8 ± 0.2b	12.8 ± 0.7c	15.3 ± 0.8d	22 ± 1e
2,3-Butanedione	4.0 ± 0.2a	36 ± 2b	126 ± 6c	70 ± 4d	63 ± 3d
∑ ***Carboxylic acids***	***61 ± 3a***	***87 ± 4b***	***141 ± 7c***	***114 ± 6d***	***40 ± 2e***
Hexanoic acid	6.2 ± 0.3a	10.2 ± 0.5b	10.7 ± 0.5b	10.5 ± 0.5b	13.6 ± 0.7c
Octanoic acid	35 ± 2a	58 ± 3b	100 ± 5c	80 ± 4d	16.0 ± 0.8e
Decanoic acid	20 ± 1a	18.4 ± 0.9a	30 ± 2b	23 ± 1c	10.4 ± 0.5d
∑ ***Esters and lactones***	***63 ± 3a***	***443 ± 23b***	***1163 ± 59c***	***2914 ± 149d***	***978 ± 50c***
Ethyl lactate	7.3 ± 0.4a	28 ± 1a	67 ± 3b	187 ± 10c	488 ± 25d
Diethyl malate	nd	0.31 ± 0.02a	0.90 ± 0.05b	4.3 ± 0.2c	6.2 ± 0.3d
4-Butyrolactone	41 ± 2a	314 ± 16b	723 ± 37c	2464 ± 126d	403 ± 21b
4-Nonalactone	15.3 ± 0.8a	101 ± 5b	372 ± 19c	258 ± 13d	81 ± 4b
∑ ***Acetals***	***67 ± 3a***	***811 ± 41b***	***1887 ± 96c***	***1512 ± 77d***	***1524 ± 78d***
1,1-Diethoxyethane	67 ± 3a	811 ± 41b	1887 ± 96c	1512 ± 77d	1524 ± 78d

Different letters show significant differences at 95% confidence level; nd, not detected.

The increase in pH and buffering capacity of the must (Peinado *et al.*, 2009) suggests an increase in the degree of neutralization of tartaric acid (Moreno and Peinado, 2012).

Although the *aroma compounds* specifically present in must are known to depend on the particular grape variety, off-vine drying introduces other influential variables, such as the time during which the grapes are dehydrated (Franco *et al.*, 2004; Moreno *et al.*, 2008; Ruiz *et al.*, 2010) and enzyme activity during anaerobic metabolism of the grape (Constantini *et al.*, 2006; Cirilli *et al.*, 2012).

Musts from dehydrated grapes exhibit a substantial increase in the contents of some aroma compounds that cannot be exclusively ascribed to the loss of water from the berries. As can be seen from Table 16.1, the specific aroma compounds most markedly increasing in concentration through the effect of soleo are: acetoin (6.6 times at 11% dehydration), 2-phenylethanol (27 times at 19%), 4-butyrolactone (60 times at 21%), and isobutyl and isoamyl alcohols (60 and 44 times, respectively, at 19% dehydration). This allows aroma compounds in must to be classified into two large categories according to whether their contents change significantly (1) from 0 to 19% dehydration or (2) above 19% dehydration:

1. Aroma compounds increasing their concentration from 0 to 19% dehydration:
   Higher alcohols: isobutanol, isoamyl alcohols, butanol, pentanol, 1-hexanol, 2-phenylethanol
   Carbonyl compounds: acetoin, 2-octanone, 2,3-butanedione
   Medium-chain carboxylic acids: hexanoic and octanoic acids
   Lactones: 4-nonalactone
   Acetals: 1,1-diethoxyethane
   Vinylphenols: 2-methoxyvinylphenol.
2. Aroma compounds increasing their concentration above 19% dehydration:
   Furfurals: furfural and 5-hydroxymethylfurfural
   Esters: ethyl lactate and ethyl malate
   Lactones: 4-butyrolactone.

Costantini *et al.* (2006) distinguished two consecutive stages during the off-vine drying of Malvasia grapes under controlled temperature and humidity conditions. The first stage spanned a weight loss from 0 to 10–12%, and the second, a weight loss exceeding 19%. The former involved accumulation of abscisic acid (ABA), the amino acid proline and the enzyme lipoxygenase (LOX); the latter involved an increase of alcohol dehydrogenase (ADH) and ethanol content. According to Chkaiban *et al.* (2007), water stress in Gewürztraminer grapes under controlled dehydration is delayed with respect to grapes of the same variety dehydrated in the traditional way under uncontrolled ambient conditions; also, the former respond to water stress at a lower water loss level. Therefore, each grape variety seemingly responds at a different rate to water stress caused by dehydration, albeit in an identical manner (viz. with an increase in LOX and ADH activity, and changes in volatile aroma compounds).

In addition to the concentration effect of water loss in the grapes, *phenolic compounds* undergo enzymatic oxidation reactions involving polyphenol oxidases (PPO) and peroxidases (POX) (Macheix *et al.*, 1991). Also, the hydrolysis of some polyphenol oligomers increases the contents in the resulting monomers and the condensation of some monomers raises the levels of the resulting oligomers or polyphenols with a high molecular mass. The most marked concentration changes in the different polyphenol families occur after the third or fourth day of off-vine drying (Serratosa *et al.*, 2008; Peinado *et al.*, 2009, 2011), at about 19% dehydration. Based on the data of these authors, polyphenols in Pedro Ximénez grapes change in specific ways during the grape drying process, namely:

1. *Phenolic acids*: gallic, protocatechuic and *p*-hydroxybenzoic acids increase with drying. Gallic acid increases to a smaller extent than expected from the water loss, which suggests that it takes part in oxidation and condensation reactions. By contrast, *p*-hydroxybenzoic acid increases in proportion to the water loss, and protocatechuic acid to an even greater extent.
2. *Cinnamic acids*: caftaric, coutaric and feftaric acids increase in lower proportions than expected from dehydration of the grape berry, especially in their *cis* forms. Therefore, they seemingly take part in other reactions.
3. *Flavonols*: the contents in the glycosides quercetin-3-glucoside and kaempherol-3-glucoside are three times higher than those in quercetin-3-rutinoside, quercetin-3-galactoside, kaempherol-3-rutinoside and isorhamnetin-3-glucoside, throughout the grape drying process. Also, all increase to a smaller extent than expected from the water loss. These compounds take part in reactions favoured by UV light, and also in oxidation reactions (Spayd *et al.*, 2002; Makris *et al.*, 2006).
4. *Flavan-3-ol isomers and dimers:* catechin, epicatechin and procyanidin B1 increase in lower proportions than expected from the water loss in the grapes. These compounds undergo degradation or conversion to other phenols during the raisining of Thompson Seedless grapes (Karadeniz *et al.*, 2000). Finally, procyanidin increases threefold during this process.

A comparative study of ripe white grapes of different varieties conducted by Castillo-Muñoz *et al.* (2010), showed Pedro Ximénez grapes to have a very low content in flavonols (7.9 μmol kg^{-1}) relative to Muscat of Alexandria grapes (65.1 μmol kg^{-1}) and small-berry Muscat grapes (153.3 μmol kg^{-1}). However, the percent distribution of flavonols was very similar in Pedro Ximénez and Muscat of Alexandria grapes, the sole substantial difference being the presence of a fivefold amount of isorhamnetin 3-glycosides in the former.

Musts and wines from white grapes have lower Trolox Antioxidant Activity (TAA) than red wines owing to the absence of anthocyanins from the former. Thus, Moreno *et al.* (2007) obtained a Trolox equivalent of 4.3 mM in must from ripe grapes of Pedro Ximénez; however, TAA increased to 11.2 ± 0.4 mM – which is comparable to the typical values for red wines – during the off-vine drying process. The increase was ascribed to the concentration effect of water loss during drying, as well as to reactions converting some phenols into others, browning and polymerization reactions and the formation of brown compounds via the Maillard reaction. This reaction is started by the sugars glucose and fructose and amino acids, and is favoured by the high concentration of sugars and temperatures reached during off-vine drying. According to Peinado *et al.* (2009), the polymers of procyanidin, catechins and phenolic acids account for 60, 25 and 15%, respectively, of TAA in off-vine dried Pedro Ximénez grapes. These fractions are additionally responsible for the dark brown colour of the resulting musts; procyanidins contribute by 52% to the absorbance increase at 420 nm, followed by catechin polymers (31%) and phenolic acids (6%), the remainder being due to coloured compounds formed in the Maillard reaction.

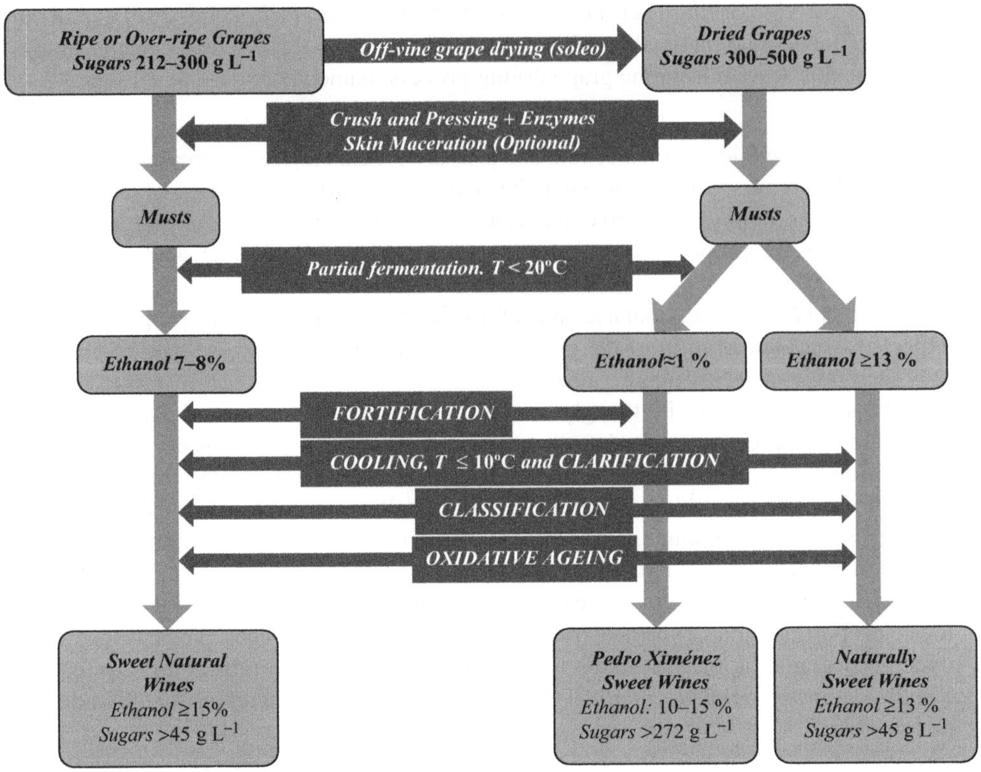

Figure 16.2 Sweet wines obtained from Pedro Ximénez or Muscat of Alexandria grape varieties.

16.4 VINIFICATION

Malaga sweet wines are obtained from Pedro Ximénez and Muscat grapes subjected to the soleo process to a variable extent. The Regulation of the Malaga DOP distinguishes between liquor wines (spirituous wines) and naturally sweet wines. The former are fortified wines obtained by adding wine alcohol at some point during the winemaking process, whereas all ethanol and sugars in naturally sweet wines come from the grapes (Figure 16.2).

According to the Montilla–Moriles DOP regulations, Pedro Ximénez wine is obtained from grapes of the Pedro Ximénez variety partially dehydrated by the soleo process to reach a sugar content below 450 g L^{-1}. The obtained must is fortified with ethanol of wine origin to 9% v/v and this mixture ('mistela') is mixed with other wines made in the same DOP, with a view to obtaining a wine with a minimum sugars content of 272 g L^{-1}. This wine can be commercialized without ageing or it can be subjected to oxidative ageing by means of the traditional 'criaderas and solera' system. Natural Sweet Wines from Moscatel or Pedro Ximénez grape cultivars and other fortified sweet wines, such as Pale Cream, Cream and Medium, are also produced in this zone and the Jerez de la Frontera zone.

16.4.1 Obtaining the must

The sugar content, colour and aroma of the must depend on the degree of soleo of the grapes (Figure 16.3). Monitoring this process by measuring the sugar content in the must and using

Figure 16.3 Musts obtained from Pedro Ximénez grapes in three different degrees of off-wine grape drying (soleo) process. (For a colour version of this figure, please see the colour plate section.)

non-destructive analytical tools such as an electronic nose allows its optimum duration with a view to obtaining the type of wine wanted (López de Lerma *et al.*, 2012). Once samples confirm that grapes have reached their optimum drying time and the sugar content in the resulting must exceeds 300 g L^{-1}, the bunches are collected and those containing rotten grapes are discarded. Only healthy grapes are stored in 20 kg boxes that are in turn placed in plastic containers holding about 1000 kg for immediate transfer to the winery.

On arrival, dry stalks are removed with a destemmer–crusher to avoid straw and wood odours, and bitter, astringent tastes in the must. The resulting product is a mixture of crushed grapes and must. Traditionally, the mixture is supplied with a small amount of SO_2 at this point and pressed. Lately, however, this procedure is being replaced with dynamic maceration techniques involving the addition of enzymes under controlled temperature conditions, avoiding in this case the addition of SO_2. The enzyme preparations most commonly used for this purpose are obtained from *Aspergillus niger* and *A. aculeatus*. Some preparations have a mainly poligalacturonase activity and are indicated for the extraction of aroma compounds and polyphenols from the solid portion of the berry. Other enzymatic preparations have pectinase (pectinesterase and polygalacturonase) activity in addition to cellulase and hemicellulase activity, and are recommended for use in aromatic grape varieties and when musts are difficult to clarify. The experiments carried out by Espejo and Armada (2010), with dynamic maceration of 3000 kg of crushed Pedro Ximénez grapes and two commercial enzyme preparations in a stainless steel tank at room temperature for 3 hours, provided an increased value of Brix degree and must volume, especially with the preparation additionally having hydrolytic activity on cellulose. The total polyphenol index (TPI) of the musts was unaltered and the sweet wines obtained were given better sensory scores than their traditionally made counterparts.

The mixture of must and solid grape parts provided by maceration is transferred to a horizontal pneumatic press to obtain the initial must. The resulting paste is placed on vertical presses delivering up to 300 atm to obtain a second, darker must. The must yielded from dehydrated grapes is about 0.3–0.4 L kg^{-1} grapes.

Racking must removes suspended substances, solids and plant debris accompanying it. Must is usually racked for 24–48 hours, with or without the addition of enzymes and clarifying agents (gelatin, bentonite, silica gel) intended to improve stability and filterability in the resulting wine. Racking must leads to the formation of fine lees during the fermentation process and provides wine with a more balanced and pleasant aroma. Dynamic racking techniques (e.g. centrifugation, flotation and rotary filtration) are scarcely used with sweet wines – only a few winemakers occasionally use rotary filters. Racked musts are corrected in their acidity and supplied with SO_2 before partial or total fermentation, or fortified with ethanol from wine distillation, depending on the particular wine to be obtained.

16.4.2 Fermentation of high-sugar grape musts

The use of *S. cerevisiae* to ferment high-sugar musts is confronted with two major technical problems. One is that the fermentation start is complicated by an initial lag phase that is rather sluggish and delays the process or even causes it to stop prematurely. The other is the difficulty of stopping the fermentation at the optimum time for obtaining a particular type of wine. These problems in the production of special sweet wines have traditionally been avoided in some winemaking areas by adding wine alcohol to musts from dehydrated grapes. The difficulty of using yeasts to ferment high-sugar musts is discussed in depth in Chapter 10. In Pedro Ximenez grapes, the osmotolerant *S. cerevisiae* strains X4 (CECT 13014) and X5 (CECT 13015) were used to conduct the fermentation of must from dehydrated grapes of the same variety towards the production of natural sweet wine (García-Martínez *et al.*, 2007). Must with a sugar content of 375 g L^{-1} was subjected to two types of fermentation. In one, the must was allowed to ferment to 12% (v/v) ethanol; in the other, the process was stopped at 8% (v/v) ethanol. In both cases, the resulting wine was fortified to 15.5% (v/v) ethanol. Based on the results, strain X4 exhibited slightly slower fermentation kinetics, and produced less volatile acidity but higher alcohols, carbonyl compounds and polyols, than did strain X5. Also, an electronic nose detected differences between partially fermented and unfermented musts, but not among the wines fermented by each yeast strain (García-Martínez *et al.*, 2011). The high content of glycerol and acetic acid found in the partially fermented musts relative to dry wines obtained from ripe and undehydrated grapes of the same variety is consistent with the effect of osmotic stress on yeasts under such drastic conditions. Also acetaldehyde, acetoin, methanol, propanol, diethyl succinate and 2,3-butanediol showed increased concentrations.

In summary, sweet wines such as Pedro Ximénez, obtained by partial fermentation of must from dehydrated grapes, possess a more complex aroma than traditionally made, unfermented wines due to the effect of the grapes contributing specific varietal aromas and the yeasts synthesizing some aroma compounds during fermentation.

At present, some winemakers in Malaga and Montilla–Moriles are producing special sweet wines by fermenting musts from Muscat or Pedro Ximenez grapes partially dehydrated by the off-vine grape drying process, to reach a minimum sugar content of 360 g L^{-1} in the resulting must. The fermentation process is conducted in large stainless steel tanks at a controlled temperature of 18 ± 2°C, using yeasts selected in terms of their resistance to high sugar contents and their low production of volatile acidity up to an ethanol content of 13% (v/v) in the fermenting must. At this point, the must is supplied with SO_2 to a total concentration of 100 mg L^{-1} in the medium, the temperature is lowered to 10° C and enological adjuvants such as bentonite and gelatine are added to facilitate sedimentation of the yeasts. This operation requires keeping large volumes of wine at a low temperature over long periods and hence uses large amounts of energy. Periodic transfer of the wine to other tanks at the same temperature reduces the risk of back-fermentation. The sweet wines thus obtained are highly appreciated by virtue of their high quality; also, all ethanol and sugar come from the grapes and their acetic acid content rarely exceeds 1 g L^{-1}.

16.4.3 Types of wines

The classification of commercial Malaga wines is very confusing owing to the wide variety of wines and mixtures defined by the Malaga DOP Regulatory Council, and also the overlap among their definitions. As noted earlier, there are two broad categories of sweet wines, namely: (i) wines fortified with wine alcohol and, in some cases, supplied with must concentrate; and (ii) naturally sweet wines where ethanol and sugar come entirely from the grapes.

The fortified wines labelled as Pedro Ximen (or Pedro Ximénez) and Moscatel must be made from at least 70% of the respective grape variety; the remainder can be made up with wines obtained from the authorized Lairén, Doradilla and Romé grapes. Naturally sweet wines labelled with the name of the grape variety must be made from at least 85% of the respective grape variety. The best-known sweet wine from the Montilla–Moriles zone is Pedro Ximénez, which is exclusively made from dehydrated grapes of this variety. This region additionally produces highly appreciated natural sweet wines from Muscat grapes. Both the Malaga region and the Montilla–Moriles region produce interesting mixed wines such as those called 'Pale Cream', 'Cream' and 'Sweet'.

Traditionally, the Malaga region has produced three different types of sweet fortified wine, namely:

1. *Vino maestro (master wine)*, which is made from must with a sugar content above 228 g L^{-1} obtained from very ripe or partially off-vine dehydrated grapes. This must is fortified with ethanol up to 8% and allowed to ferment slowly to 15–16% (v/v) ethanol, at which point more than 100 g sugar L^{-1} remains unfermented. Master wines are no longer produced in Malaga; instead, freshly extracted must is fortified with ethanol to a 10–20% (v/v) content to obtain mistela.
2. *Vino tierno (tender wine)*, which is obtained from off-vine dehydrated grapes providing a sugar content above 350 g L^{-1} in the must. The must is allowed to ferment spontaneously to a minimum ethanol content of 1% and then supplied with wine alcohol up to 9–15% depending on the intended use. Tender wine is typically used to soften other wines, but never drunk as such.
3. *Vino Dulce Natural (VDN)*. The Natural Sweet Wines are produced by fermenting Pedro Ximénez or Muscat grape-musts containing more than 212 g sugar L^{-1} to reach ethanol content above 7% (v/v). In order to increase their sugar and ethanol contents, the Malaga

DOP regulations allow the addition of the following products to obtain the commercial fortified wines:

1. *Arrope*. This product is made from must obtained in the last grape pressings, which is supplied with about 2 g SO_2 L^{-1}. The must is heated directly on fire or in a water bath to reduce its volume by one-half or two-thirds. Arrope has a brown to dark-brown colour and a sweet taste of cooked must which evolves favourably during storage and ageing of the wine. Arrope quality is measured in terms of its contents in hydroxymethylfurfural (HMF) and sugars. According to Valencia (1990), a good Pedro Ximénez arrope should contain at least 3.8 g HMF L^{-1}, 393 g glucose L^{-1} and 354 g fructose L^{-1}. This product is mainly used to colour wine and usually supplied with ethanol or wine to 13% (v/v) for storage. Concentrating arrope by 50% gives so-called 'pantomima'.
2. *Vino borracho (drunkard wine)*. This product is a mixture of equal amounts of wine and alcohol coming from the wine distillation that is used to fortify wines of the same variety in order to avoid the typical clouding effect of the addition of pure wine alcohol.
3. *Grape musts*. Concentrated, rectified concentrated or partially fermented grape must obtained from on-vine over-ripened or off-vine dried grapes and alcohol-fortified must can be added in the production of fortified wines. All these products may be supplied from the Montilla–Moriles DOP.
4. *Concentrated musts and dry wines* obtained from Doradilla, Airén and Romé grapes in proportions not exceeding 30% in combination in the final product.

Authorized wines for use in the production of mixed wines include naturally sweet wine, natural sweet wine, master wine and tender wine. They must jointly account for at least 50% of the product and the amount of ethanol coming from alcoholic fermentation in the final wine must exceed 4% (v/v).

Arrope supplies to the wines the melanoidins formed by means of the Maillard reaction that has taken place during its production. These brown pigments are responsible for the colour and astringency of sweet wines (Rivero-Perez *et al.*, 2002) and evolve favourably during wine ageing. There are five types of wine depending on colour and the proportion of arrope added, namely: *Golden*, which contains no arrope, so the colour is the exclusive result of ageing; and *Rot gold*, *Brown*, *Color* and *Dunkel*, which contain 5%, 5–10%, 10–15% and $> 15\%$ arrope, respectively.

According to their sugars content, the Malaga sweet wines are known as Pale Dry, Pale Cream, Cream and Sweet. Pale Dry is the only wine without added arrope and has a sugar content below 45 g L^{-1}; all the remaining wines show a sugars content from 45–100; from 100–140, and higher than 140 g L^{-1}, respectively. The best known Malaga sweet wine is 'Malaga Dunkel', a fortified wine obtained largely from Pedro Ximénez and Muscat grapes. Also widely appreciated are 'Pajarete' and 'Lacrima'. Pajarete is an amber to dark-amber liquor wine containing 45–140 g sugar L^{-1} and aged without the addition of arrope. Lacrima is obtained from the free-run must of grapes subjected to no mechanical pressing, and 'Lacrimae Christi' corresponds to a Lacrima wine aged for at least 2 years.

Finally, Naturally Sweet Wines are produced from the musts of off-vine dehydrated Pedro Ximénez or Muscat grapes having a natural sugar concentration exceeding 300 g L^{-1}. The ethanol content of these wines must be at least 13% (v/v) and derive entirely from fermentation of the must; residual sugars in the wine also come from the grapes. Therefore, Naturally Sweet Wines are not fortified wines.

16.4.4 Ageing: description of the 'criaderas' system

The sweet wines made in the Malaga and Montilla–Moriles DOPs are subjected to oxidative ageing in oak wood casks with a maximum volume of 1000 L, which have been previously conditioned with wine. Malaga wines are named differently according to ageing time. Thus, wines aged for 0–6 months, 6–24 months, 2–3 years, 3–5 years and more than 5 years are called 'Malaga Pálido', 'Malaga', 'Malaga Noble', 'Malaga Añejo' and 'Malaga Transañejo', respectively. The Pedro Ximénez sweet wines made in the Montilla–Moriles zone can be commercialized without ageing or after oxidative ageing under the dynamic or static system.

The ageing system used in both regions is of the dynamic type and known as 'criaderas and solera'. However, a number of small winemakers use a static system. In a criaderas and solera system, each criadera consists of a row of American oak casks containing wine aged for an identical time. Each system consists of several rows that are numbered from the floor to the top (Figure 16.4). The first row, lying directly on the floor of the ageing cellar, is called the 'solera'; the row on top of it is the 'first criadera' and those above it are the second, third, and so on. The year's harvest wine is loaded on the 'sobretabla' (the uppermost row). In a system consisting of three criaderas and the solera, the youngest wine is stored in the sobretabla; this lies on the third criadera, which contains older wine, and the third lies on the second, which contains even older wine, and so on down to the solera, containing the oldest wine. The regulations allow that up to 40% of the amount of wine stored in the solera can be withdrawn for bottling each year.

The volume of wine withdrawn from the solera is replenished with an identical volume from the casks in the first criadera, which is in turn replaced with wine from the second, and so on; the rows in the sobretabla are replenished with harvest wine. The wine withdrawn

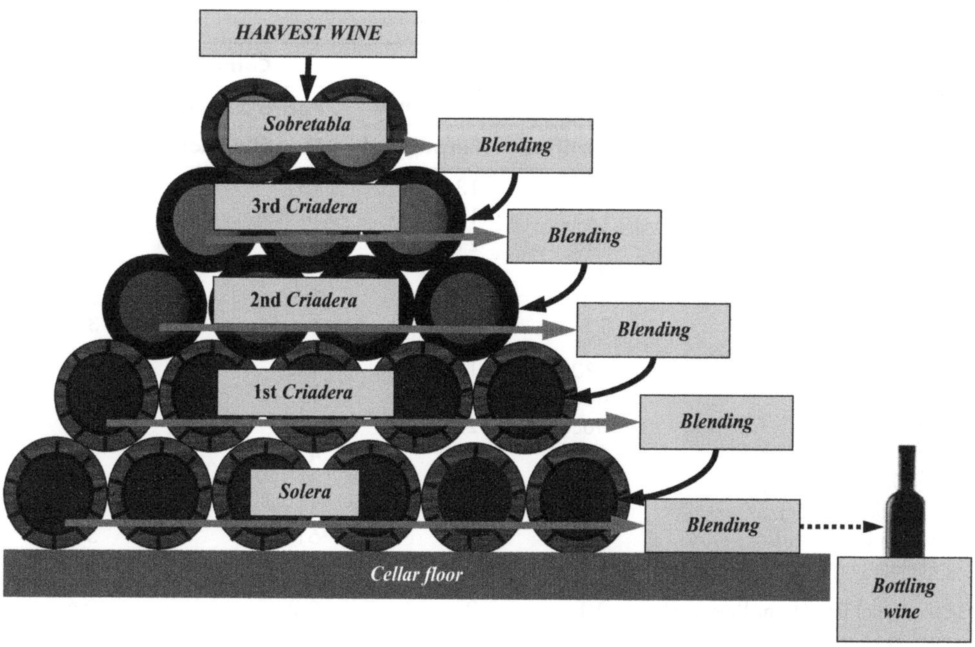

Figure 16.4 Oxidative dynamic ageing of sweet wines carried out using the criaderas system.

from the barrels in each row is mixed in steel tanks prior to transfer to the next criadera. These operations ensure uniformity in the ageing process within each row and provide a complex, homogeneous mixture of wines from different harvests in the solera. As a result, the bottled wine marketed each year possesses very similar sensory properties irrespective of the particular year that its grapes were harvested.

Unlike statically aged wines, for which the harvest year is obviously known and they can thus be classified as vintage wines, dynamically fermented wines cannot. However, the age of wine in each row, and hence the average age of that withdrawn from the solera, can be calculated from an appropriate mathematical equation including the number of criaderas, the number of barrels in each criadera and in the solera, the annual number of transfers and the volume of liquid withdrawn from each row. Valencia (1990) calculated the average age of a system consisting of three criaderas plus the solera from which wine was withdrawn by 15% once each year for 65 years to be: 6.1 years in the third criadera, 12.8 years in the second, 19.4 years in the first, and 21.6 years in the solera. Jackson (2008) found similar ages for a traditional ageing system for Sherry wines.

16.5 ANALYTICAL AND SENSORY PROPERTIES OF COMMERCIAL WINES

The regulation of each denomination of origin states the minimum contents in some components of commercial sweet wines. By way of example, Table 16.2 shows the results obtained by analysing samples of Pedro Ximénez wines from the Montilla–Moriles region, Color

Table 16.2 Analytical characteristics of non-fermented (fortified) and fermented musts (naturally sweet wines) from Pedro Ximénez grapes subjected to an off-vine drying process and commercial sweet wines produced in the Montilla–Moriles and Malaga regions.

			Commercial wines	
	Pedro Ximénez must		Pedro Ximénez	Malaga
Fractions or compounds	Fortified	Fermented	Montilla–Moriles	color
Ethanol (% v/v)	10	13	15	15
Density (kg L^{-1})	1150	1053	1140	1060
Glucose + Fructose (g L^{-1})	430	215	320	190
Volatile acidity (g acetic acid L^{-1})	0.3	1.2	0.10	0.7
Titratable acidity (g tartaric acid L^{-1})	4.1	7.1	3.36	5.45
pH	3.8	3.8	4.4	3.60
Methanol (mg L^{-1})	120	165	113	63
1-Propanol (mg L^{-1})	Traces	57	19	20
Isobutanol (mg L^{-1})	Traces	29	4	15
Isoamyl alcohols (mg L^{-1})	4	150	4	100
2-Phenylethanol (mg L^{-1})	13	28	3	20
Acetaldehyde (mg L^{-1})	10	76	31	120
Acetoin (mg L^{-1})	120	80	61	56
Ethyl acetate (mg L^{-1})	19	30	52	60
2,3-Butanediol (*m*+*l* forms)(mg L^{-1})	0.1	2	0.2	0.6
Glycerol (g L^{-1})	2	12	2	5.6

Sweet Wine from Malaga, unfermented must from Pedro Ximénez dehydrated grapes fortified with ethanol up to 10% (v/v) that has been fermented with a view to obtaining Naturally Sweet Wine containing 13% ethanol and a large amount of sugars. The differences in composition between these four wines can be interpreted in terms of the byproducts formed during alcoholic fermentation of the corresponding musts (Moreno and Peinado, 2012). Also, an electronic nose allows discrimination between unfermented wines and partially fermented wines (García-Martínez *et al.*, 2011).

The particular grape variety used and the must ageing time introduce qualitative and quantitative differences between commercial sweet wines. Marquez *et al.* (2008) analysed minor aroma compounds in 20 samples of sweet wines made from Pedro Ximénez and Muscat grapes in Andalusia, and found significant differences in 20 compounds (especially in limonene, linalool, α-terpineol, isoamylic alcohols, 5-methylfuraldehide, 4-methyl-2-pentanol acetate, ethyl decanoate, hexanoic acid and decanoic acid, with $P < 0.01$) between the two types. Campo *et al.* (2007) analysed wine distillates and various types of special wines aged for a long time and found ethyl 2-, 3- and 4-methylpentanoate and ethyl cyclohexanoate, to be present at concentrations 85–350 times their perception threshold in Pedro Ximénez wines from the Montilla–Moriles DOP, these compounds being the major contributors to the wines' sweet and fruity odours. According to Chaves *et al.* (2007), changes in colour parameters and active odorants compounds in these wines are not linearly related to ageing time since they occur largely within the first 1.3 years of ageing and are much smaller up to the seventh. Linalool, 2,3-butanedione and decanal exhibit the greatest changes in this respect during the early stages of wine ageing.

Traditionally, the denomination 'Malaga' has been associated to quality sweet wines and 'Pedro Ximénez' to wines from off-vine drying grapes of this grape variety. As a rule, these wines possess a strong, highly pleasant aroma and a colour ranging from amber to black. Also, they span a wide range of sweetness. Expert tasters have described Malaga and Pedro Ximénez wines from the Montilla–Moriles DOP in the following terms:

- *Muscat Natural Sweet Wine*: widely variable in colour from yellow to dark amber (with mahogany touches), limpid and bright; dense look; strong aroma of Muscat grapes; sweet, unctuous and persistent in the mouth.
- *Muscat Fortified Wine*: dark amber, limpid, bright colour; floral aroma of ripe fruits with wood notes, complex and mineral; powerful in the mouth; good sweetness–acidity balance, with a long, persistent aftertaste.
- *Muscat Naturally Sweet Wine*: pale yellow with light greenish hues, limpid and bright; floral aroma with honey and citric touches; fresh grape taste with smoked notes; good sweetness–acidity balance.
- *Pedro Ximénez Fortified Wine*: noble sweet liquor wine exclusively made from strongly off-vine dried Pedro Ximénez grapes with notes of ageing; dark ruby to mahogany colour with amber reflections; moderately strong aroma of dried fruits (grapes, figs, dates) and a wood and toasted notes (coffee and cocoa); velvety, unctuous, very sweet but not sickeningly so, and with a highly persistent taste in the mouth.
- *Mixed Pedro Ximénez Wine*: wine from Pedro Ximénez grapes and tender wines from off-vine dried grapes, natural sweet, naturally sweet and old oloroso wines; dark mahogany with iodine touches, limpid and bright in colour; ageing aroma reflecting the grape variety used in their production; tasty, sweet, elegant in the mouth.

ACKNOWLEDGEMENTS

The authors are grateful to the winemakers Dr F. Valencia (Larios, S.A.), F. Medina (Bodegas Málaga Virgen) and R. López-Marín for their assistance with the Malaga wines; and to Cooperativa San Acacio (Montemayor), Cooperativa La Aurora (Montilla) and B. Lucena (winemaker, Bodegas Alvear, Montilla) for their cooperation in relation to the Montilla–Moriles wines.

This work was part of a research project funded by Spain's Ministry of Economy and Competitiveness (INIA-CCAA) and FEDER (RTA2011-00020-C02-02).

REFERENCES

Bellincontro, A., D. De Santis, R. Botondi, *et al.* 2004. Different postharvest dehydration rates affect quality characteristics and volatile compounds of Malvasia, Trebbiano and Sangiovese grapes for wine production. *Journal of Food Science and Agriculture*, 84:1791–1800.

Boletín Oficial de la Junta de Andalucía. 2011. BOJA, 23 diciembre 2011, N° 249:114–199.

Campo, E., J. Cacho, and V. Ferreira. 2007. Solid phase extraction, multidimensional gas- chromatography mass-spectrometry determination of four novel aroma powerful ethyl esters. Assessment of their occurrence and importance in wine and other alcoholic beverages. *Journal of Chromatography A*, 1140:180–188.

Castillo-Muñoz, N., S. Gómez-Alonso, E. García-Romero, *et al.* 2010. Flavonol profiles of *Vitis vinifera* white grape cultivars. *Journal of Food Composition and Analysis*, 23:699–705.

Chaves, M., L. Zea, L. Moyano, *et al.* 2007. Changes in color and odorant compounds during oxidative aging of Pedro Ximénez sweet wine. *Journal of Agricultural and Food Chemistry*, 55:3592–3598.

Chkaiban, FL., R. Botondi, A. Bellincontro, *et al.* 2007. Influence of postharvest water stress on lipoxygenase and alcohol dehydrogenase activities, and on the composition of some volatile compounds of Gewürztraminer grapes dehydrated under controlled and uncontrolled thermohygrometric conditions. *Australian Journal of Grape and Wine Research*, 13:142–149.

Cirilli M., A. Bellincontro, D. De Santis, *et al.* 2012. Temperature and water loss affect ADH activity and gene expression in grape berry during postharvest dehydration. *Food Chemistry*, 132:447–454.

Costantini, V., A. Bellincontro, D. De Santis, *et al.* 2006. Metabolic changes of Malvasia grapes for wine production during postharvest drying. *Journal of Agriculture and Food Chemistry*, 54:3334–3340.

Espejo, F. and S. Armada. 2010. Effect of enzyme addition in the making of Pedro Ximénez sweet wines using dynamic pre-fermentative maceration. *South African Journal for Enology & Viticulture*, 31: 133–141.

Flanzy, C. 2000. Metabolismo anaerobio de la uva. In *Enología: Fundamentos científicos y tecnológicos* (ed. C. Flanzy), Ed. AMV-Mundi Prensa, Madrid, Spain.

Franco, M., R.A. Peinado, M. Medina, *et al.* 2004. Off-vine grape drying effect on volatile compounds and aromatic series in must from Pedro Ximénez grape variety. *Journal of Agricultural and Food Chemistry*, 52:3905–3910.

García-Martinez, T., A. Bellincontro, M.N. López de Lerma, *et al.* 2011. Discrimination of sweet wines partially fermented by two osmo-ethanol-tolerant yeasts by gas chromatographic analysis and electronic nose. *Food Chemistry*, 127:1391–1396.

García-Martínez, T., O. Maestre, R.A. Peinado, *et al.* 2007. Aislamiento y caracterización de levaduras tolerantes a altas concentraciones de azúcares. In *Avances en Ciencias y Técnicas Enológicas*, Ed. Junta de Extremadura, pp. 102–104.

Jackson, R.S. 2008. Specific and distinctive wine styles. In *Wine Sciences*, Elsevier/Academic Press, San Diego, CA, pp. 520–587.

Junta de Andalucía. 2011. http://www.juntadeandalucia.es/agriculturaypesca/ifapa/ria/ (last accessed 31 December 2012).

Karadeniz, F., R.W. Durst, and R.E. Wrolstad. 2000. Polyphenolic composition of raisin. *Journal of Agricultural Food Chemistry*, 48:5343–5350.

López-Alejandre, M.M. 2000. *Los vinos de Montilla–Moriles*, Consejería de Agricultura y Pesca, Junta de Andalucía, Sevilla, Spain.

López, P.A., J. Moreno, and M. Medina. 1988. Caracterización del envero y madurez en la uva de *Vitis Vinifera* Pedro Ximénez cultivada en la zona Montilla–Moriles. *Revista Agroquímica y Tecnología Alimentos*, 28:274–284.

López de Lerma, M.N., A. Bellincontro, F. Mencarelli, *et al.* 2012. Use of electronic nose, validated by GC–MS, to establish the optimum off-vine dehydration time of wine grapes. *Food Chemistry*, 130:447–452.

Macheix, J.J., J.C. Sapis, and A. Fleuriet. 1991. Phenolic compounds and polyphenol-oxidase in relation to browning in grapes and wines. *Critical Reviews in Food Science and Nutrition*, 30:441–486.

Makris, D.P., S. Kallithraka, and P. Kefalas. 2006. Flavonols in grapes, grape products and wines: burden, profile and influential parameters. *Journal of Food Composition Analysis*, 19:396–404.

Marquez, R., R. Castro, R. Natera, *et al.* 2008. Characterisation of the volatile fraction of Andalusian sweet wines. *European Food Research and Technology*, 226:1479–1484.

Moreno, J. and R.A. Peinado. 2012. *Enological Chemistry*, Elsevier/Academic Press, San Diego, CA.

Moreno, J., J. Peinado, and R.A. Peinado. 2007. Antioxidant activity of musts from Pedro Ximénez grapes subjected to off-vine grape drying process. *Food Chemistry*, 104:224–228.

Moreno, J.J., F. Cerpa-Calderon, S.D. Cohen, *et al.* 2008. Effect of postharvest dehydration on the composition of pinot noir grapes (*Vitis vinifera* L.) and wine. *Food Chemistry*, 109:755–762.

Peinado, J., M.N. López de Lerma, and R.A. Peinado. 2011. Synergistic antioxidant interaction between sugars and polyphenolics from a sweet wine. *European Food Research Technology*, 231:363–370.

Peinado, J., M.N. López de Lerma, J. Moreno, *et al.* 2009. Antioxidant activity of different phenolics fraction isolated in must from Pedro Ximénez grapes at different stages of the off-vine drying process. *Food Chemistry*, 114:1050–1055.

Rancho de la Merced. 2011. http://www.variedadesdevid.com/variedad/66/moscatel-de-alejandria/; http://www.variedadesdevid.com/variedad/71/pedro-ximenez/ (last accessed 31 December 2012).

Ribéreau-Gayon, P., D. Dubordieu, B. Donèche, *et al.* 2000. Harvest and pre-fermentation treatments. In *Handbook of Enology. Vol I. The Microbiology of Wine and Vinifications*, John Wiley & Sons, Ltd, Chichester, pp. 299–326.

Rivero-Perez, M.D., S. Perez-Magariño, and M.L. Gonzalez-San José. 2002. Role of melanoidins in sweet wines. *Analytica Chimica Acta*, 458:169–175.

Ruiz, M.J., L. Zea, L. Moyano, *et al.* 2010. Aroma active compounds during the drying of grapes cv. Pedro Ximénez destined to the production of sweet Sherry wines. *European Food Research and Technology*, 230:429–435.

Serratosa, M.P., A. Lopez-Toledano, J. Merida, *et al.* 2008. Changes in color and phenolic compounds during the raisining of grape cv. Pedro Ximénez. *Journal of Agricultural and Food Chemistry*, 56:2810–2816.

Spayd, S.E., J.M. Tarara, D.L. Mee, *et al.* 2002. Separation of sunlight and temperature effects on the composition of *Vitis Vinifera* cv. Merlot berries. *American Journal of Enology and Viticulture*, 53:171–182.

Valencia, F. 1990. *Monografía sobre los vinos de Málaga*, Larios, S.A., Málaga, Spain.

Valero, A., S. Marin, J.A. Ramos, *et al.* 2005. Ochratoxin A-producing species in grapes and sun-dried grapes and their relation to ecophysiological factors. *Letters in Applied Microbiology*, 41:196–201.

17 Tokaj

Zoltán Kerényi

Corvinus University of Budapest, Institute Vitivinicole Research Station Kecskemét, Kecskemét, Hungary

17.1 INTRODUCTION

One of the most famous natural sweet dessert wines, or 'Vin Doux Naturel' (VDN), is Tokaj *aszú*, produced in Hungary, in East-Central Europe. The VDN is a category of dessert wines determined by the fact that the alcohol and sugar content of the end product originate exclusively from the grapes, differing in this respect from other famous dessert wines, such as Port and Sherry. The history of the aszú vinification goes back to the fifteenth century, and the first written mention of Tokaj wines is dated 1495 (*Regni Ung Historia* by Istvánffy, 1724).

The typical, unique character of Tokaj wines derives from four co-existing factors:

- ecological conditions (soil, climate, grape varieties, etc.);
- *Botrytis cinerea*, noble rot;
- special grapevine processing, vinification;
- long ageing in small barrels at low temperatures (10–12°C).

The main features of the Tokaj winemaking technology are outlined in the following sections.

17.2 TOKAJ WINE TYPES

Mention of Tokaj wines generally evokes thoughts of sweet wine, although there are many different Tokaj wines. Undoubtedly, the most famous is Tokaji aszú, but some other products are worthy of note, such as the Tokaj varietal wines, Furmint and Hárslevelű (cuvées), the Tokaji szamorodni (dry and sweet), Tokaji fordítás and máslás, and Tokaji esszencia. It is important to clarify from the outset that the expression 'Tokaj**i**' (meaning 'of Tokaj'), is an attributive Hungarian word for the wine's geographical origin. Tokaj is the name of a town and also a wine district in Hungary. The official appellation of its wines is listed in the European wine register as Tokaj (E-Bacchus database, 2010), but the traditional, local name Tokaji can be officially used as well. The old English spelling 'Tokay' should be avoided, as this name currently refers to wines other than Tokaji, for example dry wines of Californian (Flame of Tokay,), or French (Tokay d'Alsace) origin.

Sweet, Reinforced and Fortified Wines: Grape Biochemistry, Technology and Vinification, First Edition.
Edited by Fabio Mencarelli and Pietro Tonutti.

17.2.1 Tokaj varietal wines

These wines are made from the six legally permitted grape varieties with the name Tokaj: Furmint, Hárslevelű (Lindenleaf), Yellow Muscat (Goldmuskateller), Kövérszőlő, Zéta and Kabar. The Furmint is of the greatest importance, because it yields nearly 70% of the whole vintage and serves predominantly as base wine for aszú production. The characteristic wine made of this variety is suitable for long ageing. Hárslevelű – grown on 20% of the region – with its silken linden-flower–honey varietal character and light acidity – is a perfect partner for blending. Some of the four other varieties are traditional and the more recently bred varieties are of lesser importance. Tokaj Yellow Muscat (falsely named as Muscat Lunel) is one of the most popular Tokaj varietal wines, due to its characteristic Muscat bouquet and fine acids. Yellow Muscat is used for aszú basic wines as well.

17.2.2 Tokaji szamorodni

Tokaj varietal wines are processed from healthy grapes that mostly contain less then 2% botrytized berries; *szamorodni* is the wine type proceeding from clusters infected by a higher percentage of noble rot (more than 10%). In the case of szamorodni, the noble-rotted berries of the grape are not selected from the cluster, but are processed together with the healthy grapes to make Tokaj szamorodni dry or sweet wines. The word 'szamorodni' is of Polish origin, and means 'as it comes', meaning that the clusters are picked in their original state, as nature created them. Depending on the proportions of healthy and noble-rotted berries in the clusters, dry or sweet szamorodni can be fermented from the grapes. The higher the percentage of aszú berries, the greater the sugar content and the probability that residual sugar will remain after fermentation. Szamorodni is a difficult wine to categorize because of its similarity to the aszú winemaking process in terms of ageing and costs, although the market positions these wines below the aszú wines. In spite of the uncertainties, the szamorodni wines have an important role in the palette of gastronomy. Some dry szamorodni wines can be compared with dry flor sherry (fino) types, which have the same characteristic aldehydes and acetals.

17.2.3 Tokaji aszú wines

Categorization of Tokaj aszú wines is based on the amount of aszú berries used for one unit volume. The quantity of the aszú berries is measured using a traditional unit, the *puttony* (hod). The higher the puttony number, the higher the sugar concentration and the quality of the aszú. Traditionally, the noble-rotted aszú berries, picked out from the clusters, are collected in small wooden tubs (puttony) that contain approximately 20–24 kg of aszú berries; they are then added to a typical Tokaj barrel, named Gönczi (136 L capacity), of base wine. The number of puttony added per barrel gives the final *puttonyos* level of the wine, which is printed on the label. At the end of the fermentation period, the remaining sugar content will be the basis for classification. According to Hungarian legislation, the lowest puttony number is three, and the highest is six. The classification of all Tokaj wines is outlined in Table 17.1 (Botos and Marczinkó, 2006).

17.2.4 Tokaji fordítás and máslás

Both types are unique to the Tokaj wine region and are produced from two byproducts of the winemaking process, namely marc and lees. Both byproducts, which have a high percentage of sugar, are mixed with dry wine giving a refermentable base material.

Table 17.1 Tokaj wine categories (Adapted from Botos and Marczinkó (2006)).

		Szamorodni		Aszú					
	Quality wine, varietal and cuvée	**Dry**	**Sweet**	**3 puttonyos**	**4 puttonyos**	**5 puttonyos**	**6 puttonyos**	**Aszúeszencia**	**Essencia**
Harvest	October	October-November	October-November	October-November	October-November	October-November	October-November	October-November	October-November
Grape berries	not infected	botrytized	botrytized	botrytized	botrytized	botrytized	botrytized	botrytized	botrytized
Harvest method	normal	together	together	selected	selected	selected	selected	selected	selected
Grape and wine processing						aszú berries + base wine or must			
Wine origin	same	same	same			practically different			
Sugar content ($g\ L^{-1}$)		0–10	10–60	60–90	90–120	120–150	150–180	>180	>450
Min sugar-free extract ($g\ L^{-1}$)		23	23	25	30	35	40	45	50
Min. alcohol content (% v/v)		12	12						1.2–8.0
Min. ageing (in barrels)		2 (1) year	2 (1) year	3 (2) year	3 (2) year	3 (2) year	3 (2) year	3 (2) year	

17.2.5 Tokaji esszencia

Tokaji esszencia is the nectar exuded by the aszú berries under their own weight during storage. In reality, it cannot be qualified as a wine, due to its low (2.0–7.0 % v/v) alcohol content, but it is still the most expensive winery product, because of its extremely rich flavour. It is nectar and a high-value wine combined, with a sugar content of more than 600 g L^{-1}. It is so thick that it is for tasting only, and is therefore rarely consumed. It is called 'no' wine, because there is no pressing, no fermentation, no sulfur addition, no clarification, and it consists of native features only. The esszencia has a role in aszú winemaking technology: it can be added to aszú wines to adjust sugar content after fermentation.

17.3 ASZÚ WINEMAKING

Aszú winemaking is a unique process combining the extraction of incomparable compounds from a special base material (aszú berries) with a particular method (small barrels and low temperature). The whole process is lengthy and a lot of the work is done by hand, which justifies the high price of aged aszú wines. The development of noble rot on the berries depends on the characteristics of the weather conditions in the vintage year. If the berries are not ripe enough at the time of *Botrytis* attack, they will be totally destroyed, the sugar content will diminish, and the berry skin will disintegrate due to the pectinase activity of the grey mould. If the sugar concentration is more than 16–17%, *Botrytis* can only penetrate the berry but is not able to consume sugars, only organic acids. During this metabolism, the *Botrytis* produces a lot of oxidized, mostly hydroxi- and oxo, compounds. One of the oxidization products is the gluconic acid that develops from glucose in high concentrations (3–10 g L^{-1}).

Harvest in Tokaj starts in October to mid-November at the latest, but some shrivelled and botrytized berries can also be harvested from the earlier ripening grapes in September. First the mature (aszú) berries are hand-picked; often women perform this duty, as they are generally more meticulous than men. The grape is harvested in a number of passes through the vineyard. With each pass the pickers will pluck only the suitable affected bunches, with the first pass generally regarded as the one that yields the highest quality botrytized grapes. Then the less advanced plots are harvested by mixing the botrytized and ripe berries. The collection of these berries can be performed in the vineyard, or at the winery, on selecting tables. The selected aszú berries can be classified both by sugar content and by appearance. The aszú berries should have no heavy mycelia growth on the surface, and the skins should be oily brown. Figure 17.1 illustrates the differences between first-class aszú berries and the less suitable, conidia-covered berries.

After selection from the bunch, the aszú berries are stored for many weeks before processing. Storage is carried out in open containers, vats or tanks and the grapes are treated with SO_2 to protect against acetic acid bacteria and wine flies. The quality and chemical composition of the aszú berries can be influenced during storage by mould and yeast biota harbouring on the skin, which also have a sensory impact on the aszú wine character. The digestion of the berry skin begins already in the vineyard, and the *Botrytis* facilitates the growth of other saprophytic micro-organisms (other moulds, yeast, bacteria), which can also influence the complexion of the wine. The digested berries are easy to extract, to release the aszú features before fermentation.

The juice of the desiccated grapes cannot be extracted by pressing, and a special maceration method has traditionally been applied, which makes the vinification technology of Tokaji

Figure 17.1 Different types of botrytized grains for Tokaji wine. (For a colour version of this figure, please see the colour plate section.)

aszú unique (Alkonyi, 2000). The extraction of noble-rotted berries is a crucial factor, among others, during Tokaj aszú wine production. The time span of soaking has a basic impact on sugar content and other volatile and non-volatile compound concentration in the base aszú wines or musts. The extracting media is mainly young wine with a higher alcohol content (up to 14% v/v) or must with the same potential alcohol. The extracting must is sometimes already in fermentation, which accelerates the process. The higher the ratio of aszú berries with mould other than *Botrytis* on the surface, the lower the fermentation soaking time. The impact of soaking time on aszú wine quality was investigated in pilot experiments for volatile compounds in wine (Kerényi *et al.*, 2007). Figure 17.2 illustrates the changes in the concentration of some volatile compounds typical for aszú wines. The general soaking time of the aszú berries varies between 12 and 48 hours, depending on the vintage year. Increased extraction time favours the development of volatile acidity but higher sugar concentration as well. The aszú berries extract has to be pressed to obtain juice for aszú fermentation. The presses used during this step are gentle pneumatic presses to avoid excessive disintegration of the skin remains and the seeds. More recently, pressing occurs under a 'nitrogen cushion' to avoid excessive oxidization of the grains.

The core of the fermentation is the microbiology, the biota of the unprocessed juice and the temperature. Depending on the initial sugar concentration, the fermentation length can extend to several months, and up to one year is not exceptional. The musts of botrytized grapes have a complex yeast biota, and display a higher level of biodiversity than normal wines. The nutritional and environmental conditions for yeasts are extremely difficult. Many investigations have shown that non-*Saccharomyces* species, mainly *Candida stellata* (probably *C. zemplinina*) may contribute significantly to the fermentation of botrytized wines. In many cases, *C. zemplinina* strains dominate fermentation during the first weeks. This strain was later identified as *C. zemplinina* by Sipiczki (2003). *Saccharomyces* strains dominate fermentation sooner or later, as in non-botrytized wine. The speciality of the aszú fermentations is that, beside various *S. cerevisiae* strains, *S. uvarum* (formerly known as *S. bayanus* var. *uvarum*) can be isolated from these wines (Magyar *et al.*, 2008). These species are well

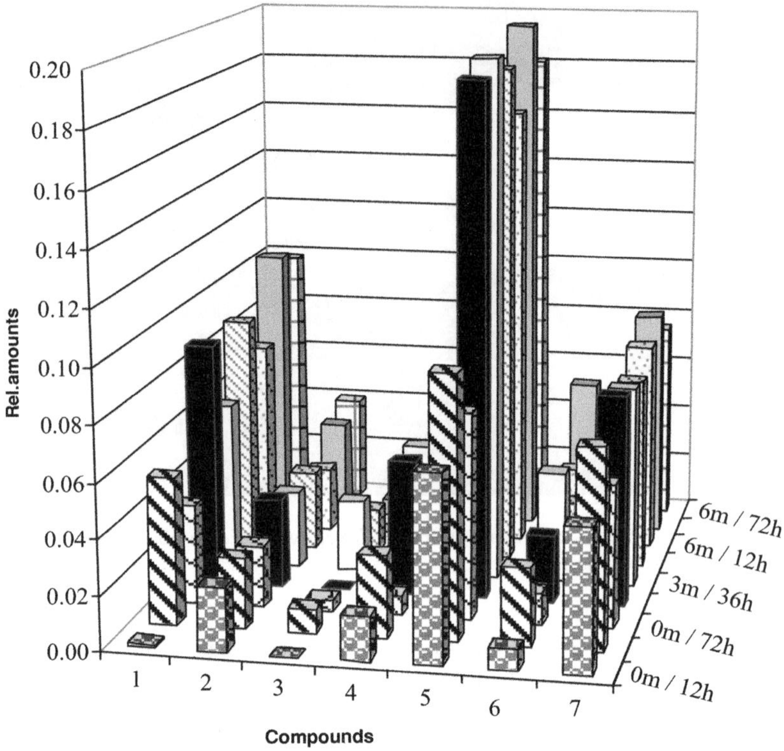

Compounds	**No.**
γ-butirolactone/10*	1
2(3H)-furanone, 5-ethyldihydro	2
2H-pyran-2-one,-tetrahydro	3
2H-pyran-2-one, tetrahydro-4-methyl-	4
2(3H)-furanone, dihydro-3-OH, 4,4-dimethyl +	5
2(3H)-furanone, dihydro-5-pentyl	
2H-pyran-2-one, tetrahydro-4-hydroxy-4-methyl	6
2H-pyran-2-one, tetrahydro-4-hydroxy-6-pentyl	7

Figure 17.2 The amount of lactones measured by GC-MS (0–6 months ageing, 12–72 hours soaking time).

adapted to aszú wine fermentations. *Saccharomyces uvarum* is a cryotolerant yeast, showing good growth and fermentation rate at low temperatures (8–12°C), typical for Tokaj cellars. In Tokaj, most small wineries rely on spontaneous fermentation, although the use of starter cultures is spreading, and is standard in most new estates. Beside commercial starters, some wineries use local *S. cerevisiae* (var. *bayanus*) strains selected from the winery microbiota.

Botrytized must possess a unique chemical composition that greatly impacts the products and byproducts of alcoholic fermentation, as well as subsequent reactions (Kerényi, 1977). The high sugar content of the juice significantly reduces the growth, fermentation rate, and survival of yeasts (Lafon-Lafourcade, 1983), and also affects secondary metabolism. Botrytized wines show higher glycerol and acetic acid content, due to the stimulating effect of high sugar concentration. The elevated volatile acidity of aszú wines is generated by acetic acid bacteria proliferating on the grape and detectable in the barrel-aged wine as well. This may reach in some cases 2 g L^{-1}, the level that is permitted by law. The production of carbonyl compounds (such as acetaldehyde, pyruvic acid and 2-ketoglutaric acid) increases dramatically as a consequence of the thiamine depletion in botrytized grapes. These compounds are responsible for the high SO_2-binding capacity of aszú wines. Nitrogen deficiency is another cause of the enhanced production of acetic acid. The fermentation temperature has a significant impact on the chemical composition of aszú wines, in addition to the well-known effect of low temperatures on the compounds arising during fermentation.

17.4 AGEING, TREATMENT AND STABILIZATION OF ASZÚ WINES

A two-year barrel-ageing period is compulsory for aszú wines in Tokaj, at least. The duration of the barrel ageing is debated; many wineries keep to the tradition of ageing aszú wines for as many years as the puttony number. The long ageing not only has economic implications, but also raises some technological problems. The botrytized grapes have a yeast biota that is able to ferment at higher alcohol and sugar concentrations, so the ageing period can sometimes lead to refermentation of the desired sugar level. The inhibition of this refermentation by sulfuring is disadvantageous, because aszú wines are intentionally exposed to slow oxidation. The chemical changes associated with barrel ageing are mainly oxidation of alcohols, aldehydes, phenolics, and the formation of esters, acetals and lactones. The longer the ageing, the more hints of dried fruits, chocolate, bread and coffee are perceptible and add to the primary notes of apricot, quince, honey and mushroom-like *Botrytis*. The cold temperature in Tokaj cellars can slow the proliferation of osmotolerant yeasts, but the sugar level has to be checked constantly.

There is a special factor in ageing aszú wines: the fungus *Cladosporium cellare* (*C. cellare*). The thick grey-black coat of this mould, consisting of hyphae and fruiting bodies, is a characteristic feature of walls and equipment in Tokaj cellars and the fungus has an important role in clearing and buffering the air, especially through humidity regulation. *C. cellare* utilizes only volatile compounds that are present in the airspace of the cellar. The volatiles that originate from the barrels do not always provide the carbon source for the mould. Since *C. cellare* cannot tolerate ethanol contents above 2% (v/v), it never grows directly on the surface of wine. It has no direct impact on the quality of wine, although it beneficially influences the purity and humidity of the air in the cellar (Magyar, 2006, 2010).

Over the last few years, a growing diversity of wine maturation concepts and methods has been observed in Tokaj. Wines ranging from short maturation to ageing in new barriques for several years, are offered by the wineries. Subsequent bottle ageing is typical. Before bottling, the wines normally undergo stabilization, including bentonite fining for protein removal and cold stabilization to avoid tartrate salt crystallization. A unique feature of botrytized wines is the formation and precipitation of calcium mucate crystals, a salt of mucic acid (Dittrich and Grossmann, 2011). Mucic acid salts are not found in normal wines, but its presence has been observed in the crystalline deposits formed throughout the ageing of sweet white wines made from grapes with noble rot, such as Tokaji aszú. Because supersaturated solutions remain stable for long periods, cooling is ineffective in maintaining stability. Reducing the calcium content of wine with DL-tartaric acid addition may reduce the risk for crystalline instability.

There is another unique technological problem caused by *Botrytis*: the presence of glucans, which can inhibit clarification. These β-D-glucans, even at low concentrations (2–3 mg L^{-1}), significantly reduce the filterability of the wine. Concentrations at 50 mg L^{-1} make filtration impossible. The breakdown of these glucans is possible by enzymes.

Botrytized wines need extreme care and sterility during bottling, since the risk of in-bottle refermentation is very significant in spite of the high sugar content. Hot filling at 50–55°C would be beneficial in terms of microbiological stability but has not been used because of quality considerations. Fine filtration, including membrane filtration, is widely used, followed by sterile filling and corking. DMDC (dimethyl dicarbonate) and potassium sorbate are frequently added for microbial stabilization, in addition to sulfur dioxide.

Author's note: I have used the word 'Tokaj' to refer to the place and 'Tokaji' to refer to the wine produced in the region.

REFERENCES

Alkonyi, L. 2000. *Tokaj – The Wine of Freedom*, Spread BT, Budapest, pp. 82–89, 128–161.

Botos, E. and F. Marczinkó. 2006. *Tokaj Wine Atlas*, Bor-Kép, Budapest, pp.18–26, 59–65.

Dittrich, H.H. and H. Grossmann. 2011. *Mikrobiologie des Weines*, 4th edn, Verlag Ulmer, Stuttgart.

E-Bacchus database. 2010. Register of designations of origin and geographical indications protected in the EU. European Commission at http://ec.europa.eu/agriculture/markets/ wine/e-bacchus/ (accessed 18 December 2012).

Kerényi, Z. 1977. Tokaji borkülönlegességek aromaanyagainak gázkromatográfiás vizsgálata. III. Nem illó és nehezen illó aromakomponensek GC-analízise. *Borgazdaság*, 25:26–29.

Kerényi, Z., É. Miklósy, M. Verczné Hodos, *et al.* 2007. Impact of soaking time of aszú berries on the volatiles and sensory value in young aszu wines. *XXXth OIV World Congress of Vine and Wine, Budapest*, 10–16 June.

Lafon-Lafourcade, S. 1983. Wine and brandy. In *Biotechnology*, Vol. 5 (eds H.J. Rehm and G. Reed), Verlag Chemie, Weinheim, pp. 81–163.

Magyar, I. 2006. *Microbiological Aspects of Winemaking*, Corvinus University of Budapest, Faculty of Food Science, Budapest.

Magyar, I. 2010. *Borászati mikrobiológia*, Mezőgazda Kiadó, Budapest.

Magyar, I., T. Tóth, and A. Pomázi. 2008. Oenological characterization of indigenous yeasts involved in fermentation of Tokaji aszú. *Bulletin de l'OIV*, 81:35–43.

Sipiczki, M. 2003. Candida zemplinina sp. nov., an osmotolerant and psychrotolerant yeast that ferments sweet botrytized wines. *International Journal of Systematic and Evolutionary Microbiology*, 53:2079–2083.

18 Vin de Paille

Pierre Louis Teissedre, Bernard Donèche and Kleopatra Chira

Faculty of Oenology, Université Bordeaux Segalen, Institut des Sciences de la Vigne et du Vin, Villenave d'Ornon, France

18.1 INTRODUCTION

Vin de paille is the French definition of 'straw wine', made only in the ripest vintages. It is a wine made from grapes that have been previously dehydrated in order to concentrate their juice. Vin de paille was first produced in Queyssac-les-Vignes in the Corrèze department, where some farmers decided to recreate the famous wine known as 'miel des muses' made by the Romans, and given by Saint Éloi to King Dagobert in 622 AD. The red grape varieties used to make this wine were Cabernet Franc and Cabernet Sauvignon, and the white varieties were Chardonnay and Sauvignon.

18.2 VIN DE PAILLE OF JURA

The best-known example of Vin de paille is produced in the Côtes du Jura. The Jura wine region is located in the Jura Mountains, in the foothills of the Alps, in the extreme east-central part of France, not far from the Swiss border. The Jura wine region has a surface of 2150 ha, of which 1750 ha are recognized as French AOCs (Appellation d'origine controlée). The lands of the Jura date back to the secondary or Jurassic era, which began about 150 million years ago, and they occupy two-thirds of the department. The vineyards are localized on slopes with an altitude that varies between 200 and 400 metres. The vineyard soils are generally composed of clay on the lower flat lands, and are more limestone-based on the higher elevations. Deposits of marl are scattered throughout the region where some of the area's most highly regarded vineyards are to be found (Chaudat, 2004). Vin de paille of Jura is produced either in the famous Chateau-Chalon area or in one of three other recognized French AOCs: Cotes du Jura, Arbois and Etoile. The predominant white grape varieties are: Chardonnay, also called *melon* in Arbois, which represents almost 48% of the total planted surface; and Savagnin, also named *naturé* in Arbois (circa 17%). The most important red varieties are: Pinot Noir, also called *noirin* in Arbois (circa 11%); Poulsard or Ploussard (circa 18%); and Trousseau, representing almost 5% of the total varieties in the north Arbois appellation (Boidron *et al.*, 1995; Chaudat, 1997). The latter two are native varieties. All the varieties, with the exception of Pinot Noir, are allowed to be used for Vin de paille production. Pinot Noir was excluded in 1990 due to the soil and climatic conditions, it does not produce wines of comparable quality to those of Burgundy.

Sweet, Reinforced and Fortified Wines: Grape Biochemistry, Technology and Vinification, First Edition.
Edited by Fabio Mencarelli and Pietro Tonutti.

18.2.1 Characterization and identification of grape cultivars for Vin de paille

Among the *V. vinifera* varieties, Chardonnay, Savagnin and Poulsard are the ones mostly used for Vin de paille of Jura (Boidron *et al.*, 1995; Chaudat, 1997). Savagnin is the predominant Jura grape variety and it has been studied and described by several authors. Savagnin is morphologically close to Traminer, while Pinot Noir is considered to be the product of a cross between Riesling Renano and Traminer (Regner *et al.*, 2000). The Savagnin variety is synonymous with grape cv. 'Alvarinho' (Loureiro *et al.*, 1998). Applying three different approaches, simple sequence repeats (SSR), amplified fragment length polymorphism (AFLP), and methyl-sensitive amplified length polymorphism (MSAP), it was revealed that Savagnin and Alvarinho are genetically very similar to the Traminer group (on average, a similarity of 92.9%) (Imazio *et al.*, 2002). Even though these two cultivars have very similar shoots parameters, they have different cluster grapes and leaf shapes. Alvarinho has two clusters per fruiting branch, while Savagnin Blanc generally has only one. Alvarinho leaves are pentagonal whereas Savagnin Blanc leaves are between pentagonal and round. Alvarinho leaves are light-to-medium green, while Savagnin Blanc leaves are darker. Savagnin Blanc produces the largest berries but they have the same seeds (Santiago *et al.*, 2007).

Using nine microsatellite markers, Imazio *et al.* (2002) were not able to distinguish Gewürztraminer accessions from Savagnin, suggesting that these genotypes are in fact clones derived from the same mother plant. This was confirmed by a genetic analysis study based on microsatellite markers. In the same study, the analyses of wines, musts and berries showed that linalool and geraniol were present only at very low concentrations in Savagnin Rose and Chardonnay 76, both considered as non-aromatic cultivars (Duchêne *et al.*, 2009). Savagnin Rose and Gewürztraminer are two pink berry cultivars that are considered as two varieties in the French official catalogue and are grown to produce neutral or aromatic wines, respectively. These two cultivars, as well as the white grape variety, Savagnin Blanc, are classified in the varietal group known as Traminer or Savagnin (Pacottet, 1903). Traminer and Traminer-related cultivars represent the second important gene pool of the Middle Ages (Bassermann-Jordan, 1975). In a simplified way, they can be seen as a combination of quality and quantity. Under favourable climatic conditions they are important vines that produce high-quality wine. We may assume, for this reason, that the Savagnin variety constitutes a great percentage of Vin de paille production.

Poulsard is one of the 96 cultivars of *V. vinifera* that have the potential for the parentage of Pinot (Regner *et al.*, 2000). Poulsard, red variety, gives elegant, aromatic wines with fruity and animal notes, and light structure and colour. Poulsard skin is very thin and produces very pale-coloured red wines, even after extended maceration; it can also be used to produce white wines. Because of this, Poulsard is often blended with other red-skin varieties or used to produce lightly coloured rosé wines.

The Trousseau variety, covers only a small percentage of the Jura region and is not usually used for Vin de paille production. It gives highly coloured wines with significant tannin levels. Using standard but well-described methods involving paper chromatography, partial hydrolysis, alkaline degradation and spectral characterization, the red pigments of *V. vinifera* variety 'Trousseau' were identified. At 21.5° Brix the berries had 850 mg of total anthocyanin per kg, which was 49.8% malvidin-3-glucoside, 36.9% peonidin-3-glucoside, 8.75% cyanidin-3-glucoside, and 4.55 % petunidin-3-glucoside. The wine can potentially age for 15–20 years (Carreno-Diaz and Luh, 1969).

Figure 18.1 Grapes on straw mats before being pressed. (For a colour version of this figure, please see the colour plate section.)

18.2.2 Production of Vin de paille in the Jura

Vin de paille production is performed in four stages: grape picking, grape dehydration, grape pressing, and wine fining. The healthiest and most mature grapes are selected from three Jura grape varieties: Chardonnay, Savagnin and Poulsard for the Vin de paille of Jura.

After harvest, traditionally the grapes were placed on straw mats for at least 6 weeks. The AOC protocol obliges the producer to follow a dehydration procedure of 6 weeks minimum. Grapes usually stay on the straw mats until they are pressed at between 3 and 5 months (Figure 18.1). Nowadays, grapes are generally hung up (Figure 18.2), and the straw mats may be replaced by modern racks in dry, ventilated, unheated rooms. Producers rarely use straw as it must be organic and at the end of the dehydration procedure it has to be manually separated from the grape clusters. Grapes are inspected on a weekly basis to remove spoilt grapes by hand. The basic yield of Vin de paille is fixed at 20 hL ha^{-1} (Chaudat, 2004).

At the end of the dehydration procedure, normally between end-December and end-February, and after having requested permission by registered letter from the INAO to press the grapes in a particular area and at a specific time, the pressing procedure for Vin de paille production begins. INAO stands for 'Institut National de l'Origine et de la Qualité'

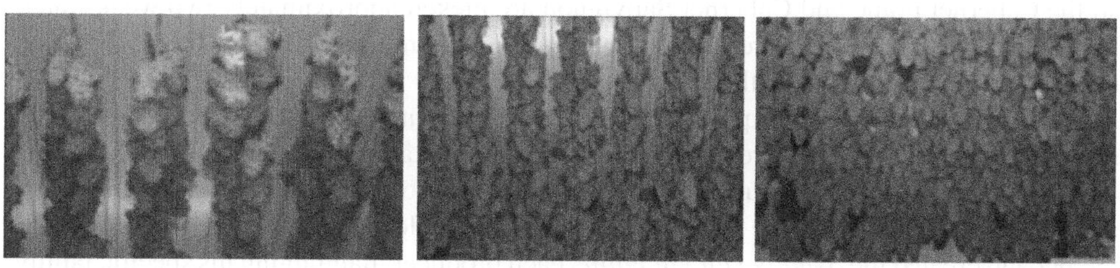

Figure 18.2 Grapes hanging up before being pressed. (For a colour version of this figure, please see the colour plate section.)

(previously known as the *Institut National des Appellations d'Origine*) and is the French organization, within the Ministry of Agriculture, that is charged with regulating French agricultural products under protected designations of origin (PDOs). Some producers blend different varieties; the proportion used depends on the vintage, but 10–20% of Savagnin is always added for acidity. At the moment of pressing, the grapes should present a concentration of sugars greater than 306 g L^{-1} of must. After pressing the dehydrated berries (with high sugar levels), 15–18 L of must is obtained per 100 kg of grapes. Skin contact and pressing not only affect the extraction of varietal aromas but also the phenolic compounds (hydroxycinnamic acids and some flavonoids) (Maggu *et al.*, 2007). The fermentation in barrels is long as it starts in winter and the sugars concentration is high. The slow fermentation, which ends naturally, gives a naturally soft wine with between 14.5° and 18° alcohol (as imposed by the AOC protocol). It will then age for three years in small barrels to develop its aromas of caramelized fruits, prune, honey, caramel, or caramelized orange. After oak ageing, wines are transferred to 375 mL bottles labelled 'Vin de paille' and indicating the vintage.

18.3 VIN DE PAILLE IN HERMITAGE

The Tain l'Hermitage winery began to produce Vin de paille again in 1985. The size of the Hermitage appellation is strictly limited by its local geography. In a very small vineyard (30 ha), the best grapes of the Marsanne and Roussanne varieties are harvested at full maturity, or even later if the weather permits. Marsanne grapes come from two very old vines, one of which is over a century old. The Vin de paille production is similar to that of the Jura: each grape cluster is cleaned carefully to remove damaged grains and then it is gently stored on wooden racks. There are fewer than 2 kg of grapes per box so that clusters are not in contact with each other. They remain there for five weeks up to two months, and then the grapes are pressed in order to extract the remaining juice before fermentation takes place. The wine is matured for two years before being bottled in 0.5 L bottles.

18.4 VIN PAILLÉ OF CORRÈZE

Vin Paillé is produced in the Domaine du Bas Queyssac, which is situated on the border between the Corrèze and the Lot, on the hillside of Queyssac-les-Vignes. Vin Paillé has an old tradition among wine growers of the region of Beaulieu and Meyssac. The best grapes are selected and picked by hand and set to rest for two months on wooden racks. Particularly, Chardonnay and Sauvignon Blanc grapes are hand-picked in mid-September, and the Cabernet Franc and Cabernet Sauvignon are picked approximately two weeks later. The wooden racks are left to rest for about two months in a well-aerated barn. The wine production is similar to that of the Jura.

About 7 kg of grapes are required to produce 1 L of Vin Paillé. The grapes are strongly pressed in order to extract the remaining juice and fermentation takes place. The Vin Paillé is then matured for two years, under the strict supervision of the wine grower, before being bottled. Towards the end of the twentieth century, several local producers of Vin Paillé got together and created the *Syndicat du Vin Paillé*, each producer maintaining his specific family tradition and knowledge. The Syndicate comprises about 20 producers today, spread over 26 municipalities in the cantons of Beaulieu in the Dordogne, and Meyssac.

18.5 VIN DE PAILLE DEHYDRATION PROCESS

The skin is the part of the fruit with the highest amount of volatile and polyphenolic compounds. Skins also contain a large amount of enzymes, such as polyphenols oxidase (Fraignier *et al.*, 1995). These specific enzymes cause the oxidation of the phenolic compounds, and thus can influence the browning process. Skins generate great interest in the production of Vin de paille because they provide aromas. During the withering process, grapes designed for Vin de paille are dehydrated until their sugar level is extremely high. However, at the same time, such postharvest treatment can produce adverse effects in the grapes, increasing, for example, fungal development and ochratoxin A (OTA). OTA is a fungal metabolite dangerous for human health as it can have nephrotoxic, immunotoxic, mutagenic, teratogenic and carcinogenic effects (Walker, 2002; Bellí *et al.*, 2004; Gambuti *et al.*, 2005). During the water loss period, grapes are subjected to dehydration until they acquire the desired sugar content and the a_w drops from 0.95 to 0.75. In these conditions, only a few micro-organisms such as black *Aspergilli* are still able to develop (Brera *et al.*, 2008). Among the black *Aspergilli*, *A. carbonarius* is the most important source of OTA in non-fortified wines (such as Vin de paille). Abarca *et al.* (2003) demonstrated that dehydrated grapes were contaminated by OTA as a consequence of *Aspergillus spp.* infections. In order to determine the origin of OTA, the authors carried out a mycological study on dried grape samples. Fungal contamination was detected in 98% of the samples investigated and black *Aspergilli* were isolated from all the positive samples. *A. niger var. niger* was found in 98% of the samples while *A. carbonarius* was found in 58% of the total. For this reason, they concluded that black *Aspergilli* were the dominant fungal species and, among these, *A. carbonarius* had shown a consistent ability to produce OTA and was the main source of this mycotoxin in these substrates.

Wines such as Vin de paille that are made from grapes dehydrated in cool-dry chambers contain higher OTA levels (average 1.35 mg L^{-1} and 50% positive samples) than those made from grapes dried in hot-dry chambers (average 0.58 mg L^{-1} and 33% positive samples). This can depend on the longer period of time needed to dehydrate the grapes in the cool-dry chamber (2 to 6 months for Vin de paille) and on the environmental conditions, which are more favourable for fungal development and OTA synthesis. Gambuti *et al.* (2005) confirmed these findings by leaving the grapes dehydrating in an aerated room for two months, and showed that wine made from these grapes had OTA levels twice as high as wines obtained from grapes at full maturity. In the end, we can affirm that the variations in the OTA content depend on the dehydrating process, with levels decreasing in wines obtained from grapes dried in sunlight, in cool-dry chambers (such as Vin de paille) and in hot-dry chambers, respectively.

18.6 SENSORY CHARACTERISTICS OF VIN DE PAILLE

The aromatic profile of Vin de paille has not been investigated. Guichard *et al.* (1993) were the first to identify and quantify the compound sotolon (4,5-dimethyl-3-hydroxy-2(5H)-furanone) in Vin de paille made from Savagnin grapes. The highest levels (between 120 and 260 μg L^{-1}) are found in the French 'Vin jaune' and Tokay wines, and the lowest ones in Vin de paille (between 6 and 51 μg L^{-1}). Vin jaune is made using matured grapes from the same Savagnin vine; however, during its ageing in oak barrels, French Jura flor yeasts (mainly *S. cerevisiae*) develop, while one third of the volume evaporates. Obviously, in these wines, as well as in Tokay wines, the presence of sotolon is related to the use of grapes infested by mould. Tokay wines are partly grown under a yeast film like Vin jaune. Sotolon is the key compound in the French flor-sherry Vin jaune, due to its low threshold (10 $\mu g\,L^{-1}$ in wine) (Dubois *et al.*, 1976; Thuy *et al.*, 1995). Its odour is described as 'nutty' at lower concentrations and as 'curry' at higher ones. Sotolon contributes a great deal to the aromas of various foods. Takahishi *et al.* (1976) found that sotolon gives the burnt note to aged Saké (140–430 $\mu g\,L^{-1}$), as well as giving a curry aroma to fenugreek seeds (Girardon *et al.*, 1986). This molecule is also found in dried tobacco, soy sauce (more than 50 $\mu g\,L^{-1}$) (Nunomora *et al.*, 1976), and in sugar molasses (0.01–0.1 $mg\,L^{-1}$) (Otteneder and Majerus, 2000). Consequently, the sotolon amount is correlated with the food's organoleptic quality. It would thus be of great interest to monitor changes in the sotolon content of Vin de paille during barrel ageing.

As mentioned earlier, no in-depth research has been carried out on Vin de paille aromas. However, in the knowledge that Vin de paille is a sweet wine obtained by postharvest natural grape dehydration, and based on previous works on sweet wines (Schneider *et al.*, 1998; Miklósy *et al.*, 2000, 2004; Cliff *et al.*, 2002; Bellincontro *et al.*, 2004; Franco *et al.*, 2004; Miklosky and Kerényi, 2004; Guarrera *et al.*, 2005; Bailly *et al.*, 2006, 2009; Perestrelo *et al.*, 2006; Sarrazin *et al.*, 2007; Moreno *et al.*, 2008; Ortega-Heras and González-Sanjosé, 2009; Ruiz *et al.*, 2009), the following observations are made: the use of different sugar concentration methods allows the production of different types of sweet wines; however, their sensory descriptors are only dependent on the particular method used, especially when compared to ageing, which is a more critical variable (González-Álvarez *et al.*, 2012).

REFERENCES

Abarca, M.L., F. Accensi, M.R. Bragulat, *et al.* 2003. *Aspergillus carbonarius* as the main source of ochratoxin A contamination in dried vine fruits from the Spanish market. *Journal of Food Protection*, 66:504–506.

Bailly, S., V. Jerkovic, J. Marchand-Bryanet, *et al.* 2006. Aroma extraction dilution analysis of Sauternes wines. Key role of polyfunctional thiols. *Journal of Agricultural Food Chemistry*, 54:7227–7234.

Bailly, S., V. Jerkovic, A. Meurée, *et al.* 2009. Fate of key odorants in Sauternes wines through aging. *Journal of Agricultural Food Chemistry*, 57:8557–8563.

Bassermann-Jordan, F. 1975. *Geschichte des Weinbaus*, Pfalzische Verlagsanstalt GmbH, Neustadt an der Weinstrasse.

Bellí, N., S. Marin, A. Duaigües, *et al.* 2004. Ochratoxin A in wines, musts and grapes juices from Spain. *Journal of the Science of Food and Agriculture*, 84:591–594.

Bellincontro, A., D. De Santis, R. Botondi, *et al.* 2004. Different postharvest dehydration rates affect quality characteristics and volatile compounds of Malvasia, Trebbiano and Sangiovese grapes for wine production. *Journal of the Science of Food and Agriculture*, 84:1791–1800.

Boidron, R., J.M. Boursiquot, J.P. Doazan, *et al.* 1995. *Catalogue des variétés et clones de vigne cultivés en France*, Ed. ENTAV.

Brera, C., F. Debegnach, V. Minardi, *et al.* 2008. Ochratoxin A contamination in Italian wine samples and evaluation of the exposure in the Italian population. *Journal of Agricultural Food Chemistry*, 56:10611–10618.

Carreno-Diaz, R. and B.S. Luh. 1969. Anthocyanin pigments in Trousseau grapes. *Journal of Food Science*, 34:415–419.

Chaudat, P. 1997. *De la cuve au palais; le système d'approvisionnement des vins d'Arbois*. Thèse de Doctorat, Paris V-René Descartes.

Chaudat, P. 2004. *Les mondes du vin. Ethnologie des vignerons d'Arbois*, L'Harmattan, Paris.

Cliff, M., D. Yuksel, B. Girard, *et al.* 2002. Characterization of Canadian ice wines by sensory and compositional analyses. *American Journal of Enology and Viticulture*, 53:46–53.

Dubois, P., J. Rigaud, and J. Dekimpe. 1976. Identification of 4,5-dimethyltetrahydrofuran-4,5-dione in a flor sherry wine. *Lebensmittel-Wissenschaft & Technologiev*, 12:366–368.

Duchêne, E., J.L. Legras, F. Karst, *et al.* 2009. Variation of linalool and geraniol content within two pairs of aromatic and non-aromatic grapevine clones. *Australian Journal of Grape Wine Research*, 15: 120–130.

Fraignier, M-P., L. Marques, A. Fleuriet, *et al.* 1995. Biochemical and immunochemical characteristics of polyphenol oxidases from different fruits of Prunus. *Journal of Agricultural and Food Chemistry*, 43:2375–2380.

Franco, M., R.A. Peinado, M. Medina, *et al.* 2004. Off-vine grape drying effect on volatile compounds and aromatic series in must from Pedro Ximénez grape variety. *Journal of Agricultural and Food Chemistry*, 52:3905–3910.

Gambuti, A., D. Strollo, A. Genovese, *et al.* 2005. Influence of enological practices on ochratoxin A concentration in wine. *American Journal of Enology and Viticulture*, 56:155–162.

Girardon, P., Y. Sauvaire, J-C. Baccou, *et al.* 1986. Identification de la 3-hydroxy-4,5-dimethyl-2(5H)-furanone dans l'arôme des graines de fenugrec (*Trigonella foenum graecum* L.). *Lebensmittel-Wissenschaft & Technologiev*, 19:44–46.

Gonzalez-Álvarez, M., R. Noguerol-Pato, C. Gonzalez-Barreiro, *et al.* 2012. Sensory quality control of young vs. aged sweet wines obtained by the techniques of both postharvest natural grape dehydration and fortification with spirits during vinification. *Food Analytical Methods*, May:1–12.

Guarrera, N., S. Campisi, and C.N. Asmundo. 2005. Identification of the odorants of two passito wines by gas chromatography-olfactometry and sensory analysis. *American Journal of Enology and Viticulture*, 56:394–399.

Guichard, E., T.T. Pham, and P. Etievant. 1993. Quantitative determination of sotolon in wines by high-performance liquid chromatography. *Chromatographia*, 37:539–542.

Imazio, S., M. Labra, F. Grassi, *et al.* 2002. Molecular tools for clone identification: The case of the grapevine cultivar 'Traminer'. *Plant Breeding*, 121:531–535.

Loureiro, M.D., M.C. Martinez, J.M. Boursiquot, *et al.* 1998. Molecular marker analysis of *Vitis vinifera* 'Albarino' and some similar grapevine cultivars. *Journal of the American Society of Horticultural Science*, 123:842–848.

Maggu, M., R. Winz, P.A. Kilmartin, *et al.* 2007. Effect of skin contact and pressure on the composition of Sauvignon Blanc must. *Journal of Agricultural and Food Chemistry*, 55:10281–10288.

Miklosy, E. and Z. Kerényi. 2004. Comparison of the volatile aroma components in noble rotted grape berries from two different locations of the Tokaj wine district in Hungary. *Analytica Chimica Acta*, 513: 177–181.

Miklosy, É., Z. Kalmar, and Z. Kerényi. 2004. Identification of some characteristic aroma compounds in noble rotted grape berries and Aszú wines from Tokaj by GC-MS. *Acta Alimentaria*, 33:215–226.

Miklosy, É., Z. Kalmàr, V. Pölös, *et al.* 2000. Study of volatile aroma components in young Tokaji Aszu wines by GC-MS. *Chromatographia*, 51:S305–S308.

Moreno, J.J., F. Cerpa-Calderon, S.D. Cohen, *et al.* 2008. Effect of postharvest dehydration on the composition of pinot noir grapes (*Vitis vinifera* L.) and wine. *Food Chemistry*, 109:755–762.

Nunomora, N., M. Sasaki, Y. Asao, *et al.* 1976. Isolation and identification of 4-hydroxy-2(or 5)-ethyl-5(or 2)-methyl-3(2H)-furanone as a flavor component in Shoyu (soy sauce). *Agricultural and Biological Chemistry*, 40:491–495.

Ortega-Heras, M. and M.L. Gonzàlez-Sanjosé. 2009. Binding capacity of brown pigments present in special Spanish sweet wines. *LWT – Food Science and Technology*, 42:1729–1737.

Otteneder, H. and P. Majerus. 2000. Occurrence of ochratoxin A (OTA) in wine: influence of the type of wine and its geographical origin. *Food Additives & Contaminants*, 17:793–798.

Pacottet, P. 1903. Savagnins. In *Ampélographie* (eds P. Viala and V. Vermorel), Masson, Paris.

Perestrelo, R., A. Fernandes, F.F. Albuquerque, *et al.* 2006. Analytical characterization of the aroma of Tinta Negra Mole red wine: Identification of the main odorants compounds. *Analytica Chimica Acta*, 563:154–164.

Regner, F., A. Stadlbauer, C. Eisenheld, *et al.* 2000. Genetic relationships among Pinots and related cultivars. *American Journal of Enology and Viticulture*, 51:7–14.

Ruiz, M.J., L. Zea, L. Moyano, *et al.* 2009. Aroma active compounds during the drying of grapes cv. Pedro Ximénez destined to the production of sweet Sherry wine. *European Food Research and Technology*, 230:429–435.

Santiago, J.L., S. Boso, P. Gago, *et al.* 2007. Molecular and ampelographic characterisation of *Vitis vinifera* L. 'Albariño', 'Savagnin Blanc' and 'Caíño Blanco' shows that they are different cultivars. *Spanish Journal of Agricultural Research*, 5:333–340.

Sarrazin, E., S. Shinkaruk, T. Tominaga, *et al.* 2007. Odorous impact of volatile thiols on the aroma of young botrytized sweet wines: Identification and quantification of new sulfanyl alcohols. *Journal of Agricultural and Food Chemistry*, 55:1437–1444.

Schneider, R., R. Baume, C. Bayonove, *et al.* 1998. Volatile compounds involved in the aroma of sweet fortified wines (Vins Doux Naturels) from Grenache Noir. *Journal of Agricultural and Food Chemistry*, 46:3230–3237.

Takahashi, R., M. Tadenuma, and S. Sato. 1976. 3-Hydroxy-4,5-dimethyl-2(5H)-furanone, a burnt flavoring compound from aged sake. *Agricultural and Biological Chemistry*, 40:325–333.

Thuy, P.T., E.G. Guichard, S. Pascal, *et al.* 1995. Optimal conditions for the formation of sotolon from .alpha.-Ketobutyric acid in the French 'Vin Jaune'. *Journal of Agricultural and Food Chemistry*, 43:2616–2619.

Walker, R. 2002. Risk assessment of ochratoxin: current views of the European Scientific Committee on Food, the JEFCA and the Codex Committee on Food Additives and Contaminants. *Advances in Experimental Medicine and Biology*, 504:249–255.

19 Botrytized Wines: Sauternes, German Wines

Pierre Louis Teissedre and Bernard Donèche

Faculty of Oenology, Université Bordeaux Segalen, Institut des Sciences de la Vigne et du Vin, Villenave d'Ornon, France

19.1 INTRODUCTION

There are several legends surrounding the origin of botrytized wine, which is made from *Botrytis*-infected grapes. In France, it is said that their famous Sauternes wines are due to a château owner who once told his workers not to pick the grapes until he returned from his trip. By the time he got back, the grapes had been infected by a fungus that had caused them to shrivel. Despite their disgusting appearance, the grapes were picked and turned into wine, and the taste was so exquisite that the owner declared his grapes would thereafter always be picked after the fungus had arrived. The Germans have their own story for the invention of late harvest wines. In 1775, a messenger was sent to Schloss Johannisberg in Germany's Rheingau region to give the official order to start harvesting the grapes. He was robbed on the way and delayed. By the time he got to the wine estate, the grapes had begun to raisin on the vine. They were picked anyway and produced astonishingly delicious sweet wine. However, despite the legends, there is evidence that sweet late harvest wines were already in production throughout Europe in the seventeenth century.

The legendary friendly fungus is *Botrytis cinerea* Pers.:Fr., known affectionately as the 'noble rot'. It dehydrates the grape, leaving a more concentrated sweet juice. A range of grapes can benefit from the positive effects of noble rot: Sauvignon Blanc, Semillon and Riesling. *Botrytis cinerea* is a ubiquitous, filamentous and necrotrophic fungus that attacks over 200 different plant species (Jarvis, 1977). It is one of the principal causes of quantitative and qualitative degradation in numerous crops. In many vineyards around the world, this pathogen is responsible for one of the most serious diseases affecting grapevine, namely *Botrytis* bunch rot and/or grey mould. However, in a few particular areas, climatic, cultural and *terroir* conditions allow the pathogen to cause noble rot on (over-)mature grapes (Ribéreau-Gayon *et al.*, 1998). The infection process results in increased sugar concentration and improved quality due to the acquisition of characteristic aromas that permit the production of sweet white wines, such as Sauternes-Barsac, Côteaux du Layon, Tokay and some German wines.

When *Botrytis* affects a grape, it forms circular patches on the outer film. The fungus punctures the fruit, developing an intercellular mycelium under the cuticle, generally penetrating the first five to eight layers of cells to a depth of approximately 1 mm. These patches gradually extend until the surface of the grape is completely covered. The film is digested,

Sweet, Reinforced and Fortified Wines: Grape Biochemistry, Technology and Vinification, First Edition.
Edited by Fabio Mencarelli and Pietro Tonutti.

Figure 19.1 Stages of noble rot development in berries. (For a colour version of this figure, please see the colour plate section.)

becoming thinner and more fragile, and purple-brown in colour. At this stage, the grapes are called 'fully rotten'. When the action of the mould is prolonged, the grapes become completely dessicated and wrinkled, and are described as 'roasted' (*rôti*). This is the optimal state for harvesting grapes for botrytized wines (Figure 19.1).

The interaction pathogen–host is described in Chapter 11.

19.2 SAUTERNES

The Sauternes area is situated in southwest France, approximately 40 km to the south of Bordeaux (Figure 19.2). It is crossed from north to south by the Ciron, a small tributary of the River Garonne, and is bordered in the south by the forested massif of the Gascony moors. The Sauternes appellation extends over just five communes: Sauternes, Barsac, Preignac, Bommes and Fargues. This region counts approximately 240 producers and 26 *Grand Crus* (vintage wines). Sauternes wines represent less than 1% (2000 ha) of the total Bordeaux vineyard area. Grand crus vineyards represent 45% of this surface but produce only 30% of the total quantity.

The Sauternes area sits on an alluvial plain consisting of diverse soil types: calcareous in the north and clayey in the south. The vineyards are cultivated on grounds comprising gravel layers on a subsoil of clay; a mixture of deep translucent sands and pure gravel with some banks of red clay and limestone marl, and limestone layers with oyster shell deposits, or red and white clays.

The climate is moderated by the proximity of the ocean. The maritime climate extends over the five communes of the Sauternes region, and also those of Cérons and Pujols-sur-Ciron where *Botrytis* also proliferates. Pine forests protect the region from the oceanic winds. Sauternes benefits from a microclimate that is essential to the characteristics of its wine. The spring is wet and warm with an annual average rainfall of 860 mm. However, the spring frosts and abundant rainfall can provoke 'coulure' (premature dropping of flowers or young fruit), which can reduce or even annihilate the harvest. The summer is warm, with an average temperature in July of 20°C. In autumn, morning mists settle over the vineyards, followed by a radiant sun. This seasonal combination of humidity and warmth determines the resulting wine as it favours the proliferation of *Botrytis cinerea*, the origin of the famous noble rot. The mists and heavy night-dews are a consequence of significant evaporation from the valley floor, which widens where it meets the Ciron in Sauternes, compounded by the scarcity of woodland along the river banks. The sunny periods slow the mushroom's development allowing the sugars to concentrate inside the fruit. The wind also plays an important part by drying the vineyard, which further contributes to sugar concentration inside the berries.

The rich and fruity Semillon grape is very sensitive to *Botrytis* attack (Figure 19.3). Its early budding is weakened by spring frosts. Vine leaves are quite large and thick, with five

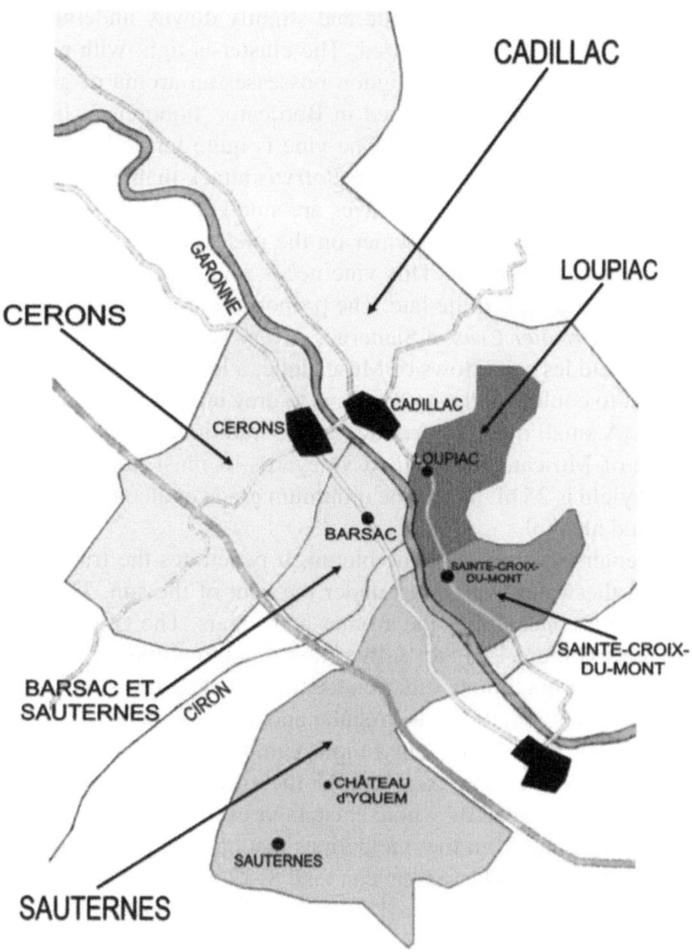

Figure 19.2 Map of the winegrowing regions of Sauternes. (For a colour version of this figure, please see the colour plate section.)

Figure 19.3 Semillon grape variety botrytized (noble rot). (For a colour version of this figure, please see the colour plate section.)

lobes, a pale-green colour on the upper side and slightly downy underneath. Vine shoots are slightly flattened and mahogany-coloured. The cluster is tight with round berries, and has a golden tint and delicate taste. Sauvignon possesses an aroma of great delicacy and a characteristic flavour; however, when used in Bordeaux 'liquoreux' (liquor-like) wines, the astringent first taste can seem unusual. The vine is quite fertile but can be susceptible to coulure. Sauvignon grapes are less prone to *Botrytis* attack than Semillon. However, the grapes are very sensitive to evaporation. Leaves are small and thick, with three lobes, and are a beautiful dark green above and downier on the under side. Vine shoots are solid, of cinnamon colour stained with brown. This vine needs rocky, clayey and light soils, and a calcareous base. Its maturation is quite late. The proportion of white Sauvignon varies from 2 to 45% among the 11 *Premier Crus* of Sauternes. Every vineyard decides on its own grape variety, which often includes some rows of Muscadelle, a less noble but very aromatic vine. Muscadelle is subject to coulure and very sensitive to grey mould, which means that *Botrytis* does not thrive well. A small quantity is added to the traditional grape variety of liquoreux wines to add a hint of Muscat. In classified vineyards, it must not exceed 3 to 5%. The authorized maximal yield is 25 hL ha^{-1}. The minimum grade of alcohol is 13% vol. of which 12.5% vol. is acquired alcohol.

Botrytis cinerea enters the berry via the bloom. It penetrates the fruit and creates small holes through which the water evaporates under the heat of the sun. The fungus develops inside the berry where it concentrates the aromas and sugars. The grape passes through the stages of over-ripening (Figure 19.1) to 'fully rotten', and finally it wrinkles and dessicates to become 'confit' with a maximum sugar concentration. The invasion of *Botrytis*, which depends on general weather conditions, is irregular and unpredictable in the same plot of land but also on the same foot and cluster. The fungus can develop differently according to the berries. Grape pickers must therefore pass through the vines several times to pick the grapes as they become ready. This can be by whole clusters or even by individual berries. The best Sauternes wines usually come from low-yield vines that have been hand-picked, on average between 5 to 12 separate times. Harvesting can take 8–12 weeks because winemakers want only the most perfectly botrytized grapes. This is the principal reason why Sauternes are so coveted – and expensive.

The *B. cinerea* fungus contributes a desirable honeyed and complex nature to both the aroma and flavour. The resulting classic Sauternes wine is rich and sweet. Thomas Jefferson visited the Bordeaux vineyards in May 1787. His detailed notes on vineyard hierarchy and descriptions of specific areas, such as Sauternes, Preignac and Barsac, and notably Yquem, evidence his great interest in the vineyards and the wine. The Sauternes châteaux were ranked in the Official Classification of 1855. Château D'Yquem was elevated to a class of its own, which is known variously as Premier Grand Cru, Grand Premier Cru or Premier Cru Superieur. The Château undisputedly makes the best wines in the area and some consider them to be the finest in Bordeaux. Among the other châteaux that produce extraordinary wines are Château Rieussec, Château Suduiraut and Château Climens. The classification also named 11 Premiers Crus (La Tour-Blanche, Château Coutet, Château Climens, Château Lafaurie-Peyraguey, Château Guiraud, Clos-Haut-Peyraguey, Château Rayne-Vigneau, Château Rabaud-Promis, Château Rieussec, Château Sigalas-Rabaud, Château Suduiraut), and 15 Deuxième Crus (Château D'Arche, Château Broustet, Château Le Caillou, Château Doisy-Däene, Château Doisy-Dubroca, Château Doisy-Védrines, Château Filhot, Château Lamothe Guignard, Château De Myrat, Château De Malle, Château Nairac, Château Romer du Hayot, and Château Suau).

19.3 BOTRYTIZED GERMAN WINES

German wines are primarily produced in western Germany, along the banks of the River Rhine. Approximately 60% of German wine production is situated in the federal state of Rhineland-Palatinate, where 6 of the 13 wine-producing regions (*Anbaugebiete*) known for quality wine are situated (Figure 19.4).

Riesling has been documented since 1435 (close to Rheingau where this variety now accounts for nearly 80% of production). Rheingau follows the Rhine from Hochheim to Lorch, within the federal state of Hesse. The 12.5 miles (20 km) from Wulluf to Bingen are the only point at which the Rhine flows in an east–west direction for any significant distance. It is this deviation from its usual northerly journey, around the Taunus hills, that is the heart of the Rheingau and provides the steep, south-facing slopes that are home to the region's vineyards. Rheingau has approximately 7750 acres (3100 ha) of vines. Climatically, the Rheingau is cool-continental with an annual average temperature of 10°C, 1643 hours of sunshine per year, and only 536 mm of annual rainfall (conditions similar to those in the Napa Valley, California). Geology is varied with slate, quartzite, sandstone, gravel and loess (a type of sedimentary soil formed by the continuous deposition of sand and clay, often loosely held together by limestone). The majority of this soil is quartz (up to 80%), with small

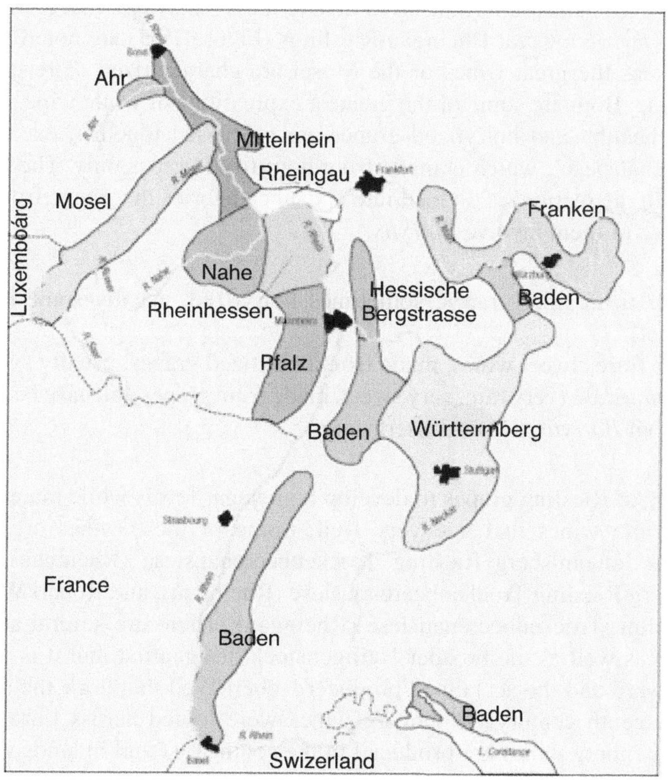

Figure 19.4 Map of the winegrowing regions in Germany. (For a colour version of this figure, please see the colour plate section.)

Figure 19.5 Riesling variety botrytized grapes. (For a colour version of this figure, please see the colour plate section.)

amounts of feldspath, mica and other minerals, all of which influence the characteristics of the wines. Soils rich in loess are porous and well drained, and generally very fertile. This *terroir* means that although the Rheingau makes up only 3% of Germany's total vineyard area, it accounts for 44 first-growth sites and nearly one-fifth of Germany's 244 VDP-classified *Erste Lage* (first-class) sites (a term used to denote those vineyards that are recognized for their exceptional *terroir*). Great Rheingau Rieslings (Figure 19.5) are noted for their depth and power, whereas the great wines of the Mosel are characterized more by their superb elegance and poise. Both are some of the greatest expressions of white wine.

In Germany, healthy and botrytized grapes are processed together, except in the case of 'Trockenbeerenauslese', which is made from botrytized grapes only. The QmP category (quality wine with distinction) is divided into six subcategories that range from the driest to the sweetest; three of them involve *Botrytis*:

- *Auslese* (made from riper grapes, sometimes botrytized, usually some residual sugar, generally sweet)
- *Beerenauslese* (rare, sweet wines, made from botrytized grapes, usually sweet)
- *Trockenbeerenauslese* (very rare, very sweet, made from grapes that have been fully 'dried' on the vine from *Botrytis*, sweet-dessert style).

It is the ability of Riesling grapes to develop high sugar levels while maintaining acidity that produces white wines that age very well. Some of these wines are very famous, such as: Schloss Johannisberg Riesling Trockenbeerenauslese (Rheingau), Robert Weil Kiedrich Turmberg Riesling Trockenbeerenauslese (Rheingau), and Robert Weil Kiedricher Grafenberg Riesling Trockenbeerenauslese (Rheingau). There are several single vineyard sites in Kiedrich as well as the broader Heiligenstock designation, but it is the Erste Lage Gräfenberg vineyard and the acclaimed producer Robert Weil that draw the most attention. By the late nineteenth century Gräfenberg wines were touted across Europe as German Grand Crus. Gräfenberg wines are produced in the south-west and in lands with a gradient of up to 60%. The soils are rocky and well hydrated with prevailing winds that allow for an extended maturation time, resulting in elegant Riesling akin to that of the Mosel. Bereich Johannisberg is the only official subregion of Rheingau and is contiguous with the entire Rheingau region. It stretches from the mouth of the Main river where it flows into the Rhine just south of Wiesbaden, following the Rhine first north, and then westward, until the river again swings north at Bingen and on to Lorch, where the Mittelrhein begins. The Erste Lage Schloss Johannisberg vineyard does not have to put a village name on its wine labels. This is due in part to its status as a monopole estate, but more notably because it was the first site to deliberately cultivate spätlese (late harvest) grapes in 1775. Thomas Jefferson declared the producer to be 'the best in the Rhine, incomparable to anyone else'. Riesling can be produced in a range of different styles, with associated diverse tastes. The key to all wines is balance, which is something that good Riesling producers excel at. When grown in cool climates, Riesling has the ability to reach levels of acidity that can match any degree of sweetness. Even the most honeyed and *Botrytis*-affected sweet wines can still taste fresh, and the most mouth-puckering dry wines can show delicate floral notes of lime blossom and perfume.

19.4 COMPOSITION OF BOTRYTIZED GRAPES AND WINES

The chemical indicator of *B. cinerea* infection on grapes is the presence of gluconic acid. In particular the *Gluconobacter oxydans* bacteria is active in the synthesis of acetic acid (acetic acid bacteria invades botrytized grapes) as well as gluconic acid and ethyl acetate (Sponholz and Dittrich, 1985). Gluconic acid is not known to be of sensory significance in wine as the sweet taste of the cyclic ester is too slight to be perceived. It has been shown that intramolecular gluconic acid esterification reactions led to the formation of two lactone γ- and δ-gluconolactone (respectively glucono-1,4-lactone and glucono-1,5-lactone) in botrytized must or wine. These lactones are in equilibrium with gluconic acid, γ- and δ-gluconolactone representing, respectively, 5.8 and 4.1% of the acid level. Correlations between the must's SO_2 binding power, gluconic acid and consequently its lactones, have been shown. The SO_2 affinity of a mixture containing this acid and γ- and δ-gluconolactone was determined, and gluconic acid appeared to be indirectly responsible for 8% of the bindable SO_2 in musts from botrytized grapes (Barbe *et al.*, 2002). The significant changes in composition of the grapes can be attributed to byproducts of *B. cinerea* metabolism and the juice concentration from berries drying (Table 19.1). An evaporative water loss is noticeable (the weight of the fresh berries is reduced by 50%) with an increase in sugars content (16–30%) and gluconic acid (between 1 to 2 g L^{-1}), and a high reduction in the concentrations of tartaric acid (close to 2.8-fold), citric acid (30–70%) and ammonia (close to sevenfold) in *Botrytis-infected* berries (Charpentié, 1954). Galactose and arabinose have been found to be higher in wines made from *B. cinerea*-infected grapes than in those made from healthy grapes (Dittrich and Barth, 1992). Glucose can in some cases be metabolized, which explains the higher fructose-to-glucose ratio in botrytized grapes and wines. A creation and accumulation of glycerol and others polyols (sorbitol, xylitol, arabitol, erythyritol, mannitol, meso-inositol) occurs during noble rot development. High levels of glycerol (14.6 to 24.7 g L^{-1}) have been reported in botrytized German Trockeenberenauslese and French Sauternes wines (Amerine, 1954). Glycerol is typically found at concentrations of 4–10 g L^{-1} in dry wine and in the case of the noble rot and late harvest wines, levels in excess of 20 g L^{-1} are not uncommon (Ribéreau-Gayon *et al.*, 1998). In order to have an effect on the viscosity of a wine (Noble and Bursick, 1984), the glycerol concentration must be of 25.8 g L^{-1} and botrytized wines can reach that level. Erythritol and sorbitol can each reach a level between 200 and 250 mg L^{-1} in botrytized wines (Liu and Davis, 1994). During infection by *Botrytis*, the synthesis of

Table 19.1 Chemical composition for Sauternes and German botrytized grapes and wines.

	1000 berries[a]	
	Healthy berries	**Infected roasted rotten berries**
Fresh weight (g)	2020	980
Must volume (mL)	1190	450
Sugars (g)	294	143
Total acidity (meq)	146	50
pH	3.33	3.62
Tartaric acid (meq)	85	15
Malic acid (meq)	96	54
Citric acid (meq)	3.2	1.6
Acetic acid (meq)	6.4	2.4
Gluconic acid (meq)	0	4.8
Ammonia (mg)	101	25
Amino acids (mg)	1526	638
Proteins (mg)	3350	1708

Wines	BE1[b]	BE2[b]	BE3[b]	TR1[b]	TR2[b]	SA1[c]	SA2[c]
Total extract ($g\ L^{-1}$)	163	152	119	299	300		
Sugar level ($g\ L^{-1}$)	74	103	78	224	194	96	186
Alcohol (% vol.)	7.9	6.3	7.9	5.3	6.4	11.55	15.05
Total acidity ($g\ L^{-1}$)	8.7	9.4	7.9	11.4	10.5	4.5	5
Glycerol ($g\ L^{-1}$)	12.0	13.6	10.9	13.0	40.0	16	25
Acetic acid ($g\ L^{-1}$)						0.53	1.32
Acetaldehyde ($mg\ L^{-1}$)	73	62	139	56	163	22	48
Tannin ($mg\ L^{-1}$)	250	390	390	291	446		
pH	3.2	3.2	3.0	3.6	3.5	3.64	4.1

BE, Beerenauslese; TR, Trockenbeerenauslese; SA, Sauternes
Sources: [a] Ribéreau-Gayon *et al.* (1998)
[b] Watanabe and Shimazu (1976)
[c] Divol (2004) vintage 2003

macromolecular polysaccharides compounds occurs in grapes. Two groups of compounds can be distinguished: one includes the polymers of mannose and galactose, with small amounts of rhamnose and glucose (20 000 to 50 000 daltons), resulting in an increase in glycerol and acetic acid during alcoholic fermentation (Dubourdieu, 1981). A second family consist of ß-glucans, polymers of glucose (100 000 to 1 000 000 daltons), which when present even in small quantities can block filters during the clarification process. This is one of the reasons why ß-glucanase can be used during winemaking to hydrolyse these compounds in order to faciliate filtration and clarification. The pressing of botrytized grapes after harvesting is generally light in order to minimize the liberation of ß-glucans into the juice (ß-glucans are localized under the berries skins with the fungal cells). *Botrytis* can also induce calcium salt of mucic acid, by producing an enzyme that can oxidize galacturonic acid to mucid acid (galactaric acid). Mucid acid binds slowly with calcium and forms an insoluble salt of mucate calcium with small white clumps of precipitate particles, which occasionally form in bottles when wine contains more than 0.1 $mg\ L^{-1}$ mucid acid (the level reached with 10–25% of *B. cinerea*-infected grapes) (Wuerdig, 1977). The laccases family is a group of enzymes produced by *B. cinerea*; they are suspected to inactivate anti-fungal phenolics such as stilbenes. Different laccases are induced by grape juice, and pectin may increase laccase synthesis in the presence of phenolics (Marbach *et al.*, 1985). Laccases can oxidize different types of phenols (para, ortho and meta-diphenols, diquinones, tannins), including quinone glutathione complex (Salgues *et al.*, 1986) and non-phenols (ascorbic acid). Laccase activity is 30 times higher than tyrosinase and it is stable at juice pH and is not inhibited by SO_2 (concentrations of 50 mg L^{-1} SO_2 at pH 3.4 are necessary in wine to inhibit laccase action but about 125 mg L^{-1} SO_2 is needed in most cases; Kovac, 1979). Unlike grape polyphenol oxidase (PPO), fungal laccases (LAC) are only slightly inhibited by oxidized phenolics (Dubernet, 1974). More severe browning occurs in juices obtained from botrytized grapes compared to juices from sound grapes. Two types of oxidizing enzymes are present in botrytized white grapes and wines: laccase and glucose oxidase (GOX) have similar evolution both during the over-ripening of grapes and during winemaking. Yet, PPO is severely inhibited by the addition of SO_2 following the alcoholic fermentation, and shows a marked instability in both the must and wine environments. GOX, however, remains free and active in solution and helps develop the main characteristics of the wine. In particular, as is to be expected from its activity, GOX oxidizes tartaric acid, ethanol and glycerol (the major components of must and wine), respectively to glyoxylic acid, acetaldehyde and glyceraldehyde. Then, by nucleophilic additions under acidic conditions, these products react with catechins and proanthocyanidins to form several new compounds, some of which appear in a coloured form. These reactions can have an impact on the visual quality of the wine (Vivas *et al.*, 2010). Alcoholic fermentation is stopped by SO_2 addition, an operation called 'mutage'. These wines present an alcohol level ranging between 13 and 15% vol. and often contain more than 60 g L^{-1} residual sugars. These wines, which are not filtered, are conserved in barrels for 12–24 months or more. Musts from botrytized grapes have levels of SO_2-binding compounds higher than those from healthy grapes (Asvany, 1985). In wine, according to Ribéreau-Gayon (1972), these compounds come either from alcoholic fermentation or from the oxidation of grape sugars. Sulfur-dioxide-binding compounds such as galacturonic, glucuronic, and 2- and 5-oxogluconic acids (Sponholz and Dittrich, 1984), xylosone and 5-oxofructose (Burroughs and Sparks, 1964), and R-dicarbonyl compounds such as glyoxal and methylglyoxal (Barbe *et al.*, 2000), have already been described. Some of the carbonyl compounds found in musts from botrytized grapes, such as glyoxal and methylglyoxal, may explain a considerable part of bindable SO_2. Others, such as 2- and 5-oxogluconic acids, produced by gluconic acid oxidation in proportions respectively of 2.5 and 1, play an interesting role as SO_2-binding indicators. The levels of some compounds such as dihydroxyacetone, 5-oxofructose, and ∂-gluconolactone in balance with gluconic acid are well correlated with SO_2 binding powers and also explain a large part of the bindable SO_2 in musts. During alcoholic fermentation, among these three compounds, only dihydroxyacetone is metabolized by yeast. Thus, two compounds present in grapes, ∂-gluconolactone and 5-oxofructose, with three yeast SO_2-binding byproducts, ethanal, pyruvic, and 2-oxoglutaric acids, explain much of the SO_2-binding power in wines from botrytized grapes (Barbe *et al.*, 2000). In earlier studies, hydroxypropanedial has been isolated and identified from musts and wines produced from botrytized grapes. This triose reductone was possibly characterized by a large SO_2 combination power and redox properties. It is normally present in very low concentrations in dry white or red wines, but this carbonyl compound occurs in huge levels if B. cinerea and other damaging moulds with associated rotten and acetic characters are present in spoiled berries (Guillou *et al.*, 1997). Hydroxypropanedial (a reductone) has been found in sweet botrytized white wines (vins liquoreux, mainly Sauternes) between 25 to 612 mg L^{-1}. Reductones may react with amino acids and form browning compounds (Yamagushi, 1969; Shimohara *et al.*, 1974). Products of the browning reaction are also observed in most dicarbonyl compounds such as glyoxal, methylglyoxal and diacetyl (Velisek and Davidek, 1978). During noble rot development, the evolution of stilbene monomers and oligomers for Sauvignon and Semillon grapes from the Sauternes appellation area has been observed (Landrault *et al.*, 2002). For Semillon grapes, the levels of trans-astringin, trans-piceid, trans-resveratrol and pallidol increase, reaching an optimum level at fully rotten and roasted rotten stages. In the case of Sauvignon grapes, the levels of trans-resveratrol appear to decrease during all stages of noble rot development, but levels of trans-astringin, trans-piceid and pallidol are highest for the spotted grape stage. However, for both varieties, the levels of *trans*-astringin, *trans*-resveratrol and *trans*-piceid are quite low (<0.13 mg kg^{-1} for each compound), reaching a maximum of 0.5 mg kg^{-1} for pallidol, during the different stages of noble rot development. These results are in agreement with the ability of pathogenic *B. cinerea* strains to biotransform resveratrol into a variety of unidentified oxidized metabolites as a means of reducing the antifungal effects of resveratrol, thus facilitating invasion of the fungus into host-plant tissues. For viniferin, the highest level is obtained during stage 3 of noble rot development, for both Semillon and Sauvignon grapes, with levels close to 2 mg kg^{-1}. Fully rotten and roasted rotten grape stages seem to affect the level in this molecule for both varieties in that the levels are reduced to almost 50%. During the last stages of noble rot development viniferin could be used by the plant to act against *Botrytis* development or could be combined with other phenolic compounds to protect some vegetal cell structures. Levels of astilbin during noble rot development appear almost stable for Semillon but increase considerably for Sauvignon to between +300 and +400% to reach concentrations between 25 and 30 mg kg^{-1} for speckled and spotted grape stages, falling to 7.11 mg kg^{-1} at the roasted rotten grape stage. Like viniferin, astilbin could be used by the plant to act against *Botrytis* development. In botrytized Sauternes wines (vintages 1986 to 1999), concentrations of viniferin were found to be between 0.08 mg L^{-1} and 0.17 mg L^{-1}, while those for astilbin were between 0.84 mg L^{-1} and 5.86 mg L^{-1} (Landrault *et al.*, 2002).

The concentrations of resveratrol-derivatives and their antioxidative capacity in wines made from botrytized grapes have also been measured (Pour Nikfardjam *et al.*, 2006). Compared to white wines of normal quality (resveratrol 0.5–4.4 mg L^{-1}; mean: 2.1 mg L^{-1}), the wines made from German botrytized grapes had lower concentrations of resveratrol and piceid (from <0.003 to 6.3 mg L^{-1}; mean: 0.9 mg L^{-1}). This might be due to the fact, that the stilbene oxidase produced by *Botrytis* oxidizes most of the resveratrol to its corresponding metabolites. As reported by Jeandet *et al.* (1995) and Roldan *et al.* (2003), extensive fungus pressure leads to a reduction of resveratrol production in the grapes. Although the shrivelling of the grapes – caused by *Botrytis* – might concentrate the contents of the berries, at the same time the stilbene oxidase decreases resveratrol and piceid concentrations, in which oxidation is obviously the predominant process. In total polyphenol content, the wines from German botrytized grapes (vintages 1990–1999) show higher concentrations than dry white wines, due to the high amounts of botrytized berries. Values ranged from 248 to 747 mg L^{-1} with a mean of 441 mg L^{-1}. However, prolonged berry skin contact with the mash is generally avoided in Germany to protect the wine from a bitter taste, 'negative' browning reactions, and astringency. The antioxidative capacities for German wines ranged from 0.6 to 2.8 mmol L^{-1} (mean: 1.4 mmol L^{-1}) (Pour Nikfardjam, 2006). However, in recent work (Franc *et al.*, 2012), nine oligostilbenes (2 monomers, 3 dimers, 1 trimer and 3 tetramers) and three flavonols, previously characterized in grapes, were identified and quantified. Results of samples of Semillon and Sauvignon grape pomaces from 2009 processed at different stages of *B. cinerea* are shown in Table 19.2.

Table 19.2 Stilbenes concentration (mg 100 g^{-1} dw) of pomace extracts collected in 2009 at different stages of *B. cinerea* development (Franc *et al.*, 2012).

	Semillon			Sauvignon	
	Healthy berries	**Speckled berries**	**Roasted rotten berries**	**Healthy berries**	**Roasted rotten berries**
Ampelopsin A	0.7 ± 0.7	4.8 ± 4.5	9.1 ± 5.2	0.4 ± 0.0	30.3 ± 2.6
Hopeaphenol	65.2 ± 20.6	13.6 ± 13.6	19.1 ± 7.3	101.6 ± 0.1	68.9 ± 12.9
Piceatannol	12.2 ± 2.4	21.5 ± 6.9	38.3 ± 29.2	14.3 ± 7.5	35.0 ± 3.7
Resveratrol	135.6 ± 19.5	110.3 ± 91.1	1093.8 ± 70.6	265.7 ± 37.1	1483.4 ± 46.8
R-2-viniferin	28.1 ± 31.2	112.2 ± 121.9	160.6 ± 152.1	193.7 ± 76.4	75.8 ± 26.7
Miyabenol C	174.7 ± 17.8	211.9 ± 57.5	842.2 ± 87.7	521.2 ± 54.4	841.1 ± 86.0
ε-viniferin	462.7 ± 81.0	174.7 ± 48.4	723.5 ± 112.4	812.6 ± 71.0	1027.8 ± 69.2
R-viniferin	32.5 ± 8.3	4.3 ± 2.3	22.9 ± 14.1	83.7 ± 28.8	45.7 ± 17.3
iso-ε-viniferin	34.5 ± 0.8	4.1 ± 2.4	63.6 ± 21.0	34.0 ± 16.6	54.2 ± 10.1
Total stilbenes	946.3 ± 117.0	657.3 ± 313.1	2973.1 ± 297.8	2027.1 ± 145.1	3662.2 ± 112.3

19.5 DIFFERENCES BETWEEN BOTRYTIZED (SAUTERNES AND GERMAN) WINES: SPECIFIC RULES AND PRECAUTIONS

The German wines Beerenauslese and Trockenbeerenauslese are made with botrytized grape juices containing more sugar than is converted to alcohol during fermentation. The wines are correspondingly sweet and low in percentage alcohol (generally between 6 to 8% vol.). Typically, there is just one sweet style produced in Sauternes, whereas there are several botrytized styles produced in Germany.

For Sauternes, legislation dictates that the grapes must reach a minimum level of 221 g L^{-1} of sugars to be sufficiently mature and that the finished wine must have at least 15% vol. of natural alcohol and 12% vol. of potential alcohol. The rate of residual sugar must be greater than or equal to 45 g L^{-1}. The total alcohol level cannot exceed 21%. In Germany the sugar level is defined in Oechsle degrees. On the Oechsle scale, one degree Oechsle (°Oe) corresponds to one gram of the difference between the mass of one litre of must at 20°C and 1 kg (the mass of 1 litre of water); for example, must with a specific mass of 1084 g L^{-1} has 84°Oe. Auslese means literally 'selective harvest'. It is the highest Prädikat besides the rare Beerenauslese and Trockenbeerenauslese and is normally 'nobly' sweet, though dry and off-dry versions also exist. Auslesen are usually made from grapes affected by *Botrytis*. On the Mosel, the minimum must weight for Auslese is 88°Oe. Before the 1971 German Wine Law, with its corresponding Prädikat system, was introduced, Auslese wines were made from specially selected grape bunches, but not necessarily botrytized. Beerenauslese, the second highest Prädikat in the must-weight hierarchy, are noble sweet wines made from predominantly botrytized grapes, with perhaps a small amount of the ripest yellow berries added. Required must weights on the Mosel are at least 110°Oe. Formerly, Beerenauslese wines were defined as Auslese of a particularly fine quality. Trockenbeerenauslese ('dry berry selection'), has a Prädikat level for wines made from selected shrivelled grapes affected by noble rot. Wines produced in the Mosel region must have at least 150°Oe. The best Trockenbeerenauslese are amongst the sweetest (sometimes over 300 g natural sugar) and most expensive dessert wines, often containing only 6–8% alcohol.

However, cautious concern exists over the residual sugars content in botrytized wines because it can make the wine unstable with respect to micro-organisms. Measures have to be taken to avoid possible microbial spoilage. These include: cooling the wine by refrigeration to place yeasts in unfavourable growth conditions (+5°C); sterile filtration systems to remove yeasts; and use of SO_2 to stop the yeasts multiplying (to avoid off-flavours in the wine, a target level of 40–60 mg L^{-1} free SO_2, adjusted according to the pH, is the rule, but the maximum limit of 300 mg L^{-1} total SO_2 in liquoreux wines must be respected).

Sorbic acid can also be used (authorized maximum dose of 200 mg L^{-1} or 20 g hL^{-1}) as it possesses antifungal properties and can block yeast activity; it is therefore useful in avoiding refermentations of sweet wines. Sorbic acid (approved for treatment as potassium sorbate with a maximum residue in wine of 200 mg L^{-1}) is an effective preservative against yeasts when applied at a concentration of 200 mg L^{-1} in combination with 100 mg L^{-1} of SO_2. Special precautions should be taken when using *Saccharomyces bayanus* as pure culture as this strain is resistant to sorbic acid. Sorbic acid is not lethal to yeast cells but only inhibits activity. A sterile filtration system should be used for sweet wine when employing sorbic acid as a preservative (Tromp and Agenbach, 1981). Dimethyl dicarbonate (DMDC) also has an interesting fungicidal action that inactivates wine spoilage yeasts such as *Brettanomyces*. It is approved for use in the EU for wines presenting more than 5 g L^{-1} of residual sugars and the legal maximum dose is 200 mg L^{-1}. The risk in the use is the production of methanol due to DMDC degradation. In one study, (Costa *et al.*, 2008), several microbial species associated with wine were tested against increasing concentrations of DMDC. The concentration that induced complete cell death upon addition to wine was regarded as the minimum inhibitory concentration (MIC). In wines with 12% (v/v) ethanol and pH 3.50, the inactivation depended on the initial cell concentration. For an initial inoculum of 500 CFU ml^{-1}, the MIC of the yeast species *Schizosaccharomyces pombe*, *Dekkera bruxellensis*, *Saccharomyces cerevisiae* and *Pichia guilliermondii* was 100 mg L^{-1}. The most sensitive strains belong to *Zygosaccharomyces bailii*, *Zygoascus hellenicus* and *Lachancea thermotolerans*, with a MIC of 25 mg L^{-1} DMDC. For inoculation rates of about 106 CFU ml^{-1}, the maximum dose of DMDC legally authorized was not effective against the most resistant species. The addition of 100 mg L^{-1} potassium metabisulphite (PMB), equivalent to 1 mg L^{-1} molecular SO_2, increased the inactivation effect of 100 mg L^{-1} DMDC over initial yeast populations of 106 CFU mL^{-1} but did not kill *S. pombe* and *S. cerevisiae* completely. Lactic acid and acetic acid bacteria were not killed by the addition of 300 mg L^{-1} DMDC. Trials performed in wines before bottling showed that in most samples indigenous bacterial populations were not affected by 200 mg L^{-1} DMDC. Therefore, under winery practice, DMDC at the maximum dose legally permitted may be regarded as an efficient preservative to control low contamination rates of yeasts, but it is ineffective against lactic acid and acetic acid bacteria (Costa *et al.*, 2008).

Production of *B. cinerea* for potential introduction into a vineyard has been studied because dry growing regions are less likely to become naturally infected. Inoculation of grapes was performed (Akau *et al.*, 2004) and *B. cinerea* was produced in both solid-phase fermentation and liquid fermentation. Stored products were evaluated for grape colonization in grape bioassays and in field trials, and for *B. cinerea* density using colony-forming unit analyses and a nucleic-acid-based method. *Botrytis cinerea* colony-forming unit density was significantly correlated to the probability of successful grape colonization in grape bioassays (*P*-value: 0.0002). There was little difference in the rate of grape colonization in laboratory bioassays among solid-phase fermentation, liquid fermentation and plate culture products. Although the initial *B. cinerea* colonization rate of field grapes was slightly greater on vines treated with solid-phase fermentation and plate culture products compared with vines treated with product from liquid fermentation, there was no significant difference in final colonization between vines treated with solid-phase fermentation, liquid fermentation and plate culture products, and untreated vines. The lack of difference may have been due to the natural inoculum of *B. cinerea* present in the field at the time of inoculation, vegetation management, and vectoring of *B. cinerea* by insects. The environmental management of vineyards and inoculum history play an important role in the production of botrytized wines.

REFERENCES

Akau, H.L., K.M. Miller, N.C. Sabeh, *et al.* 2004. Production of *Botrytis cinerea* for potential introduction into a vineyard. *Bioresource Technology*, 92:41–48.

Amerine, M.A. 1954. Composition of wines. 1: Organic constituents. *Advances in Food Research*, 5:353–370.

Asvany, A. 1985. Les techniques de vinification permettant de diminuer les doses de SO_2. *Bulletin de l'OIV*, 652-653:621–623.

Barbe J.C., G. De Revel, and A. Bertrand. 2002. Gluconic acid, its lactones, and SO_2 binding phenomena in musts from botrytized grapes. *Journal of Agricultural and Food Chemistry*, 50:6408–6412.

Barbe, J.C., G. De Revel, A. Joyeux, *et al.* 2000. Role of carbonyl compounds in SO_2 binding phenomena in musts and wines from botrytized grapes. *Journal of Agricultural and Food Chemistry*, 48:3413–3419.

Burroughs, L.F. and A.H. Sparks. 1964. The identification of sulfur dioxide-binding compounds in apple juices and ciders. *Journal of the Science of Food and Agriculture*, 15:176–185.

Charpentié, Y. 1954. Contribution à l'étude chimique des facteurs de l'acidité des vins. Thèse d'Ingénieur-Docteur, Bordeaux.

Costa, A., A. Barata, M. Malfeito-Ferreira, and V. Loureiro. 2008. Evaluation of the inhibitory effect of dimethyl dicarbonate (DMDC) against wine microorganisms. *Food Microbiology*, 25:422–427.

Dittrich, H.H. and A. Barth. 1992. Galactose and arabinose in mosten und weinen der auslese gruppe. *Wein-Wissenschaft*, 47:129–131.

Divol, B. 2004. *La microbiologie des vins issus de raisins botrytisés au cours de l'élevage. Caractérisation des souches de* Saccharomyces cerevisiae *responsables de refermentations*. Thèse de Doctorat, Institut National Polytechnique de Toulouse, Toulouse, France.

Dubernet, M. 1974. *Recherches sur la tyrosinase de* Vitis vinifera *et la laccase de* Botrytis cinerea*: Applications technologiques*. Thèse de Doctorat, Université Bordeaux II, Bordeaux, France.

Dubourdieu, D. 1981. Les polysaccharides secrétés par *Botrytis cinerea* dans la baie de raisin, leur incidence sur le métabolisme de la levure. In *Actualités Oenologiques et Viticoles* (eds P. Ribéreau-Gayon and P. Sudraud), Dunod, Paris, pp. 224–230.

Franc, C., S. Gagné, D. Deffieux, *et al.* 2012. Oligostilbenes composition of botrytized white grape pomace. In *Oeno 2011*, Ed. Dunod, Bordeaux.

Guillou, I., A. Bertrand, G. de Revel, *et al.* 1997. Occurrence of hydroxypropanedial in certain musts and wines. *Journal of Agricultural and Food Chemistry*, 45:3382–3386.

Jarvis, W.R. 1977. *Botryotinia and Botrytis species: Taxonomy, physiology and pathogenicity. A guide to the literature*, Monograph N°15, Canada Department of Agriculture, Ottawa, p. 195.

Kovac, V. 1979. Etude de l'inactivation des oxydases du raisin, par des moyens chimiques. *Bulletin de l'OIV*, 52:809–826.

Jeandet, P., R. Bessis, M. Sbaghi, *et al.* 1995. Resveratrol content of wines of different ages: relationship with fungal disease pressure in the vineyard. *American Journal for Enology and Viticulture*, 46:1–4.

Landrault, N., F. Larronde, J.C. Delaunnay, *et al.* 2002. Levels of stilbene oligomers and astilbin in French varietal wines and in grapes during noble rot development. *Journal of Agricultural and Food Chemistry*, 50:2046–2052.

Liu, S.Q. and R.C. Davis. 1994. Analysis of wine carbohydrates using capillary gas liquid chromatography. *American Journal of Enology and Viticulture*, 45:229–234.

Marbach I., E. Harrel, and A.M. Mayer. 1985. Pectin a second inducer for laccase production by Botrytis cinerea. *Phytochemistry*, 24:2559–2561.

Noble A.C. and G.F. Bursick. 1984. The contribution of glycerol to perceived viscosity and sweetness in white wine. *American Journal of Enology and Viticulture*, 35(2):110–112.

Pour Nikfardjam, M.S., G. Laszlo, and H. Dietrich. 2006. Resveratrol-derivatives and antioxidative capacity in wines made from botrytized grapes. *Food Chemistry*, 96:74–79.

Ribéreau-Gayon, P. 1972. Connaissance de la nature des combinaisons de l'anhydride sulfureux dans les vins. *Bulletin de l'OIV*, 507:406–413.

Ribéreau-Gayon, J., Y. Glories, A. Maujean, *et al.* 1998. *Handbook of Enology. The Microbiology of Wine and Vinifications*, Vol. 2, 1st edn, John Wiley and Sons, Ltd, Chichester.

Roldan, A., V. Palacios, I. Caro, *et al.* 2003. Resveratrol content of palomino fino grapes: influence of vintage and fungal infection. *Journal of Agricultural and Food Chemistry*, 51:1464–1468.

Salgues, M., V. Cheynier, Z. Gunata, *et al.* 1986. Oxidation of grape juice 2-S-gluthationyl caffeoyl tartaric acid by *Botrytis cinerea* laccase and characterization of a new substance: 2,5-di-S-glutathionyl caffeoyl tartaric acid. *Journal of Food Science*, 51:1191–1194.

Shimohara, K., Y.K. Tseng, H. Sato, *et al.* 1974. Browning by the reaction of triose-reductone and amino-acids. *Science Bulletin of the Faculty of Agriculture, Kyushu University*, 28(3):139–144.

Sponholz, W.R. and H.H. Dittrich. 1984. Uber das Vorkommen von Galakturon- und Glucuronsaüre sowie von 2- und 5-Oxo-Gluconsaüre in Weinen, Sherries, Obst- und Dessertweinen. *Vitis*, 23:214–224.

Tromp, A. and W.A. Agenbach. 1981. Sorbic acid as preservative: its efficacy and organoleptic threshold. *South African Journal of Enology and Viticulture*, 2(1):1–5.

Velisek, J. and J. Davidek. 1978. Reactions of glyoxal with amino-acids. Formation of the brown pigments. Sb. Vys. Sk. *Chem-Technol. Praze*, E: Potraviny, E46:35–50.

Vivas, N., N. De Vivas Gaulejac, C. Vitry, *et al.* 2010. Occurrence and specificity of glucose oxidase (E.C: 1.1.3.4) in botrytized sweet white wine. Comparison with laccase (E.C: 1.10.3.2), considered as the main responsible factor for oxidation in this type of wine. *Vitis – Journal of Grapevine Research*, 49(3):113–120.

Watanabe M. and Y. Shimazu. 1976. Application of *Botrytis cinerea* for wine making. *Journal of Fermentation Technology*, 54:471–478.

Wuerdig, G. 1977. Appearance of mucic acid in musts prepared from grapes contaminated with Botrytis. *Bulletin de l'OIV*, 50:50–56.

Yamagushi, N. 1969. Studies on products of the browning reaction between reducing sugars and amino-acids. Antioxidant activities of browning reaction products and various reductones. *Journal of Food Science Technology*, 16(4):140–144.

20 Ice Wine

Nikolin Musabelliu

Winemaker, Toronto, Ontario, Canada

20.1 INTRODUCTION

Ice wine is a sweet dessert wine that is made from naturally frozen grapes left on the vine. The earliest indications of Ice wine production date from the Roman period. Later in Franconia, Germany, over 200 years ago, as a result of an unusual drop in temperature, the grapes were harvested frozen. In order not to lose the harvest, the growers and winemakers decided to squeeze the frozen grapes and in this way, discovered the liquid gold nectar called Eiswein (Ice wine).

Making Ice wine is a gamble, since winemakers must choose to leave grapes on the vine for a longer period. Climate conditions are the most important factor, and the production of Ice wine happens only if the weather cooperates. The production process must be finished before the grapes thaw. The biggest producers of Ice wine worldwide are Austria, Canada and Germany since these countries experience the cold climate conditions that are essential for the production of the wine (Table 20.1).

Nowadays, Ice wine is also produced in other countries, such as Australia, Croatia, Czech Republic, France, Hungary, Italy, Israel, Luxembourg, Moldavia, New Zealand, Slovakia, Slovenia, Switzerland and the United States.

20.2 LEGAL REQUIREMENTS

A universal agreement for the production and quality standards of Ice wine was signed in 2000 by the major Ice wine producing countries Austria, Canada and Germany; with the purpose of protecting consumers against counterfeit Ice wine.

Labelling wine produced from artificially frozen grapes as Ice wine was forbidden by the OIV (L'Office International de la Vigne et du Vin), in Paris in 2003. In order to label grapes as Ice wine, Eiswein and Vin du Glace, they should be naturally frozen on the vine. The grapes used in Ice wine production should come from the same region.

According to German law, 'Eiswein' is classified under Qualitätswein mit Prädikat (QmP). Frozen grapes for Ice wine production must be harvested at a minimum of −7°C (19.4°F) and the minimum sugar concentration level should be between 110–128° Oechsle (26–30° Brix). In Germany the minimal alcohol content is 5.5 % alcohol by volume and the

Sweet, Reinforced and Fortified Wines: Grape Biochemistry, Technology and Vinification, First Edition.
Edited by Fabio Mencarelli and Pietro Tonutti.

Table 20.1 Ice wine volumes in litres for Austria, Germany and Canada 2005–2010.

	2005	2006	2007	2008	2009	2010
Austria	N/A	N/A	N/A	52 555	91 663	113 383
Germany	1 645 900	85 700	593 200	1 139 100	N/A	N/A

Source: www.oesterreichwein.at; www.deutscheweine.de

	Canada					
	2005	**2006**	**2007**	**2008**	**2009**	**2010**
British Columbia	76 020	150 000	155 000	126 000	70 000	125 000
Ontario	264 561	406 239	564 948	1 262 574	1 001 340	910 485

Source: www. info@winebc.com; www.vqaontario.ca

maximum quantity of total SO_2 allowed is 400 mg L^{-1}. In Canada 'Icewine' is regulated by the VQA (Vintners Quality Alliance). A VQA regulation states that frozen grapes for Ice wine production must be harvested at −8°C (17.6°F) or lower after 3 consecutive days in order for grapes to fully freeze. The minimal level of sugar must be 35° Brix equivalent to 153.5° Oechsle. According to the VQA, Ice wines in Canada must have at least 7% alcohol, a concentration of acetic acid not above 2.1 g L^{-1} and residual sugar must be 125 g L^{-1} (100 g L^{-1} in British Colombia).

Austrian law dictates that 'Eiswein' must have 25° KMW (Klosterneuburger Mostwaage). Under Luxembourg legislation, 'Vin de Glace' must have 120° Oechsle (28.1° Brix).

20.3 GRAPE VARIETY

Ideally grapes for making Ice wine should be resistant to disease, of an aromatic variety, late ripening, and must have high acidity. The most common grapes used are Cabernet Franc, Riesling and Vidal. Other varieties for producing Ice wine are: Cabernet Sauvignon, Chardonnay, Chenin Blanc, Gewürztraminer, Merlot, Sangiovese, Semillon, Syrah, Sauvignon Blanc and Tempranillo.

In Austria, Eiswein is produced from Welschriesling and Scheurebe. In Canada, the hybrid grape variety Vidal is more common because its tough skin can resist harsh weather. In Germany, Riesling is typically used for Ice wine production.

20.4 VINIFICATION

20.4.1 Plundering by birds

Ice wine grapes provide food for birds and other animals during the winter. When there is no snow, there are no obstacles to prevent animals from approaching the plants. When food is scarce later in winter, the grapes attract crowds of birds which can empty a vineyard in minutes. Precautions are taken to minimize bird plundering and to maximize the protection of grapes from damage. During the autumn period, the grapes are protected using propane canons and later on, the best and the only method to protect them is by covering the vines with nets. In Canada, long nets are used for grape protection whereas in Europe, short nets are preferred.

20.4.2 Grape harvest

Grapes for Ice wine must be harvested frozen without artificial refrigeration, and they must be completely frozen before harvesting starts. The perfect harvesting period is between December and February or March.

Healthy grapes are very important and significant for the production of a great Ice wine. In fact, grapes in healthy condition provide fruit flavours and good acidity. There are two methods for harvesting frozen grapes, one is the traditional method of hand picking and the other is harvesting by machine. In hand harvesting, grape pickers should wear gloves so that the grapes will not defrost. The machine harvesting method gives cleaner juices because only the berries are collected and the stems remain on the vines. This method is also more efficient because only a small quantity of berries is lost. Harvesting is usually carried out under darkness in the very early morning hours, before sunrise.

20.4.3 Fermentation process

The sugar concentration of frozen grapes for the production of Ice wine can be calculated as follows: °Oechlse $= 21 + (17 \times$ minus temperature) (Ziraldo and Kaiser, 2007).

For example: at −9°C grapes would reach 174° Oechlse $[21 + (17 \times 9) = 174]$ 174° Oechlse = 39° Brix.

While the grapes are crushed and destemmed, the winery doors are left open so that they maintain the cold harvesting temperature. A basket press is used for grape pressing in the winemaking process because the juice obtained is cleaner compared to other pressing methods. The lower the temperature of frozen grapes, the slower the juices are released. During the pressing process, because the grapes are frozen and the temperature remains between −8°C and −12°C, only a small quantity of thick liquid is released from individual berries. Ice wine juice is not optimal for yeast growth because the sugar level is too high. The main problem is the formation of acetic acid and ethyl acetate during the fermentation of this type of must which creates hyperosmotic stress to yeast cells (Kontkanen *et al.*, 2004). Together with the formation of acetic acid and ethyl acetate, a high content of glycerol is produced (Sutherland *et al.*, 1997). For Ice wine fermentation, winemakers inoculate the must with a special yeast able to tolerate high osmotic pressure due to the high sugars content. *Saccharomyces bayanus* (of the genus *Saccharomyces*) is the most frequently used because it produces less acetic acid during fermentation and is cryotolerant. Recently two hybrids between *Saccharomyces bayanus* and *Saccharomyces cerevisiae* named AWRI1571 and AWRI1572 have been proposed with good results in terms of low volatile acidity formation and reasonable lapse of time to reach an ethanol concentration of 10° (v/v) (Yang, 2010). After the juice is clarified, a fermentation starter and nutrients are added to help the yeast during the fermentation process. The temperature should be maintained between 15 and 16°C to preserve the fruit character of the finished wine. The fermentation process is very slow and may take several months to complete. When the alcoholic concentration reaches approximately 7–13% alcohol by volume the fermentation is stopped by chilling the must. Malolactic fermentation is not allowed. In order to prevent re-fermentation Ice wine is anaerobically stored. Later, it is sulfured and because of the high residual sugar level, it is sterile filtered.

20.4.4 Ice wine taste

Ice wines from white grape varieties tend to be pale or light golden in colour when they are young and they become amber-golden as they age. They have honey flavours blended with full tropical aromas, apricot, citrus, lychee, pineapple and tangerine. The red varieties tend to produce wines that have a light pink colour, red berry flavours and fresh strawberry, cherry and raspberry aromas. The level of acidity is an essential distinguishing characteristic of Ice wine, and the key to producing an ideal Ice wine is maintaining the balance between acidity level and sugar concentration.

Ice wines are expensive and are frequently sold in smaller bottles (50 ml, 200 ml and 375 ml) than standard table wines. High concentrations of volatile acids, consisting mainly of acetic acid, are often found in Ice wine. Wine yeasts produce acetic acid as a byproduct of the hyperosmotic stress response caused by high sugar concentrations (>35° Brix) in grape must. Volatile acid levels in Ice wine often exceed legal limits (Erasmus *et al.*, 2004). An interesting study was done to characterize the volatiles of Ice wine from two different areas (Cliff *et al.*, 2002). Eight volatile compounds differed significantly between British Columbia and Ontario Ice wines. Ontario Ice wines had the highest fruity and floral aromas and a golden copper colour while German Ice wines had higher acidity with a nutty/oily character. British Columbia Ice wines were sweeter with more body/viscosity, and a more persistent taste.

In recent years Canada has started the production of sparkling Ice wine.

ACKNOWLEDGEMENT

The author would like to thank Eranda Troqe for her linguistic advice.

REFERENCES

Cliff, M., D. Yuksel, B. Girard, *et al.* 2002. Characterization of Canadian Icewines by sensory and compositional analyses. *American Journal of Enology and Viticulture*, 53:46–53.

Erasmus, D.J., M. Cliff, and J.J. Van Vuuren. 2004. Impact of yeast strain on the production of acetic acid, glycerol, and the sensory attributes of Icewine. *American Journal of Enology and Viticulture*, 55:371–378.

Kontkanen, D., D. Inglis, G. Pickering, *et al.* 2004. Effect of yeast inoculation rate, acclimatization, and nutrient addition in Icewine fermentation. *American Journal of Enology and Viticulture*, 55:363–370.

Sutherland, F.C., F. Lagos, C. Lucas, *et al.* 1997. Characteristics of Fps1-dependent and -independent glycerol transport in *Saccharomyces cerevisiae*. *Journal of Bacteriology*, 179:7790–7795.

Yang, F. 2010. *Study of New Yeast Strains as Novel Starter Cultures for Riesling Icewine Production*. Master's Thesis, Brock University, Ontario, Canada.

Ziraldo, D. and K. Kaiser. 2007. *Icewine: Extreme Winemaking*, Key Porter Books.

21 Port

Tim Hogg

Escola Superior de Biotecnologia, Universidade Católica Portuguesa, Porto, Portugal

21.1 INTRODUCTION

Port is a sweet fortified wine, vinified in the Douro demarcated region from grapes grown in that region (IVDP, 2012). Legally it falls under the category of naturally sweet fortified wine and is thus defined as having been produced by the addition of wine-derived spirit to a partially fermented, or unfermented grape must. European Union regulations for fortified wines stipulate an acquired alcohol content by volume of between 15 and 22 % (European Council, 1987). In the case of Port, this is generally between 18 and 20%. The name 'Port' comes from the city of Porto, Portugal's second city and the capital of the northern region. Although always known as Porto wine (*Vinho do Porto*), the ageing, blending and bottling of the Port does not in fact take place within the city, but either in a defined bonded area in the city of Vila Nova da Gaia, which occupies the southern margin of the Douro estuary across the river from Porto, or indeed within the Douro demarcated region itself.

The unique characteristics of Ports as they are presented in bottled, commercial examples, are due to a combination of the attributes of the grape varieties employed, the agro-climatic conditions under which the grapes are cultivated, the vinification methods used and the types of ageing regimes applied. A further key element is the role of the professional who blends the wine at the various stages of production, this role being of particular significance in the production of Port. Although Port is defined by its geographical origin, products with similar production processes and gross final composition exist in many parts of the wine-producing world (Voss, 1989). In the case of many non-European producers these wines were often initially designed to imitate Port, which is not to say that many have not developed into highly credible products in their own right (Voss, 1989). Other classic, sweet, fortified wines of Europe that share some elements of production with Port include: Madeira and Moscatel de Setubal (Portugal), Marsala (Italy), Pineau de Charentes and the Vin Doux Naturel of Languedoc-Roussillon (France).

Another facet of Port, which is relevant to introduce here as it impacts on a discussion of the science and technology of production, is the concept of category. Port has a number of legally defined and controlled categories which refer both to the style of wine – of which there are a number – and also its relative quality level. This official classification is administered by the *Instituto dos Vinhos do Douro do Porto* (IVDP, 2012) and reflects an official acceptance of a wine as having been made in a certain manner and as meeting certain, largely measurable, quality parameters. Indeed the specific legislation regarding all aspects of Port is highly

Sweet, Reinforced and Fortified Wines: Grape Biochemistry, Technology and Vinification, First Edition.
Edited by Fabio Mencarelli and Pietro Tonutti.

complex and dynamic, so much so that any global written treatment of it will probably be obsolete before it is published. Readers interested in this aspect of Port are directed to the official web site of the IVDP (www.ivdp.pt). Similarly, numerous books and articles exist that cover many aspects of the history, economics and culture of Port and the reader interested in these aspects is referred to these (e.g. Bradford, 1978; Fonseca *et al.*, 1987; Voss, 1989; Mayson, 2005).

The focus of this chapter is very clearly on those technical and scientific aspects relevant to Port from the Douro valley; elements of the law and business are only introduced where they frame specific aspects of technical relevance. One example of a legal alteration which has changed the technical landscape is the removal, in the 1970s, of the legal obligation for Port to pass some of its ageing process in the bonded area of Vila Nova de Gaia (the city on the southern margin of the mouth of the Douro, opposite Porto). This change led to the liberalization of production and commercialization, permitting new companies, grape growers and cooperatives to bottle and sell wine directly from the region in which the grapes were grown. A more recent business-based development, which has also altered the production landscape, involves the dramatic increase in demand for the unfortified wines from the region. The relevance of this is considerable as most producers of Port are now also producers of table wines, which are much less tightly legislated as to the quantities produced (Mayson, 2005). The technical expertise, winemaking practices and specialist equipment for producing quality table wines are in many cases quite different to those required for Port.

The intention in this chapter is to present descriptions of how the major categories of Port are produced and, where they are known, the underlying scientific phenomena that are specific to these categories. Viticultural aspects will only be discussed when they relate directly to the specificity of the products. Likewise the descriptions of the generic winemaking processes will not be presented unless they are unique to, or particularly relevant to Port.

21.2 STYLES OF PORT

There is a range of legally defined categories under which Ports are presented for sale; it is not within the scope of this chapter to describe these in detail although those interested in this aspect can find up-to-date and detailed information on the IVDP web site (IVDP, 2012). However, it is necessary to describe some of the categories in order to frame the production steps which define the underlying styles. Red Ports can be classified into one of three basic categories, although each one requires some qualification. Ruby Ports are those which are aged in a manner that accentuates their fruity characteristics, full-bodied structure and red colour; this means ageing in large wooden vessels for between 3 to 5 years before blending and bottling. Tawny Ports on the other hand are aged in smaller wooden vessels and handled in a manner that accentuates the brown and yellow hues, lighter body and spicy and nutty aromas. The third basic category is Vintage Port, which is distinct from the other categories in that it is made from selected grapes and from a single year's harvest. Wines of this category are aged in wood for between 2 and 3 years before bottling. Vintage Ports are classically made for extended ageing in bottle (often many decades) although some markets appreciate these wines whilst relatively young (Figure 21.1).

Particularly in the case of Ruby and Tawny Ports, there is a considerable range of quality levels which are legally defined and controlled by the IVDP. These include categories such as Late Bottle Vintage, which is a very high-quality Ruby Port from a single year's harvest, bottled after 4 to 5 years of wood ageing. Within the Tawny Ports there are those that are

Figure 21.1 Different styles of Port. From the left, the first two wines are a young White Port and an aged White Port, respectively. The following three wines are Tawnies of decreasing age. The penultimate wine is a Ruby and the one on the far right is a Vintage Port. (Source: Instituto dos Vinhos do Douro e Porto, 2012.) (For a colour version of this figure, please see the colour plate section.)

sold with an indication of their age, that is, 10, 20, 30 or more than 40 years old. These are very high-quality wines which have been blended from Tawny Ports to give an average age equivalent to that of the category. Any of these categories will have a strict definition of the specifications that must be complied with in order to receive authorization from the IVDP to commercialize the Ports under that specific category. A large part of these specifications is qualitative and relates to the composition of the wine as it is presented. Many parameters are measurable by analytical chemistry methods and are thus entirely objective, but all wines must be accepted by a tasting panel before being authorized to be sold under any particular category.

21.3 PORT PRODUCTION

21.3.1 Geography

Port is made from grapes grown in the Douro demarcated region in northern Portugal. The region covers 250 000 ha, of which about 45 000 ha are under vine; it is the world's oldest demarcated wine area, the original boundaries dating from 1761 (Fonseca *et al.*, 1987). The region is centred on the valley of the River Douro and in particular the stretch that runs from the village of Barqueiros, 65 km inland from Porto, to the Spanish border near Freixo de Espada à Cinta, a further 100 km east. As mentioned earlier, in order to qualify for a Certificate of Origin from the *Instituto dos Vinhos do Douro e Porto*, wines must be vinified from fruit grown in the demarcated region and aged either within this area or the bonded area of Vila Nova de Gaia. As in many classic European wine regions, a detailed classification exists which is intended to relate the intrinsic quality of the fruit produced to geographical subdivisions within the global demarcated region. In the case of Port the major subregions are: the *Baixo Corgo*, *Cima Corgo* and the *Douro Superior* (Table 21.1).

Whereas the *Baixo Corgo* and, to a lesser extent, the *Cima Corgo* are densely planted with vines, large areas of the *Douro Superior* remain uncultivated (Reader and Dominguez, 2003). The climate thus changes from Atlantic in the lower areas of the Douro valley to Mediterranean in the *Douro Superior* (Fonseca *et al.*, 1987); this, together with viticultural practices, has an important influence on wine quality. A recent study by Santos *et al.* (2011),

Table 21.1 The subregions of the Douro demarcated region.

Subregion	Total area (ha)	%	Area under vine (ha)	% of Total area under vine
Lower Corgo	45 000	18	14 582	32.4
Upper Corgo	95 000	38	20 969	22.1
Upper Douro	110 000	44	10 175	9.2
Total	250 000		45 726	18.3

Source: Instituto dos Vinhos do Douro e Porto, 2012.

modelled the grapevine yield and possible scenarios of production in the region according to climate change predictions. The models suggest a continuing increase in yields until the end of the twenty-first century. The authors are clear about the limitations of the approach and do not claim that it describes the whole picture, as the trend appears to be in the opposite direction (Schultz and Jones, 2010).

At a more detailed and local level, land parcels for the production of grapes for Port production are subject to a complex system of classification, which allows authorization for production to be distributed on the basis of potential quality. Points are attributed to land parcels according to the grape varieties planted, slope, soil type, altitude, aspect and productivity, amongst other factors. Parcels are thus classified with a letter from A to F and the higher graded vineyards (i.e. A and B) are authorized to turn a greater proportion of grapes into Port than the lower grades (letters E and F) (Portaria n.o 413/2001). The higher ranking vineyards are those at the lower altitudes in the *Cima Corgo* and *Douro Superior* whose grapes are considered to be more suitable for premium wines. Annually around 55 million hL of must is authorized for Port, but this can be substantially reduced according to industry-wide supply-and-demand considerations. The amount produced is controlled by the IVDP by limiting the amount of fortifying spirit that is authorized for use in the harvest campaign.

21.3.2 Viticulture

The majority of varieties of *Vitis vinifera* that are planted (on hybrid rootstocks) in the Douro region and used for Port production are considered to be native to the region. When exactly the varieties now employed became native varieties is undoubtedly an interesting subject for agricultural and economic historians, but will not be dealt with further here. At present Port can only be produced from authorized varieties, although special dispensation is allowed for grapes from old vineyards where the varieties are mixed in the same plot (Portaria n.o 413/2001).

Vines planted since the late 1980s have been essentially incorporated into consolidated blocks segregated by variety and, most often, they are a specific clone of that variety (Magalhães, 2009). Modern molecular biology techniques have been applied in recent years to discriminate more objectively between specific Douro varieties and to explore the relatedness of some of the most important ones (Lopes *et al.*, 1999, 2006; Castro *et al.*, 2011). Aside from their intrinsic scientific interest, these studies might form the basis of methods to give more objectivity to certain aspects of the market in vines and grapes.

The majority of vineyards have vines planted in horizontal rows, either on hand-built stone terraces, bulldozed earth terraces, or, where the slope permits, in unterraced plots (Reader and Dominguez, 2003). Irrigation is only permitted in situations where water stress could damage the long-term health of the vines and always requires prior approval from the IVDP (IVDP, 2012).

21.3.3 Harvest

The Port harvest is largely manual with most grapes being hand-picked (cut) into small plastic boxes holding about 25 kg and either transported directly to the winery in these containers or transferred to larger steel bins for transport. Classically, grape analysis for maturity determination on picked samples employs a combination of simple measurements such as dissolved solids, pH value, titratable acidity and berry tasting. As in other premium wine-growing regions, a more systematic approach to determining the picking date is constantly being sought. More precision is facilitated by the varietal and clonal plantings (discussed earlier), which aids homogeneity. There is a tendency for the tasting component to become more systematic and there is also experimentation with instrumental methods.

Sugar levels are measured on reception at the winery, both for reasons of process management and, in those cases in which the grapes themselves are being transacted, in order to help define payments.

At the winery grapes are transferred into receival hoppers and pumped by screw-conveyer to a mechanical crusher. Conventional roller or beater crushers are used and stalks are largely removed. In wineries in other hot regions, cooling of the must is often effected by chilling the pipes that conduct the grape must from the crusher in order to diminish oxidative degradation and the growth of undesired microbes. For the same reasons, sulfur dioxide (SO_2) is added, either as potassium metabisulfite solution or as a solution of SO_2 gas, at levels of between 50 and 150 mg SO_2 per kg of grapes and pH can be adjusted by adding tartaric acid (Reader and Dominguez, 2003). The target pH at this stage will depend on the specific nature of the Port being produced.

21.3.4 Vinification

Port is made from red or white grapes that are vinified and matured separately, to produce red or white wines. Recently, rosé (or pink) styles have begun to be produced from sweet, pressed, red grapes after brief contact with skins; the resulting juice is then fermented as a white wine. Fortification occurs during primary alcoholic fermentation; the resulting ethanol concentration (circa 19% v/v) arrests yeast activity. In this way the sugar present after fortifying, and also in the final product, is entirely derived from the original grape must. The timing of spirit addition is dependent on the style of wine desired, the majority of wines being fortified after approximately half the original sugar in the must has been consumed.

It is in the fermentation of red grape musts, which make up by far the greatest part of the total, where one of the major differences to standard winemaking is observed. The relatively short time that passes between crushing and the arresting of fermentation by spirit addition, is the only period during which colour and flavour compounds can be extracted from the grape skins. This period is rarely more than 48 hours and thus much shorter than in table wine fermentations. Also relevant is the fact that the extractive medium itself never reaches the ethanol concentration of a table wine fermentation (Bakker *et al.*, 1996). Specific strategies have evolved to ensure that the extraction of desirable compounds is maximized, and these are discussed later.

The biochemistry of the alcoholic fermentation in Port is no different to that for any other wine and it will not be discussed further here. It is generally true that the alcoholic fermentation which precedes the addition of spirit in Port vinification is carried out by *Saccharomyces cerevisiae*. In the vast majority of vinifications the *S. cerevisiae* will not have been purposely added as active dry yeast, but will derive from previous ferments. This may be effected by actively leaving a small amount of the previous fermenting must in the fermenting vessel, or by simply assuming that the incoming must will become passively inoculated from the surfaces which have not been hygienized between fermentations. For specific styles – such as the recently introduced Pink Port, and others where musts have exceptionally high sugar levels – active dry *S. cerevisiae* may be employed, if not systematically. Theoretically, as fermentation is arrested by fortification before alcoholic fermentation has run its course, non-*Saccaromyces* species may be expected to play a greater role in the fermentation, and therefore in the final composition, than when fermentation proceeds to dryness (Moreira *et al.*, 2011). The classical means of fermentation is that performed in granite troughs called *lagares*; these are approximately 1 m deep and often hold more than 6000 kg of must. Maceration in *lagares* is assisted by regular treading sessions by barelegged workers. The particular combination of geometric proportions of the filled trough and the relatively soft mechanical mixing has made the foot-trodden *lagar* the most favoured method of fermenting high-quality Ports. The favourable geometry is given by the fact that the solid cap, which naturally rises to the top of the ferment in red wine fermentations, presents a much larger surface area-to-volume ratio than regular fermentation tanks (Birks and Sarmento, 1991). Unfortunately the availability and cost of the labour required for foot treading has made this an exceptional occurrence nowadays, although companies value the undeniable attractiveness of this practice to journalists and other trade visitors. At one time there was considerable doubt in the industry as to whether the foot-trodden *lagar* is actually superior to the major modern competing method, which is standard, pumped-over, tank fermentation. In the few articles that do appear in the literature, the data appears to be inconclusive. Birks and Sarmento (1991) showed a greater extraction of colour and tannins in foot-trodden *lagares* when fermenting grapes from the same batch of high-quality grapes. Apparently the major difference between the methods could be attributed to the initial mechanical mixing, as the subsequent kinetics of extraction was similar. Bakker *et al.* (1996) found no significant differences between the two configurations in a similarly controlled, experimental study. More recently, Barbosa *et al.* (2003), in another experimental study, did find a significant difference in *lagar*-fermented musts. In this case, as well as colour and tannins, the glycosidically bound terpene precursors were found to be very much enhanced in the *lagar* fermentations. This is consistent with the anecdotal observations that *lagar*-fermented Ports are more suited to post-bottling ageing, as is the case of vintage-style Ports.

Nowadays, however, the majority of Port is fermented in standard red-wine fermenting tanks with active cooling systems; the personal preferences of the winemaker determine the details of pumping over frequency, fermentation temperature, and so on, as is the case for other wine styles. However, the perceived advantages of the *lagar* system for the higher quality wines has led to a number of companies developing their own, individual robotic solutions to partially imitate the foot-trodden *lagar* systems. In all of these systems, the geometry of the *lagar* has been largely maintained, although stainless steel contact surfaces are very much preferred and active cooling systems are always employed. The mechanical interventions to imitate foot treading vary somewhat. The only model for which there is a detailed description in the literature, employs a programmable gantry which passes over a series of adjacent *lagares* and 'works' them sequentially via hydraulically operated beams (Symington *et al.*,

Figure 21.2 Robotic lagares in series, the gantry will work one ferment and then pass over to the adjacent one by means of the rails running under the tops of the lagares. Image courtesy of Symington Family Estates. (For a colour version of this figure, please see the colour plate section.)

2011) (Figure 21.2). Silicone rubber pads are the points that put downward pressure on the grape solids, imitating a human foot. The efficacy of robotic *lagares* generically has led to their adoption, in the Douro region, for the processing of grapes for high-quality table wines.

The temperature of fermentation is a decision of the individual winemaker. Bakker (1989) showed that fermentations conducted above 28°C, resulted in greater extraction of anthocyanins, and darker and (initially) sensorially preferred wines, when compared to those controlled at 22°C. The differences and preferences did not persist during subsequent maturation of the wines. Extractive enzymes have been found to increase pigment and flavour extraction in Port vinifications, these improvements persisting during further maturation (Bakker *et al.*, 1999). Enzymes are apparently sparsely employed in the industry.

Modern styles of White Port do not normally suffer prolonged maceration on the skins, and are generally pressed before fermentation becomes visible. The rest of the fermentation (which occurs at lower temperatures than for red wines), occurs in the liquid must to which fractions of the pressings may be added, according to the style intended. The grape varieties and pressing method influence the presence of specific phenolics in the wine (Ramos *et al.*, 1999); browning can be reduced and flavours enhanced by optimizing skin contact and oxygenation (Ho *et al.*, 1999b).

21.3.5 Fortification

Fortification employs a wine-derived spirit of 76–78 % ethanol by volume obtained from Portugal or elsewhere in Europe. The process is normally performed when approximately 4–5 % of ethanol (by volume) has been produced from fermentation, although this depends on the intended sweetness of the wine. The liquid fraction is drained from the solids and mixed immediately with spirit. The remaining grape pomace is pressed and the press juice can be partially mixed into the free-run liquid. The actual mixing is normally performed by simultaneously transferring sprit and must into a tank. Ethanol levels and pH values are adjusted at this stage, as needed. The role of the grape spirit on the quality of Port has long been a subject of concern. As a significant part of the volume of the final product is made up of spirit, concentrations in the latter of vestigial food safety hazards, such as urethane or heavy metals, are relevant (Reader and Dominguez, 2003). The controlling body, the IVDP, is responsible for approving fortification spirit, ensuring that these can comply with certain minimum organoleptic and analytical standards. Although the role of the non-ethanol components of the fortifying spirit in aspects such as ageing behaviour and aromatic quality, were previously empirically recognized, it is only over the past decade or so that such considerations have begun to appear in the literature. Certain specific aromatic components of fortifying spirit have been shown, in experimental wines, to contribute directly to the aromatic complexity of Ports that have not yet been subjected to extended ageing, although it is not certain that these attributes remain relevant (Rogerson and de Freitas, 2002). The aldehyde component of spirits has, again in experimental scenarios, been shown to contribute to the development of colour and tannins in Ports (Pissarra *et al.*, 2005). This latter finding is consistent with what is known of the role of carbonyl compounds, and aldehydes in particular, in the development of the phenolic component of wines during ageing.

21.3.6 Ageing

As mentioned earlier, by law, wines must be aged either in the Douro or in Vila Nova da Gaia. Ageing takes place in old oak vessels ranging in capacity from slightly less than 600 L to as much as 200 000 L (Reader and Dominguez, 2003). Old wood vessels have been shown to be very distinct as an ageing medium for Port, influencing the development of coloured and other phenolics (Ho *et al.*, 2001a, b). Racking is a very important process in Port maturation and will be periodically performed on all wines prior to them becoming part of one of the blending lotes. The size of the ageing vessel, and the aeration regime during racking, greatly influence the style that develops, and wine destined for Ruby and Vintage Ports will be predominantly aged in large vessels, whilst those for Tawnies will be aged in small casks.

21.3.6.1 Pigments and ageing

Postvinification ageing is critical to the development of Port's unique characteristics and this, in part, has resulted in it receiving the most attention from the scientific community. The most abundant anthocyanin in Port grapes and the resulting wine is malvidin, as malvidin 3-glucoside, followed by malvidin 3-*p*-coumarylglucoside and malvidin 3-acetylglucoside and their derivatives (Bakker and Timberlake, 1985a, b; Mateus *et al.*, 2002b). The actual composition varies significantly among varieties (Bakker and Timberlake, 1985a, b), the year of harvest and the altitude of planting (Mateus and de Freitas, 2001; Mateus *et al.*, 2002a). Normal practice is to allow young Ports to settle undisturbed for two to three months after the vintage. During this period, the colour increases to a maximum, a phenomenon colloquially known as 'closing up' (Reader and Dominguez, 2003). The effect is believed to be due to the formation of polymers of anthocyanins, which are bridged with other phenolic compounds via an aldehyde (principally acetaldehyde) (Bakker and Timberlake, 1986). The anthocyanin molecules can also react with pyruvic acid and produce pigmented adducts such as Vitisin A, which are relatively stable (Bakker and Timberlake, 1997). From work in model solutions, Romero and Bakker (2000) propose that pyruvic acid and acetaldehyde compete in reacting with the original anthocyanin. In the case of the pyruvic acid adducts, of the predominant Port anthocyanins listed above, these have been shown to increase for the first 100 days after fortification and then progressively diminish over time (Mateus *et al.*, 2002a), consistent with the observed 'closing up' (Reader and Dominguez, 2003). The aldehyde bridging and pyruvic acid adduct formation probably predominate over direct condensation of anthocyanins with phenolics because of the high levels of free aldehyde and pyruvate in young Port compared with red table wines (Bakker and Timberlake, 1986). It is to be expected that pyruvate will be in a higher concentration in Port than in table wines as it is known to be excreted in the early stages of fermentation and re-imported later (Mateus *et al.*, 2002a; Morata *et al.*, 2003). Aldehydes are both derived from fortification spirit (Pissara *et al.*, 2005) and produced by yeast during fermentation. The aerating mixing that occurs at fortification will also favour acetaldehyde production by oxidizing SO_2, which is bound to it, and by promoting oxidation of ethanol (Reader and Dominguez, 2003). Less abundant species of anthocyanin-based pigments are thought to have increasing roles as Ports lose their dense purple colour and gain red and orange tones (Mateus *et al.*, 2002b). A previously unrecognized family of anthocyanin-catechin-based pigments have been discovered in Ports and given the name Portsins (Oliveira, J. *et al.*, 2006). The contribution of these blueish-coloured compounds, which are unlikely to be exclusive to Port or *Douro* wines, to the overall colour is not certain but they are intrinsically stable and are thus likely to contribute more as the background pigmentation diminishes over time. As premium Port of all of the major styles is designed to age, the chemical basis of the colour changes occurring over extended ageing have been the subject of study. A previously unrecognized pyranoanthocyanin – vinylphenol (purple hue) – was identified in two-year-old Ports and presented as further evidence of the importance of anthocyanin-pyruvic acid adducts as major contributors to colour as ageing progresses (Mateus *et al.*, 2006).

The established role of co-pigmentation in the development and stabilization of wine colour has led to some studies on the abundance of non-coloured phenolics in Port grapes and Ports (Andrade *et al.*, 2001). These exploratory papers suggest that these compounds vary quantitatively between varieties of Port grapes thus providing analytical support for producers incorporating co-pigmentation into their winemaking strategies.

21.3.6.2 Flavour and ageing

Characteristic flavours of Ports are known to be derived from the grapes, the fortifying spirit and compounds derived from the ageing process of the particular style. It is nowadays possible to identify the molecular species responsible for specific aromas in wine by a combination of sensory analysis and gas chromatography olfactory detection (GCO) and gas chromatography mass spectrometry (GCMS) (Silva Ferreira *et al.*, 2003b). Variations on this approach have been applied to identify the molecules responsible for a number of positive aromas characteristic of Port grapes, in particular the premium variety, Touriga Nacional. Using one-year-old lots of monovarietal Ports of Touriga Nacional, Falqué-Lopez *et al.* (2004) characterize them as having 'plum brandy', 'mulberry', 'cherry', 'wild fruits' and 'dry raisin' aromas. Later studies by the same group identified linalool and linalyl acetate as being significant contributors to the consistently cited bergamot descriptor (de Pinho *et al.*, 2007). Rogerson *et al.* (2002) identified 1,3-dimethoxybenzene in young Ports made from Tinta Cão and Tinta Barroca grapes. Although aroma active, being described as having 'a sweet medicinal odour with hazelnut, resinous and woody notes', it was detected at concentrations just below its aroma threshold in Port, and is likely to contribute to the overall aroma but not in an identifiable way. Similarly, the norisoprenoid, 2,6,6-trimethylcyclohex-2-ene-1,4-dione (TMCHD), which is described as having sweet honey aroma was identified in one-year-old Ports of the Touriga Francesa and Touriga Nacional varieties (Rogerson *et al.*, 2001). The role of norisoprenoids in Port aroma has been given considerable attention apparently being relevant in both young and extensively aged wines (Silva Ferreia and Guedes de Pinho, 2004; Mendes-Pinto, 2009). It has been proposed that, as degradation products of grape carotenoids, certain norisoprenoids responsible for primary fruit aroma may be released from these precursors during ageing (Silva Ferreira *et al.*, 2008; Oliveira, C. *et al.*, 2006). This phenomenon is relevant for Port as carotenes can persist after Port vinification whereas in dry wines they are essentially absent (Mendes-Pinto *et al.*, 2005). The abundance of specific carotenes in Port grapes is also known to be greatly influenced by certain viticultural practices, such as bunch shading (Oliveira *et al.*, 2004), and also by plant water status (Oliveira *et al.*, 2003).

Although the composition and sensory characteristics of tawny, ruby and vintage styles are quite distinct, the underlying alterations provoked by the ageing regimes are not themselves exclusive to any one style. Certainly esterification occurs in all configurations of ageing; ethyl lactate, diethyl malate, triethyl lactate, and monoethyl and diethyl succinate have been proposed as indicators of wine age (Mendes Moreira *et al.*, 1994). Acetate and acetate esters appear to increase with time; older wines have measurably higher levels of volatile acidity which is sought after in tawny styles and not associated with microbial action (Reader and Dominguez, 2003). In wood-aged tawnies, certain heterocyclic acetals derived from glycerol and acetaldehyde have been proposed both as age indicators and contributors to the aroma of the older wines (Silva Ferreira *et al.*, 2002). A consistent age indicator and very significant aroma compound in tawnies is sotolon (3-hydroxy-4,5-dimethyl-2(5H)-furanone) (Silva Ferreira *et al.*, 2003a). Sotolon imparts a nutty, spicy (fenugreek) aroma, and its experimental addition to wines leads to these being unanimously rated as older (Silva Ferreira *et al.*, 2003a). The increases in wood-extracted phenolic compounds noted in some aged Tawny Ports (Ho *et al.*, 1999a), are likely to contribute to the aged flavour, although this has not been demonstrated in the cited study.

21.3.7 Blending

The majority of Port styles are intended to be highly consistent in terms of organoleptic quality and average age of the wines within the particular style. Only in the case of those wines that carry a harvest year (mostly vintage and late-bottled vintage) would some kind of year-to-year variation be considered acceptable (Reader and Dominguez, 2003). Hence the role of the blender is particularly relevant. Practices vary across the industry, but for those producers that present a large range of products and categories, general practice is to prepare and hold blending batches, which serve as the building blocks of the final blends. These in turn are prepared from other reserve batches, which are effectively the more base 'colours' from which the blender builds up the lotes. It is normal, when preparing a batch with fresh wine, to include some of the previous batch and thus add to the consistency. Chemical and physical analyses of all batches and their component wines will be kept, and these will help the blender to maintain consistency throughout. This is most important for elements such as colour in the large-volume products (Bakker and Arnold, 1993). However, it must be stressed that by far the largest part of defining the composition of the lotes and the final blends is down to the tasting team of the blender.

21.3.8 Fining and stabilization

The preparation of Ports for bottling differs little from that applied to table wines, where the large quantity, younger wines will require careful stabilization to remove unstable tartrates, pigments and colloids, while older and/or more premium styles will require much less treatment (Reader and Dominguez, 2003). This is because the older tawny styles are naturally more stable and because the consumer of vintage style wines would expect some sediment to form over time. It is certainly the case that fining agents and procedures, and centrifugation, filtration and cold stabilization procedures do not differ substantially to those used in table wines. All Ports are bottled within the Douro region or in the bonded area of Vila Nova de Gaia, and the majority use t-bar, stopper corks although vintage and other bottle-ageing categories are closed with a conventional pulled cork.

REFERENCES

Andrade, P.B., G. Mendes, V. Falco, *et al.* 2001. Preliminary study of flavonols in port wine grape varieties. *Food Chemistry*, 73:397–399.

Bakker, J. 1989. The effect of fermentation temperature on ports as measured by high performance liquid chromatography, spectroscopic analysis and sensory evaluation. *Actualités Oenologiques 89, Comptes Rendus du 4e Symposium International d'Oenologie* (eds P. Ribéreau-Gayon and A. Lonvaud), Dunod, Paris, pp. 403–407.

Bakker, J. and G.M. Arnold. 1993. Analysis of sensory and chemical data for color evaluation of a range of red Port wines. *American Journal of Enology and Viticulture*, 44:27–34.

Bakker, J. and C.F. Timberlake. 1985a. The distribution of anthocyanins in grape skin extracts of port wine cultivars as determined by high performance liquid chromatography. *Journal of the Science of Food and Agriculture*, 36:1315–1324.

Bakker, J. and C.F. Timberlake. 1985b. The distribution and content of anthocyanins in young port wines as determined by high performance liquid chromatography. *Journal of the Science of Food and Agriculture*, 36:1325–1333.

Bakker, J. and C.F. Timberlake. 1986. The mechanism of color changes in aging port wine. *American Journal of Enology and Viticulture* 37:288–292.

Bakker, J. and C.F. Timberlake. 1997. Isolation, identification, and characterization of new color-stable anthocyanins occurring in some red wines. *Journal of Agricultural and Food Chemistry*, 45:35–43.

Bakker, J., S.J. Bellworthy, T.A. Hogg, *et al.* 1996. Two methods of Port vinification: a comparison of changes during fermentation and of characteristics of the wines. *American Journal of Enology and Viticulture*, 47:37–41.

Bakker, J., S.J. Bellworthy, H.P. Reader, *et al.* 1999. Effect of enzymes during vinification on color and sensory properties of Port wines. *American Journal of Enology and Viticulture*, 50:271–276.

Barbosa, A., A.C. Silva Ferreira, H. Melo, *et al.* 2003. Determination of free and bound terpenes and norisoprenoids during alcoholic fermentation of Port wine. In *Proceedings of Actualités Enologiques*, Ed. Tec-Doc, pp. 589–593.

Birks, A.G. and M.M. Sarmento. 1991. Maceration for premium Port wine production. *Australian and New Zealand Wine Industry Journal*, 6:278–28.

Bradford, S. 1978. *The Englishman's Wine. The Story of Port*, Christie's Wine Publications, London.

Castro, I., J.P. Martín, J.M. Ortiz, *et al.* 2011. Varietal discrimination and genetic relationships of *Vitis vinifera* L. cultivars from two major controlled appellation (DOC) regions in Portugal. *Scientia Horticulturae*, 127:507–514.

De Pinho, P. Guedes, E. Falqué, M. Castro, *et al.* 2007. Further insights into the floral character of Touriga Nacional wines. *Journal of Food Science*, 72(6):S396–401.

Falqué-Lopez, E., A.C. Silva Ferreira, P. Guedes de Pinho, *et al.* 2004. Determination of aromatic descriptors of Touriga Nacional wines by sensory descriptive analysis. *Flavour and Fragrance Journal*, 19:298–302.

Fonseca, A.M., A. Galhano, E.S. Pimentel, *et al.* 1987. *Port Wine: Notes on Its History, Production and Technology*, Instituto do Vinho do Porto, Porto.

Ho, P., T.A. Hogg, and M.C.M. Silva. 1999a. Application of a liquid chromatographic method for the determination of phenolic compounds and furans in fortified wines. *Food Chemistry*, 64(1): 115–122.

Ho, P., M.C.M. Silva, and T.A. Hogg. 2001a. Changes to the colour and phenolic composition during the early stages of the maturation of port in wood, stainless steel and glass. *Journal of the Science of Food and Agriculture*, 81:1269–1280.

Ho, P., M.C.M. Silva, and T.A. Hogg. 2001b. Multiple imputation and maximum likelihood principle component analysis of incomplete multivariate data from a study of the ageing of port. *Chemometrics and Intelligent Laboratory Systems*, 55:1–11.

Ho, P., I. Vasconcelos, F.S.S. Rogerson, *et al.* 1999b. Effects of skin contact and oxygenation of musts on the composition of white port wines. *Sciences des Aliments*, 19:687–699.

IVDP. 2012. Instituto dos Vinhos do Douro e Porto at www.ivdp.pt (accessed 4 March 2012).

Lopes, M.S., M. Rodrigues dos Santos, J.E. Eiras Dias, *et al.* 2006. Discrimination of Portuguese grapevines based on microsatellite markers. *Journal of Biotechnology*, 127:34–44.

Lopes, M.S., K.M. Sefc, J.E. Eiras Dias, *et al.* 1999. The use of microsatellites for germplasm management in a Portuguese grapevine collection. *Theoretical and Applied Genetics*, 99:733–739.

Magalhães, N. 2009. *Tratado de Viticultura; A Videira a Vinha e o Terroir*, Chaves Ferreira Publicações.

Mateus, N. and V. de Freitas. 2001. Evolution and stability of anthocyanin-derived pigments during Port wine aging. *Journal of Agricultural and Food Chemistry*, 49:5217–5222.

Mateus, N., J.M. Machado, and V. de Freitas. 2002a. Development changes of anthocyanins in *Vitis vinifera* grapes grown in the Douro Valley and concentration in respective wines. *Journal of the Science of Food and Agriculture*, 82:1689–1695.

Mateus, N., J. Oliveira, J. Pissarra, *et al.* 2006. A new vinylpyranoanthocyanin pigment occurring in aged red wine. *Food Chemistry*, 97:689–695.

Mateus N., S. Pascual-Teresa, J.C. Rivas-Gonzalo, *et al.* 2002b. Structural diversity of anthocyanin-derived pigments in Port wines. *Food Chemistry*, 76:335–342.

Mayson, R. 2005. *Port and the Douro*, 2nd edn, Octopus Books.

Mendes Moreira, C., T. Hogg, and G. de Revel. 1994. Recherche de composés volatils témoins du viellissement dans le vin de Porto, type tawny. *Proceedings of I Congresso Internacional de la Vitivinicultura Atlântica.*

Mendes-Pinto, M.M. 2009. Carotenoid breakdown products – the norisoprenoids – in wine aroma. *Archives of Biochemistry and Biophysics*, 483:236–245.

Mendes-Pinto, M.M., A.C. Silva Ferreira, C. Caris-Veyrat, *et al.* 2005. Carotenoid, chlorophyll, and chlorophyll-derived compounds in grapes and Port wines. *Journal of Agricultural and Food Chemistry*, 53:10034–10041.

Morata, A., M.C. Gómez-Cordovés, B. Colomo, *et al.* 2003. Pyruvic acid and acetaldehyde production by different strains of *Saccharomyces cerevisiae*: relationship with Vitisin A and B formation in red wines. *Journal of Agricultural and Food Chemistry*, 51:7402–7409.

Moreira, N., C. Pina, F. Mendes, *et al.* 2011. Volatile compounds contribution of *Hanseniaspora guilliermondii* and *Hanseniaspora uvarum* during red wine vinifications. *Food Control*, 22:662–667.

Oliveira, C., A. Barbosa, A.C. Ferreira, *et al.* 2006. Carotenoid profile in grapes related to aromatic compounds in wines from Douro region. *Journal of Food Science*, 71:S1–S7.

Oliveira, C., A.C. Ferreira, P. Costa, *et al.* 2004. Effect of some viticultural parameters on the grape carotenoid profile. *Journal of Agricultural and Food Chemistry*, 52:4178–4184.

Oliveira, C., A.C. Silva Ferreira, M.M. Mendes Pinto, *et al.* 2003. Carotenoid compounds found in grapes and their relationship to plant water status. *Journal of Agricultural and Food Chemistry*, 51:5967–5971.

Oliveira, J., C. Santos-Buelga, A.M.S. Silva, *et al.* 2006. Chromatic and structural features of blue anthocyanin-derived pigments present in Port wine. *Analytica Chimica Acta*, 563:2–9.

Pissarra, J., S. Lourenço, J.M. Machado, *et al.* 2005. Contribution and importance of wine spirit to the port wine final quality; an initial approach. *Journal of the Science of Food and Agriculture*, 85:1091–1097.

Portaria n.o 413/2001. 2001. Regulamento da Classificação das Parcelas com Cultura de Vinha para a Produção de Vinho Susceptível de Obtenção da Denominação de Origem Porto. (Portuguese legal diploma regulating the classification of land parcels and defining authorized grape varieties for the Port production.)

Ramos, R., P.B. Andrade, R.M. Seabra, *et al.* 1999. A preliminary study of non-coloured phenolics in wines of varietal white grapes (códega, gouveio and malvasia fina): effects of grape variety, grape maturation and technology of winemaking. *Food Chemistry*, 67:39–44.

Reader, H.P. and M. Dominguez. 2003. Fortified wines: Sherry, Port and Madeira. In *Fermented Beverage Production*, 2nd edn (eds A.G.H. Lea and J.R. Piggot), Springer.

Rogerson, F.S.S. and V.A.P. de Freitas. 2002. Fortification spirit, a contributor to the aroma complexity of port. *Journal of Food Science*, 67:1564–1569.

Rogerson, F.S.S., S. Azevedo, N. Fortunato, *et al.* 2002. 1,3-Dimethoxybenzene, a newly identified component of port wine. *Journal of the Science of Food and Agriculture*, 82:1287–1292.

Rogerson, F.S.S., H. Castro, N. Fortunato, *et al.* 2001. Chemicals with sweet aroma descriptors found in Portuguese wines from the Douro region: 2,6,6-trimethylcyclohex-2-2,6,6-trimethylcyclohex-2-ene-1,4-dione and diacetyl. *Journal of Agricultural and Food Chemistry*, 49:263–269.

Romero, C. and J. Bakker. 2000. Effect of acetaldehyde and several acids on the formation of vitisin A in model wine anthocyanin and colour evolution. *International Journal of Food Science and Technology*, 35:129–140.

Santos, J.A., Malheiro, A.C., Karremann, M.K., *et al.* 2011. Statistical modelling of grapevine yield in the Port Wine region under present and future climate conditions. *International Journal of Biometeorology*, 55:119–131.

Schultz, H.R. and G.V. Jones. 2010. Climate induced historic and future changes in viticulture. *Journal of Wine Research*, 21:137–145.

Silva Ferreira, A.C. and P. Guedes de Pinho. 2004. Nor-isoprenoids profile during port wine ageing: influence of some technological parameters. *Analytica Chimica Acta*, 513:169–176.

Silva Ferreira, A.C., J.C. Barbe, and A. Bertrand. 2002. Heterocyclic acetals from glycerol and acetaldehyde in Port wines: Evolution with aging. *Journal of Agricultural and Food Chemistry*, 50:2560–2564.

Silva Ferreira, A.C., J.C. Barbe, and A. Bertrand. 2003a. 3-hydroxy-4,5-dimethyl-2(5*H*)-furanone: A key odorant of the typical aroma of oxidative aged Port wine. *Journal of Agricultural and Food Chemistry*, 51:4356–4363.

Silva Ferreira, A.C., T. Hogg, and P. Guedes de Pinho. 2003b. Identification of key odorants related to the typical aroma of oxidation-spoiled white wines. *Journal of Agricultural and Food Chemistry*, 51:1377–1381.

Silva Ferreira, A.C., J. Monteiro, C. Oliveira, *et al.* 2008. Study of major aromatic compounds in port wines from carotenoid degradation. *Food Chemistry*, 110:83–87.

Silva Ferreira, A.C., P. Rodrigues, T. Hogg, *et al.* 2003c. Influence of some technological parameters on the formation of dimethyl sulfide, 2-mercaptoethanol, methionol and dimethyl sulfone in Port wines. *Journal of Agricultural and Food Chemistry*, 51:727–732.

Symington, C., F. Rogerson, A. Serodio, *et al.* 2011. Constructing quality Ports with automated robotic *lagares*. *Bulletin de l'OIV*, 84(965-966-967):419–427.

Voss, R. 1989. *Pocket Guide to Fortified and Dessert Wines*, Mitchell Beazley.

22 Marsala

Andrea Zanfi[1] and Silvia Mencarelli[2]

[1] *Wine journalist, Grosseto, Italy*
[2] *Wine public relations consultant, Pisa, Italy*

22.1 THE STORY OF MARSALA WINE

A good glass of Marsala holds an entire voyage. Drinking it, one travels across its history and the land of its origins, and tastes all the beauty of the sunshine and colours of Sicily, the island for which this wine is a flagship. Located precisely at the centre of the 'sun belt', the climatic zone stretching from Portugal to Greece, Sicily has a dry and hot climate, crossed by African winds, but always tempered by the sea. The Marsala wine we taste today is the sublime synthesis of a land unique for its physical, climatic and natural characteristics, and with a long and fascinating history. It is nature and culture in its purest state.

The 'discovery' and successive industrialization of Marsala wine as we know it is due to the merchant John Woodhouse from Liverpool. In 1773, he tasted *perpetuum*, a strong and sunny wine, matured in wooden barrels, from which a small quantity was taken and replaced by a similar quantity of a younger wine. The operation was repeated after some years and then several more times. Woodhouse noticed a similarity with Madeira, a Spanish wine that was better known at the time, but which was also more expensive. He then had the notion of producing a wine similar to Madeira in the vineyards around Marsala, and selling it in his homeland, using Sicilian labour and grapes which at the time were very cheap. However, Woodhouse wanted to verify first that people would like it. In 1773, the first consignment of wine left the port of Trapani bound for Liverpool. There were fifty 412-litre pipes, each filled with two gallons of wine brandy, in order to preserve the wine as the sea trip to England lasted 40 days. The wine was an immediate success and promising enough to lead Woodhouse to invest in industrial production.

Another essential character in Marsala wine's story is Benjamin Ingham, who must be remembered for having industrialized production and for introducing precise technical criteria. Thanks to his intervention, the modern techniques of viticulture and vinification were introduced amongst growers, thus rendering Sicily's asset even more prestigious. Ingham was followed by his nephew, Joseph Whitaker, known as Pip, who continued the study of Marsala and introduced further new techniques. He is also responsible for the discovery of Motya Island and for its later purchase. The last important character in the story of Marsala is Vincenzo Florio, a native of Bagnara Calabra. He was the forefather of a family which was to gain a unique importance in Sicily's economic panorama, and he created the first large firm to produce Marsala. Thanks to his team and to the contacts he created with other

Sweet, Reinforced and Fortified Wines: Grape Biochemistry, Technology and Vinification, First Edition.
Edited by Fabio Mencarelli and Pietro Tonutti.

international ports, Marsala wine became highly successful and famous worldwide. Florio's firms introduced new working rules, responding to several requests from the working classes at that time (c.1800). These three big families not only discovered, improved and extended the reach of Marsala wine, but were also able to create a modern and flourishing entrepreneurial network, which was a rare reality not only for Sicily but also for the whole of Italy. In time, several other firms were established and developed through the production of Marsala wine, some of which are still in existence today and have become very large enterprises.

However, after its age of glory during the nineteenth century, Marsala's industry passed through a period of deep crisis, because of various reasons. First, vine diseases, such as Phylloxera, late blight and powdery mildew, spread and affected many vines seriously reducing their yield. Later, there were growing problems in trade with other countries, due to the introduction of new taxes and the limits imposed by American prohibitionism. However, during prohibitionism, Marsala wine was still exported to the United States with the label '*Hospital Size*' and a dosage indication of 'a small glassful twice a day' since doctors considered Marsala to have some therapeutic qualities.

The earliest ruling on Marsala wine is a law dated 15 October 1931; it was the first example of a law of this type in Italy. According to this legislation, finally, the territories for the production of Marsala were limited to Trapani, Palermo and the north-western part of Agrigento. With the birth of the designation 'DOC' (Denominazione di Origine Controllata), under a law of 12 July 1963 (No. 930), and other norms introduced in 1984 and 1986, the rules governing production and merchandising of Marsala become more detailed. As a result, today, only the liquorish wines respecting precise rules and with specific characteristics can be called Marsala. Among these rules is especially the place of production, which is the region of Trapani, except for Pantelleria, Favignana and Alcamo.

22.2 GRAPE VARIETIES

The varieties, flavours, ages and colours of Marsala wine are varied, and mirror its land, a diversified Sicily that leaves the traveller lost for words. Marsala was born in Trapani, in the most western part of Sicily. Here the vines are fed by chalky and clayey soils and are always influenced by sea winds. The white grapes Catarratto Lucido, Catarratto Comune, Grillo, Ansonica and Damaschino prevail; the red grapes are Nero d'Avola, Nerello Mascalese and Pignatello. Certainly the most common variety is the Catarratto, which is probably also the oldest, being at the base of the best Italian Vermouths. Based on their colour, the ampelography of the vines suggests that, on the one hand, for the Marsala Oro and Ambra wines, the grapes of Grillo and/or Catarratto and/or Ansonica and/or Damaschino can be used; for Marsala Rubino, on the other hand, the grapes of Pignatello and/or Nero d'Avola and/or Nerello Mascalese can be chosen; moreover, up to a maximum of 30% of some other white grapes can be used in the golden and amber varieties. The characteristics of the different vines represent, therefore, the first everlasting impression to which the flavours and smells of other varieties are added, mixed together with the maturation and oxidation processes.

Let us start then with the most widespread variety, the Catarratto. It is a historical variety in Sicily, where it has been cultivated for a long time and its name means 'wealth'. In 1970, Pastena distinguished four different subtypes of Catarratto: the 'comune' or 'latino', the 'lucido spargolo', the 'lucido serrato' and the 'lucidissimo'. It has a middle period of budding, and a middle-premature flowering, veraison and berry ripening. It has a semi-straight structure and good vigour. The production is rich but inconstant because of the phenomena of flower drops and shot berries. It is better to prune it short and the wine produced is the base for Vermouth and comes under the DOC of Alcamo, Contea di Sclafani, Etna, Sambuca di Sicilia, Menfi, Santa Margherita Belice, Mamertino di Milazzo, Erice and, of course, Marsala.

Grillo, a grape variety imported probably from Puglia and cultivated in Sicily, especially in the Marsala area, since the end of the nineteenth century, has a premature budding and a medium period of flowering, veraison and berry ripening. It grows up straight and with great vigour. The production is good and constant and the wine can be used alone or mixed together with other wines, but also as a base to prepare Vermouth and in other DOCs, such as those of Contea di Sclafani, Marsala, Marmertino and Erice.

Ansonica is another Mediterranean grape variety growing especially in Sicily where it is called 'Inzolia'. It has been cultivated since 1500 on the islands of Elba and Giglio and it is also present on the Argentario coast and in the territories around Livorno and Grosseto. Ansonica has strong similarities to some Greek grapes, such as Rhoditis and Sideritis. We suppose a French origin for the name *ansoria* (from the French word *sorie* meaning tawny, golden) because of the presence of Normans in Sicily and in the east Mediterranean sea. The cluster is medium-sized, has a pyramidal shape, with one or two wings, and is rather sparse. The berry is crisp and aromatically neutral. Ansonica has an intermediate budding, and a premature flourishing, veraison and medium-premature berry ripening. Production is rich and constant; vine training with shoot or mixed pruning is preferred. It adapts to the different humid and hot climates in southern Italy. Some years it is sensitive to late blight and powdery mildew. This variety is used for vinification and for direct use, either alone or mixed with other white grapes. The wine has a yellow colour with green hues; it is perfumed and has a hot and harmonic flavour but with a touch of acidity. It is one of several DOCs in Sicily and Tuscany, such as Alcamo, Ansonica Costa of Argentario, Contea di Sclafani, Contessa Entellina, Marsala, Menfi, Sambuca di Sicilia, Santa Margherita di Belice, Elba, Parrina.

The final white vine related to Marsala is Damaschino. This is a very vigorous vine, with broad drought-resistant leaves and large irregular clusters of grapes that can weigh up to 3 kg. Yield is high, and it produces grapes with a low sugar level and low acidity, from which a delicate and fine wine is obtained, with a low degree of alcohol.

With regard to red grapes, the Nero d'Avola is one of the best grape varieties in Sicily. It is also known as 'Calabrese', a name shared with several varieties in Sicily, Tuscany, Calabria and a large part of southern Italy. The bunch is medium-sized, conical in shape, with wings, and is compound and compact. It has a middle-season budding, premature flowering and a middle period of veraison and berry ripening. The structure of the plant is broad and very strong. Pruning must be short, in order to maintain a small shape, similar to a small tree or vertical trellis. It is resistant to climatic change and cryptogamous diseases, except for *Botrytis* from which it can be easily affected. The wines obtained belong to the DOC Eloro, Marsala, Cerasuolo di Vittoria, Sciacca, Santa Margherita di Belice, Contea di Sclafani, Delia Nivolelli, Sambuca di Sicilia, Contessa Entellina, Alcamo, Erice, Marmetino.

Nerello Mascalese is a typical variety of the territory around Etna where it has been cultivated for a long time. The name derives from the Contea di Mascali, a very old territory at the base of Etna, between Giarre and Mascali (around Catania), which is also the likely centre of origin or, at least, diffusion of the *cultivar*. The Nerello Mascalese probably belongs to the old group of vines 'Nigrelli', described by Sestini in 1760 (Cartabellotta, 2008). The bunch is large and compact, conical or pyramidal, long, with one or more wings, while the grape has a very frosty, thick and consistent skin. The period of budding is quite premature, and is followed by a middle period of flowering, veraison and berry ripening. Used alone or with other varieties, it is part of the DOC Contea di Sclafani, Etna, Faro, Marsala, Sambuca di Sicilia, Lamezia, Sant'Anna di Isola Capo Rizzuto.

22.3 VINIFICATION

Vinification is as normal for white wine. In the case of Grillo, the grapes are picked at an advanced stage of ripening with a sugar content around 250 g L^{-1} differing from Catarratto and Insolia, which rarely get to that stage. The grape pressing is stronger than that of a white classic vinification in order to produce more extracts and oxidizable substances. Generally the final wine has a dry extract of around 25–30 g L^{-1}. Very often in order to increase extraction, a juice extractor (drainer) and a vertical torque (basket press) are used. Must clarification is done with pectolic enzymes and is followed by flotation. Fermentation follows at 18–20°C by inoculating the alcohol-resistant yeasts (e.g. *Saccharomyces bayanus*) with a *pied de cuvée*, and by adding nitrogen, tiamin and yeast ghosts. Towards the end of the fermentation, wine racking is done and new must mass is added to fermentation lees. The wines produced with Grillo grapes have a minimum alcoholic content of 14.5% and a maximum of 15.5%, while the alcoholic content of Catarratto wines is between 13 and 13.5% alcohol.

22.3.1 The mistella

The mistella is prepared by racking the clear must from flotation and adding the filtration solid residues; the mixture is cooled down and alcohol (20% v/v) is added. Sulfur dioxide is added to another portion of must to avoid fermentation and to concentrate it by cooking (cooked must). Mistella is called *sifone* or *safuni* in Marsala's dialect and it is the product obtained by the addition of wine alcohol to the must. A good mistella is obtained either using just one grape variety (Grillo, Catarratto), as ruled by the Marsala DOC disciplinary, or by mixing several varieties. The alcoholic level of mistella must be higher than 17% but lower than 22%, and is obtained by adding alcohol, which inhibits the fermentation process, and a lot of sugar (150–180 g L^{-1}).

Sensorially, mistella presents two maturation phases:

1. during the first month after production, it has a dominant alcoholic taste that is fresh, flowery and fruity (green apple), and a smell that recalls the must and the grape from which it has been obtained;
2. after several months, the product has more mature aromas, the fruit smell develops (the green-apple aroma becomes more candy-like), and the sweet, soft and velvety sensation dominates over the alcohol.

Mistella is important in the preparation of Marsala with sugar residues, both because of the evolution of the primary aroma into tertiary aroma (typical of wines refined in wood) and because of the smoothness it gives to the wine.

22.3.2 The cooked must

Cooked must is the product resulting from the concentration of the sulfited white grape must, obtained through the traditional method of heating over direct fire and at atmospheric pressure. Nowadays, the traditional method is not much used and the cooked must is obtained through heating with a normal steam current at 100ºC coming from a conventional steam generator or boiler. Around 150–200 g HL^{-1} of SO_2 is added to the white grape cooked must. The resulting product becomes transparent through natural and static lees removal and later on, is spun to separate the substances left in suspension (colloids, mucilage, gums and proteins).

In order to improve the quality of the product a clarification with bentonite and gelatine can be used, and after flocculation a filtration can also be done. The resulting product is then ready to face the final phase of concentration. The concentration is done in a vacuum concentrator at a temperature that cannot exceed 40–45ºC, in order to obtain in only one step a must which is partially desulfurized at 22–24° Baumé. The resulting concentrate is cooled down to 20ºC and left to rest for some days to facilitate the precipitation of tartaric salts of potassium and calcium. Once this phase has been passed, the product is ready to be transformed into cooked must.

The concentrated must is kept heated at 80ºC by counter-current steam, avoiding higher temperatures which could give a burnt taste to the wine because of the caramelized sugars.

The process can be considered finished when, after a further percentage of water has been eliminated, the product develops a more amber colour and a sugar density of 26–33° Baumé.

22.3.3 The *concia*

Marsala wine can also be *conciato*, meaning that other ingredients have been added. This is the case for *Rancio* (also called *Lievito* or *Vecchione*), a very good, long-aged (up to 50 years) Marsala produced by adding the best grapes to the new Marsala mass in order to make the wine finer. In this *concia* process, Marsala wine alcohol, small quantities of cooked must and mistella, or non-fermented must by ethanol addition, are used. The *concia* is composed of two phases: sweetening, done by adding cooked must and mistella, and fortification, done either by adding ethanol from wine (higher than 96% v/v) or by adding wine brandy. By adding must or mistella, the wine reaches a sweeter flavour, closer to that of raisins, while the cooked must gives the wine a more velvety and bitter flavour and a yellow colour. The introduction of mistella and cooked must are the principal elements that define the different types of Marsala.

Similar to the solera system of blending, Marsala often goes through a *perpetuum* system, where a series of vintage blendings takes place.

22.4 THE DOC DISCIPLINARY OF MARSALA WINE PRODUCTION

Art. 1

The DOC 'Marsala' and 'Marsala wine' must be used together with the words 'Fine', 'Superiore', 'Superiore Riserva', or 'Vergine' or 'Soleras', 'Vergine Riserva' or 'Soleras Riserva', or 'Vergine Stravecchio' or 'Soleras Stravecchio', which is used for the liquorish, golden wines that have the characteristics established by the law of 28 November 1984, and those established by the production disciplinary.

The use of DOC is allowed only with specific labels indicating the ageing time, colour and sugar level, all expressed in Italian and English.

1. Marsala Fine: alcohol not lower than 17% by distillation;
   dry extract: not lower than 22 g L^{-1};
   total acidity: not lower than 3.5 g L^{-1};
   volatile acidity: not higher than 0.90 g L^{-1};
   minimum 1 year of ageing.
2. Marsala Superiore: alcohol not lower than 17% by distillation;
   dry extract: not lower than 18 g L^{-1};
   total acidity: not lower than 3.5 g L^{-1};
   volatile acidity: not higher than 1 g L^{-1};
   minimum 2 years of ageing.
3. Marsala Superiore Riserva: all the characteristics of Marsala Superiore but with:
   4 years of ageing;
   volatile acidity: not higher than 1.3 g L^{-1}.
4. Marsala Vergine or Soleras:
   alcohol not lower than 18% by distillation;
   residual sugars lower than 4%;
   dry extract: not lower than 22 g L^{-1};
   total acidity: not lower than 3.5 g L^{-1};
   volatile acidity not higher than 1.3 g L^{-1};
   minimum 5 years of ageing.
5. Marsala Vergine Stravecchio or Riserva:
   Same characteristics as Marsala Vergine but with 10 years of ageing.

The Marsala wines named 'golden' have a golden colour that can be more or less intense; the Marsala wines named 'amber' have an amber colour that can be more or less intense; the Marsala wines named 'ruby' have a ruby colour which develops golden hues as it ages.

Marsala is classified into three types, according to its level of sweetness:

1. dry: sugar level inferior to 40 g L^{-1}
2. semi-dry: sugar level superior to 40 g L^{-1}but inferior to 100 g L^{-1}
3. sweet: sugar superior to 100 g L^{-1}.

In Marsala wine to which cooked must has been added, traces of oxy-furfural are allowed as it derives from the fining and ageing processes. Ethyl alcohol or brandy can be added to Marsala wines ready for direct use without the need for further ageing, but they must be fined for at least one month before being sold.

The minimum ageing period starts at the end of the *concia* period. Marsala wines can be reprocessed several times and in such cases the product must be aged further, according to the type of Marsala desired.

The ageing must be done in wooden barrels, from oak or cherry trees, except during the first four months when ageing can be done in other containers. *Marsala Fine*, after the first four months of ageing, can be transformed into other drinks. *Marsala Vergine*, during the ageing process, can be sold as *Marsala Superiore* or *Marsala Fine*, provided the ageing for such categories has been completed and the wines have the right characteristics. *Marsala Superiore* can be sold as *Marsala Fine* if it has completed the ageing for such a category and if the wine has the right characteristics.

ACKNOWLEDGMENT

We would like to sincerely thank the winemaker Tommaso Maggio for his useful information on Marsala vinification.

REFERENCE

Cartabellotta, D. 2008. *Viticoltura ed Enologia di Sicilia. Terra, Il multimediale in Agricoltura.* www.agrinnovazione.regione.sicilia.it (accessed 23 November 2012).

23 Notes on Other Sweet Wines

Fabio Mencarelli

DIBAF, University of Tuscia, Viterbo, Italy

23.1 INTRODUCTION

In the Mediterranean basin, the production of wines from dehydrated grapes, as has been mentioned several times in this book, is an ancient and common practice. However, in addition to the most important wines described in individual chapters, there are a number of sweet wines made from dehydrated grapes which are produced in small quantities in other countries and which are characteristic of their own special regions.

23.1.1 Austria

In Austria, *Strohwein* or *Schilfwein* refers to a sweet dessert wine produced from dehydrated grapes. *Stroh* is German for straw, while *Schilf* means reed. The minimum sugars concentration of the grapes at harvest is 25° KMW (Klosterneuburger Mostwaage), equivalent to 25° Brix, which is the same as for Austrian Beerenauslese. These requirements are regulated by the Austrian wine law.

The grapes are dehydrated for a minimum of three months, either by laying the bunches on mats of straw or reeds, or by hanging them from pieces of string. After a minimum of two months, the grapes must reach a sugar content of 30º KMW (the same must weight as a Trockenbeerenauslese) before pressing. Strohwein and Schilfwein are treated as synonyms under Austrian wine law, and the choice between them depends on local naming tradition rather than on the specific material used as drying mats for a particular batch of wine.

23.1.2 Croatia

In Croatia, the most common wine made from dehydrated grapes is Prošek which is traditionally produced in the southern area of Dalmatia. The system of dehydration is the same as described for Strohwein and Schilfwein above. There are only a few commercial producers as production is typically a homemade affair.

23.1.3 Czech Republic

In the Czech Republic, Slámové víno is the term for straw wine which, under Czech wine law, is classified as a 'predicate' wine (Czech: *Jakostní víno s přívlastkem*). Czech regulations require that harvested grapes must come from a single wine subregion; the grapes must be dehydrated for at least three months either on straw or reed mats or hung in a well-ventilated

Sweet, Reinforced and Fortified Wines: Grape Biochemistry, Technology and Vinification, First Edition.
Edited by Fabio Mencarelli and Pietro Tonutti.

space, and the must weight is required to reach at least 27° NM on the Normalizovaný moštomer scale. Straw wine in the Czech Republic is typically made from white grapes that are well ripened and undamaged.

23.1.4 Cyprus

In Cyprus, the best-known sweet wine is Commandaria which takes its name from the headquarters of the Knights Templar, 'La Grande Commanderie' near Limassol, and hence claims to be the oldest named wine still in production. Commandaria is made from two indigenous grapes, the white Xynisteri and the red Mavro, an ancestor of the Négrette grape known as Pinot St-George in the United States. The grapes are left to over-ripen on the vine and when the sugar content reaches acceptable levels they are harvested. More specifically, Xynisteri is picked when it reaches around 12° Baumé and Mavro at 15–16° Baumé. The grapes are then laid out in the sun to further increase the sugar density through evaporation. When the must weight reaches 19–23° Baumé the juice is extracted through crushing and pressing. Fermentation takes place in reservoirs and will cease naturally due to the high levels of alcohol achieved at around 15%. When fermentation has been completed, at a minimum alcohol level of 10% (which is often exceeded), Commandaria may be fortified by the addition of pure 95% grape alcohol or a wine distillate of at least 70% alcohol. After this addition, the alcohol content may not exceed 20%, while its total potential alcohol (including its sugar content) must be at least 22.5% (Robinson, 2006). Thus, Commandaria may be a fortified wine, but fortification is not mandatory. The above process has to take place within the confines of 14 designated villages that lie in the Commandaria region. Commandaria, by law, is aged for at least four years in oak barrels, but this can take place outside the above designated area within Cyprus under strict control and according to the conditions laid down in Cypriot legislation.

23.1.5 Greece

Amongst the Greek islands, there is widespread production of wines from grape dehydration. Vinsanto of Santorini is made from Assyrtiko grapes, vinified after a few days of drying in the sun. The wine is then placed in barrels to mature for several years, as its capacity for ageing is measured in decades. Sweet wines of Samos Island are also made from sun-dried Muscat Blanc à Petits Grains grapes. An interesting sweet red wine is Mavrodafni from the Achaia region in Peloponnese. The dark Mavrodafni grape was named after the fruits of the Daphne bush. The wine is a fortified wine. Mavrodafni is initially vinified in large vats exposed to the sun. Once the wine reaches a certain level of alcohol, fermentation is stopped by adding distillate prepared from previous vintages. Then the Mavrodafni distillate and the wine, still containing residual sugar, is transferred to underground cellars to complete its maturation. The solera method, as described in Chapter 16 on Pedro Ximénez and Malaga, is used. Once aged, the wine is bottled and sold as a sweet wine under the 'Mavrodafni OPAP' designation (Lazarakis, 2005).

REFERENCES

Lazarakis, K. 2005. *The Wines of Greece* (Mitchell Beazley Classic Wine Library), Octopus Publishing (pp. 496). ISBN 1-84000-897-0.

Robinson, J. 2006. Commandaria. In *The Oxford Companion to Wine*, 3rd edn, Oxford University Press, Oxford (pp. 190). ISBN 0-19-860990-6.

FURTHER READING

Austrian Wine Act 1999. Bundesgesetz uber den Verkehr mit Wein and Obstwein.
Czech wine fund: Viticulture Act of 2004. Available [in Czech] at http://www.vinarskyfond.cz/legislativa/1-vinarska_legislativa_2004.pdf (accessed 30 November 2012).
For an overview of Austrian wine (facts and figures), see http://www.winesfromaustria.com (accessed 30 November 2012).

FURTHER READINGS

Part 4 Market and Marketing

24 Sweet Wine Market

Renzo Cotarella

Marchesi Antinori srl, Firenze, Italy

24.1 INTRODUCTION

When speaking of sweet wine, there is often a certain ambiguity in the definition of this wine category, which encompasses such a wide range of products.

In order to make it easier, we propose a classification according to the *manufacturing process*, which is the classification generally adopted by the majority of wine specialists.

1. Natural sweet wine: Muscats
2. Fortified wines: Port and Marsala, *Vino liquoroso* and Barolo Chinato (aromatic wine)
3. Raisins wines:
   - Natural raisins wines: Passito (i.e. Passito di Pantelleria)
   - Aged raisins wines: Vinsanto
4. Late harvest wine: Muffati, Sauternes and Ice wines.

It may also be useful to know that on the basis of the specific market area, the boundaries of these categories can be more or less flexible. In foreign markets, sweet wine is often considered as a subclass of one general group. This can complicate research making it difficult to gather details for statistical analyses.[1]

24.2 SIZE AND TREND OF THE SWEET WINE MARKET

The sweet wine market is dominated worldwide by Italy and France; between them they produce around 90 000 hL per annum.

Just as an indicator of the size of this market, one should consider that their total wine production is around 259 million hL per annum; we are therefore looking at a particular *niche market* (Table 24.1).

[1] The ICE (Italian Institute for Export Trade) does not consider sweet wine as a single category in its studies and research.

Sweet, Reinforced and Fortified Wines: Grape Biochemistry, Technology and Vinification, First Edition.
Edited by Fabio Mencarelli and Pietro Tonutti.

Table 24.1 Share of total wine production by nation in 2010.

Countries	%
France	17
Italy	16
Spain	14
United States	8
Argentina	6
Australia	4
Chile	4
South Africa	3
Germany	3
Portugal	2
Rest of the world	23

The ICE (Italian Institute for Export Trade) does not consider sweet wine as a single category in its studies and research.

24.3 PRODUCTION AND DISTRIBUTION OF NATURAL WINE IN THE ITALIAN MARKET

The Italian territory encompasses several zones with suitable *terroir*, and a huge selection of grape cultivars for the sweet wine category, especially Passito, Vinsanto and Moscato.

According to Federvini[2] research figures, for natural raisins wine alone, there are:

- 17 regional producers
- 100 general wine denominations (also used for raisins wine)
- 40 denominations and 10 subzones just for Vinsanto and Picolit.

Italy produces around 46 000 hL per annum (511 000 cases x 9L); 13% of volume (turnover €24 million) is exported, while the great majority of sales are in the domestic market and the local production area (Figure 24.1).

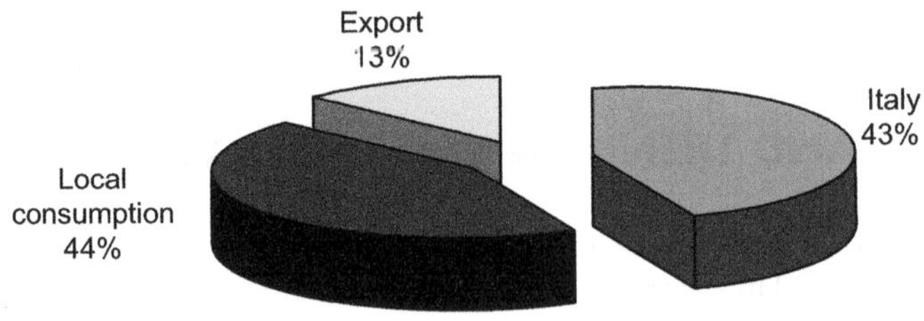

13% of volume is exported

43% of volume is consumed in the domestic market

44% of volume is consumed in the area where it is produced

Figure 24.1 Italy's export and domestic consumption share by volume. (For a colour version of this figure, please see the colour plate section.)

2 Federvini is an Italian federation of wine producers, with its headquarters in Rome.

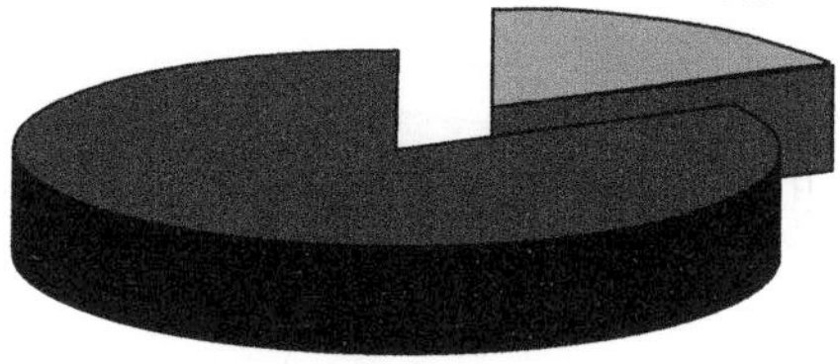

Figure 24.2 Italy's quantity and market share by channel. (For a colour version of this figure, please see the colour plate section.)

The principal import markets are: the United States, Japan, Germany, Switzerland and Australia. However, other countries such as Benelux, Canada, China, Denmark, and the United Kingdom have been increasing their import volumes.

Forty-four per cent of the total production of sweet wine is consumed in the area where it is produced. This highlights the strict relationship between the production and consumption areas, a much stronger indicator than for any other wine category. Even if sweet wine was able to match the new trends in consumption, statistics still show that the sweet wine market is experiencing a decline in sales. In 2009, sales decreased by 6.9%; less, however, than in the general wine categories. Although the official statistics are not yet available, there were negative trends in both 2010 and in 2011. In the domestic market, sweet wine is distributed mostly through *off-trade* channels with around 33 000 hL, and just 10 000 hL through the *on-trade* channel (HORECA: the Hotel/Restaurant/Café channel) (Figure 24.2).

The *major multiple* channels are 79% off-trade with a turnover of €21 million; 70% of sales are in just three types: Vinsanti, Passiti and Moscati (Figure 24.3). Supermarkets offer a vast range of wines, with a wide selection of sweet wine. According to a survey by the Industrial Research Institute (IRI; USA), there are usually 16 wines on offer: 11 sweet wines (Zibibbo, Malvasia, Vinsanto, Passito di Pantelleria, Moscato and Recioto), 4 Marsala, and 1 Port. Figure 24.4 shows the breakdown of consumption by region.

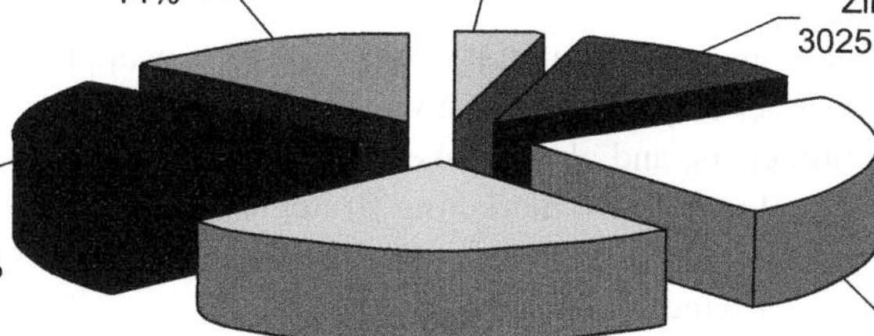

Figure 24.3 Major multiples off-trade sales and market share per wine. (For a colour version of this figure, please see the colour plate section.)

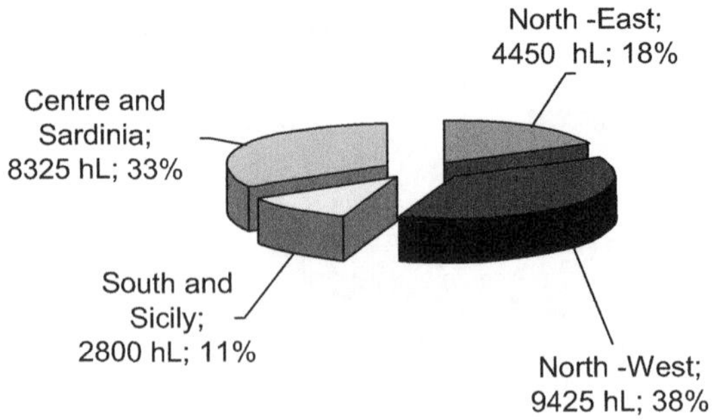

Figure 24.4 Major multiples off-trade sales and market share per region. (For a colour version of this figure, please see the colour plate section.)

Another analysis, which relates to the HORECA channel, distinguishes between *wine shops* and *restaurants*. On the one hand, their wine lists differ in size, but on the other, they both prefer to select wines from local producers.

Wine shops generally stock around 25 sweet wine labels. Wines from the Sicilian region are the most highly represented, followed by wines from Veneto, Trentino, Tuscany and Piedmont.

Restaurant wine lists offer a more limited choice with 70% offering only 5 labels at prices ranging from €20–100.

Sweet wine is almost always associated with desserts, which take a 16.7% share of a restaurant's turnover; this is a marker for the real potential of the sweet wine category.

24.4 THREATS AND OPPORTUNITIES FOR THE FUTURE OF SWEET WINE

As noted earlier, a short wine list for sweet wines can be considered as one effect of a generally negative trend. Today, we are facing several changes that are contributing to the downward trend in consumption of sweet wine. The economic crisis has reduced the average expenditure in a restaurant and it has become too expensive to order a bottle of sweet wine at the end of a meal, given that more than 50% of wine sold costs less than €12 a bottle.

In addition to these economic reasons, we cannot overlook another limit represented by the alcohol breathalyser test, a legal restriction which has had a considerable impact on the consumption of sweet wine.

Taking these considerations into account, it is vital to find new approaches to increase the sales of sweet wines. In order to promote these wines successfully, producers need to keep tight control of production costs, and also to take advantage of the limits of the sweet wine market to introduce new and innovative marketing strategies, such as '*one to one*' (blogs) to attract a new kind of consumer (targeted by sex and by age), and to discover new techniques for matching food and sweet wine.

Of particular note amongst the new marketing approaches, are some small international companies which have created 'clubs' for lovers of sweet wine as a way of promoting specially themed wine dinners, sweet wine tastings and wine courses through mailing lists and newsletters.

25 A Strategic Framework for Marketing Sweet, Reinforced and Fortified Wines

Alberto Mattiacci[1] and Costanza Nosi[2]

[1] *Department of Communication and Social Research, University of Rome, La Sapienza, Rome, Italy*
[2] *Department of Management Sciences, University Roma Tre, Rome, Italy*

25.1 INTRODUCTION

Over the last 30 years, the global wine business has become more and more crowded (Zanni *et al.*, 2005). Wineries have been spurred to increasingly adopt strategies in order to succeed or maintain their position, while distributors have enhanced their market power by changing their approach to the category. The augmented number of competitors, either as producers or retailers, enables customers to choose among a multiplicity of heterogeneous products, amplifying both *intra*- (between wines) and *inter*-competition (in the comprehensive beverage category). In addition, due to the spread of technology and production competences, the average quality of wines continues to improve in all producing countries. Hence the necessity for incumbents to defend their achieved position, and to penetrate the market with innovative products, attempting both to capture untapped demand and to conquer market shares from existing players. Market innovation within the industry is dominated by these opposing forces.

On the demand side, consumers are becoming more and more sophisticated and capable of obtaining all available information prior to making a purchase, with the aid of wine guides, Internet-based apps, wine magazines, wine rankings, and so on. Hence today, everybody is able to satisfy their personal needs (e.g. a gift purchase) and desires (e.g. discovering a new brand) in relation to wine, and according to their available budget.

The rise of new wine industries (the so-called 'New World') and the changes in living standards led by globalization, stimulate cross-cultural eating and drinking habits, leading to a increasing level of wine consumption in traditionally non-consuming countries, such as California, New Zealand, Australia and, to a lesser extent and more gradually, Asia. At the same time, per capita consumption in traditional markets such as France, Italy, Spain and Portugal has been declining in volume but, at the same time, moving towards higher quality (and higher priced) wines.

In this changing world, sweet, reinforced and fortified wines represent a very specific and circumscribed *wine class*, a subcluster of the wider category of wines – as we will explain in the following paragraphs. While basically neglected by the management and marketing literature – which focus mainly on other wine classes, such as sparkling and

Sweet, Reinforced and Fortified Wines: Grape Biochemistry, Technology and Vinification, First Edition.
Edited by Fabio Mencarelli and Pietro Tonutti.

still wines – these products constitute an interesting research topic; consumption habits, price variety, distribution policies, tradition/innovation dilemma are just a few examples of the multiple market peculiarities that make sweet, reinforced and fortified wines worthy of further investigation.

As with any other wine class, it is extremely hard to work out a univocal marketing framework and strategy for the entire category of sweet, reinforced and fortified wines. Product variety, for instance, turns into a plethora of different value offerings (from value to premium, reaching sometimes the luxury segment) and as such inhibits the development of a single effective marketing plan, suitable for all situations. As we will see later in this chapter, marketing management should be interpreted as a general tool kit that provides concepts, methods and techniques which need to be adapted from time-to-time to both product and consumer characteristics.

The first part of this contribution is dedicated to briefly illustrating the theoretical background of wine marketing and clarifying the meaning of market exchanges from a marketing point of view. The second section highlights the most relevant drivers of wine buying and consumption behaviour, while the third part focuses on the main features and critical aspects of sweet and fortified wines. The last section provides a possible framework for marketing these wines taking into account consumer purchase and consumption behaviour, and product characteristics.

25.2 WINE MARKETING: A BRIEF THEORETICAL BACKGROUND

Since its foundation, *marketing management* has been a discipline that, on a scientific basis, aims to provide business with intellectual, diagnostic and operational tools to manage market exchanges profitably (Kotler, 1967; Bagozzi, 1975; Kohli and Jarowski, 1990). Contemporary marketing originated in the United States of America around 110 years ago and has spread from there around the world, undergoing four distinct phases or 'eras' (Wilkie and Moore, 2012):

Pre-era: Pre-marketing (before 1900)
Marketing as a field of study does not exist. Empiricism prevails in the management of commerce, and market exchanges occur as simple phenomena, strongly localized at a geographical level.

Era 1: Founding the field of marketing (1900–1920)
Distribution increases: increasing industrialization of advanced economies gives rise to the technical and economic problem of serving the market efficiently and effectively. Under the influence of the field of study of economics, growing attention is devoted to market exchanges as economic institutions. 'Marketing' courses start to be offered at universities.

Era 2: Formalizing the field (1920–1950)
The discipline is formalized under the name of 'marketing fundamentals' including, in addition to distribution-related topics, pricing, selling techniques and promotion. In the United States, entities that will play a major role in the development of the discipline are founded: the American Marketing Association (AMA), the *Journal of Marketing* and the *Journal of Retailing*. Numerous conferences are organized.

Era 3: A paradigm shift in the marketing mainstream: Marketing, management and the sciences (1950–1980)
After World War II, the boosting of mass markets in the United States provides the basis for development of the discipline. Two main perspectives emerge: (i) the managerial point of view that leads marketing themes towards the so-called managerial economics; (ii) behavioural and quantitative sciences that constitute the basis for the successful advancement of marketing as a dedicated branch of learning.

Era 4: The shift intensifies: Fragmentation of the mainstream (1980–present)
The business world has to face new challenges which modify the scenario: business focus on short-term financial results, downsizing, globalization, re-engineering of organizations and structures. The marketing field develops under the influence of: (i) a 'publish-or-perish' mindset that starts to dominate academia; (ii) the diffusion and fragmentation of the conceptual framework of marketing management.

The third marketing era is the one that provides effective figures of speech able to bring marketing ideas into very different cultural and market environments: *marketing concept* (McKitterick, 1957) – customers represent the focus of the entire business activity; *market segmentation* (Smith, 1956) – it is economically convenient to target a portion of the whole potential product demand; *marketing myopia* (Levitt, 1969) – criticism of the intellectual myopia of those unable to see the market because they are too focused on the product; *marketing mix* (Borden, 1964) – companies have to invest based on a reasoned combination of levers to dominate the market and make profits; *four Ps* (McCarthy, 1960) – identification of four elements constituting the mix (price, product, place and promotion); *brand image* (Gardner and Levy, 1955) – the awareness that a trademark is something more than a sign and embodies meanings that might be valued by the customer; *marketing management* (Kotler, 1967) – the conceiving of a comprehensive intellectual framework that contains all the above-mentioned concepts and combines them into a three-phase process: analysis, planning and control.

The fourth era, the one we are experiencing today, is definitely the liveliest, both at an intellectual and an operational level. The high degree of material well-being is leading markets progressively toward a structural maturity, by increasing internal competition through the proliferation of offerings, market polarization between basic and premium, and globalization. These are a few elements of the contingent state. A *service-centred* frame of mind consolidates and spreads out (Parasuraman *et al.*, 1988); *customer satisfaction* – and the subsequent focus on customer loyalty – is a must for any organization. At a business-to-business level, customer loyalty turns immediately into the building of a long-term relationship (in the place of a one-shot market exchange) and brings the concept of 'Customer Lifetime Value' to the economic framework. Henceforth, the expansion of this concept into consumer markets is relatively fast. At the same time, digital technology is radically transforming the range of relational opportunities that can be built between the demand- and supply side. It is bidirectional (whereas before, it was unavoidably one way) and pervasive, many-to-many (whereas before, it was one-to-many). An evaluation of the structural effects of such a variable on market exchanges is still immature, since the phenomenon has just begun, but it will surely be dramatic.

From a theoretical point of view, the field of marketing management is starting to consider the specific characteristics of the wine business very late, and only during the fourth marketing era. It was not until the 1980s that Emerald Group Publishing Limited, the world's leading scholarly publisher of journals and books in business and management, founded the *International Journal of Wine Marketing* (currently published as the *International Journal of Wine Business Research*). The *Academy of Wine Business Research* was founded in

2003 to create an international society of academics and scholars devoted to the research of economic, social and management issues in the wine sector. So, it is only since 2000 that scientific journals have started to publish essays and articles devoted to the wine business. Up to then, wine marketing had essentially been investigated – when indeed it was investigated – as a specific case of the wider phenomenon of *fast-moving consumer goods* (FMCG), or of the luxury market, depending on the type of wine being considered.

This is obviously correct, both from an operational and from a theoretical point of view. The trend toward a vertical specialization of the discipline which began in the 1980s, makes 'wine marketing' worthy of attention due to some of its specific traits.

1. Wine is a *cognitive product* (Mattiacci *et al.*, 2006): value perception of the product is strongly related to consumer knowledge. As a consequence, wineries have to invest in promotional activities and not just in activities aimed at increasing the customer's ability to 'understand' a wine (tastings, cellar visits, etc.), but also in their appreciation of price as a sign of product value.
2. Consumers differ based on their geographical and cultural backgrounds (Mattiacci *et al.*, 2006): in traditional wine-producing countries, purchase and consumption follow specific rules and patterns that diverge from those where wine is only imported.
3. The pricing range of the product category is impressive. Prices range from as little as a few centimes to more than a thousand euros, giving rise to anomalous 'markets' (such as auctions, collectables) which are non-existent in the rest of the agro-food business.
4. Since wine production is positioned in a *grey area*, between agriculture and industry, it is subject to specific and strict regulation, which varies throughout the world, altering global competition. In the EU, for instance, producers have to follow the rules established by the unified European rural policy.
5. The impressive number of existing labels and brands in every market make it impossible to adopt brand awareness-based market strategies, similar to those applied in the beer and spirits markets.
6. Wine marketing communication is strongly informative rather than persuasive, and is aimed at generating product knowledge and educating consumers about the intangible values of brands. The increasing importance of health-related wine issues in various countries, mainly due to alcohol addiction problems and the spread of alcoholic beverages among young people, in fact, hinders the adoption of aggressive marketing strategies.
7. International wine distribution strategies need to be tailored either to local purchasing habits, rules and conventions, or to local laws and unwritten rules.

25.3 WINE BUYING AND CONSUMPTION BEHAVIOUR

According to the AMA, a product is 'a bundle of attributes (features, functions, benefits, and uses) capable of exchange or use; usually a mix of tangible and intangible forms. Thus a product may be an idea, a physical entity (a good), or a service, or any combination of the three. It exists for the purpose of exchange in the satisfaction of individual and organizational objectives'.[1]

---

[1] Available at http://www.marketingpower.com/_layouts/Dictionary.aspx?dLetter=P (last accessed 7 January 2013).

From the distinctive point of view of marketing management, a product is not just the output of a manufacturing process. It is something much more complex and articulated, demand- and market sensitive. In some cases, according to the *Consumer Culture Theory* (Arnould and Thompson, 2007), the product sees people as the protagonists of its ideation and implementation, not only as mere buyers and consumers (Carù and Cova, 2011). It is important to point out that, unlike other markets (e.g. soft drinks or snacks), the wine industry tends to exclude the idea of a deep involvement of the consumer in the process; market-driven products – such as *Yellow Tail* or the Italian number-one selling *Tavernello* – are not considered examples to follow, despite their excellent market performance.

Market exchange is not viewed as a 'simple' commercial transaction (i.e. money is given by purchasers in order to possess or use goods or services), but it is considered to be a more complex phenomenon, where desires, needs, culture, experiences, emotions, budgets – from a *demand-side* point of view – as well as targets, volumes, margins, rotations, scale economies, logistics – from a *supply-side* point of view – are involved. The object of the exchange is not represented by the good or the service itself, but it is the *value* that it embodies which constitutes the reason for the transaction; value is a wider construct than product, as it includes functional and symbolic, as well as experiential product performance drivers.

As shown in Figure 25.1, marketing management encompasses all aspects of the dynamic relationship between producers, distributors and final consumers. Producers – for example Seagram, Florio, etc. – must manage the exchange with the trader (*sell in*) and/or the exchange with the consumer (*sell out*). Distributors – for example Marks & Spencer, an importer – have to deal either with a business-to-business exchange, to source the products/brands to merchandise, or a business-to-consumer one, where products/brand are purchased for final consumption. The behaviour of these players is supposed to be rational. Perceived value

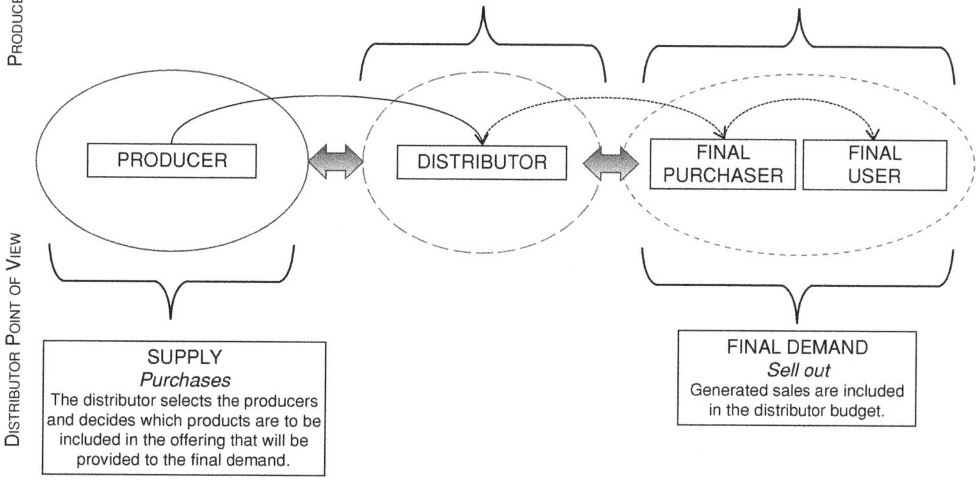

Figure 25.1 Marketing as exchange.

is at the very heart of this complex system and becomes, therefore, the key concept of contemporary marketing theory.

A great part of the international literature dedicated to wine concentrates on wine-consumer behaviour. It focuses on wine attributes influencing the purchasing and consumption process, as well as on the relevant characteristics that affect product quality evaluation by consumers after drinking (confirmation or contradiction of perceptions with regard to expectations) (Lockshin and Hall, 2003).

Quality has been found to play a major role in the wine-consumer's purchasing behaviour (Hauck, 1991). Consumers evaluate products based on intrinsic and extrinsic cues (Olson and Jacoby, 1972; Zeithaml, 1988; Grunert, 1995). While the former belong to the substance of the product and may be revealed exclusively through processes which alter its physical texture (in the case of wine, for example, sugar content), the latter do not constitute its substance, since they are physically separated from it (typically, in the case of wine, the label) (Orth and Krška, 2002).

The literature claims that consumer behaviour is deeply influenced by extrinsic cues (Lockshin and Hall, 2003). These are signs used by consumers to make their comprehensive quality judgement on goods. They influence both the expected quality – formed in the pre-purchase phase – and the perceived quality after consumption (Rao and Monroe, 1989). The usefulness of these signs lies in their informative content and the quality-predictive value they have for consumers (De Magistris, 2004). In conditions of imperfect information about product quality, consumers use extrinsic cues, mostly when intrinsic ones are difficult to ascertain, such as in the case of food products (Grunert *et al.*, 1996; Fotopoulos and Krystallis, 2003; Nosi and Zanni, 2004) and beverages, and so it is with wine (Zanni *et al.*, 2005).

According to a well-known taxonomy (Nelson, 1970), product attributes may be classified as research, experience and trust. Research attributes can be defined as informative stimuli that consumers associate with product quality, and which may be ascertained through the senses before purchase (e.g. colour, labels, price, brand of a wine). Experience attributes can be ascertained exclusively after purchase (e.g. flavour, taste). Trust attributes cannot be ascertained at all or only at high costs (e.g. chemical composition, transformations undergone during production processes, etc.).

Wine belongs, in the same measure as any other foodstuff or beverage, to the category of 'experience goods'; that is, consumers' capacity to assess quality prior to purchase is limited, since most product attributes can be assessed only during consumption. This means that, in the decision-making process of wine purchase, trust (or credence) cues prevail. Such a condition has spurred producers to turn trust attributes into search attributes to be conveyed to consumers (Zanni *et al.*, 2005). Numerous information cues play a major role in purchasing behaviour for wine: packaging (front and back labels, bottle, cask, etc.) (Gluckman, 1986; Charters *et al.*, 2000; Orth and Malkevitz, 2006; Santini *et al.*, 2006); pricing (Jenster and Jenster, 1993; Koewn and Casey, 1995) place of origin (Thode and Muskulka, 1998; O'Neil and Whatmore, 2000; Orth *et al.*, 2005; Balestrini and Gamble, 2006; d'Hauteville *et al.*, 2006); sales material (Chaney, 2000); branding (Lockshin *et al.*, 1997; Beverland, 2004, 2005; Novak *et al.*, 2006).

Furthermore, since the quality-predictive value of credence attributes is based on the credibility and reliability of the information that the consumer is able to gather before and during the purchase process, wineries endow their offerings with further external information, such as denominations, certifications, traceability labels, medals, and so on. These attributes are related to evaluation processes carried out by third parties, without a lucrative interest in the business activity of the producer. The more the consumers trust and rely on the information source, the more the information influences their quality expectations, and therefore their purchasing decision (Grunert *et al.*, 2000). Hence, control organisms on production codes, consortia, specialized press, expert jury panels, wine guides, etc. increase their importance. In the wine industry, they represent an increasingly relevant information tool for inexpert consumers who appear to be highly confused by the multiplicity of available products. Previous studies have postulated the relevance of such information cues on wine-consumer behaviour: awards (Orth and Krška, 2002); wine-guide ratings (Odorici and Corrado, 2004); denominations (Barham, 2000).

However, despite the multiplication of sources consumers can use to reduce their information asymmetry, a cognitive gap persists. The attempt of wine producers and marketers to confer multiple information contents on products can end in a confusing and discouraging purchase. Gluckman (1986) has pointed out that consumers fail to differentiate wine brands and tend to mix them up with other product attributes, such as grape variety or region. This holds true especially for the young consumer who seems to be perplexed in the face of the many cognitive aspects of wine (Mattiacci *et al.*, 2006).

25.4 WINE ACCORDING TO MARKETING THOUGHT

The basics of marketing thought (Levitt, 1969; Kotler, 1984) consider the product concept as consisting of three levels of value (Figure 25.2): the *core product* represents the reason

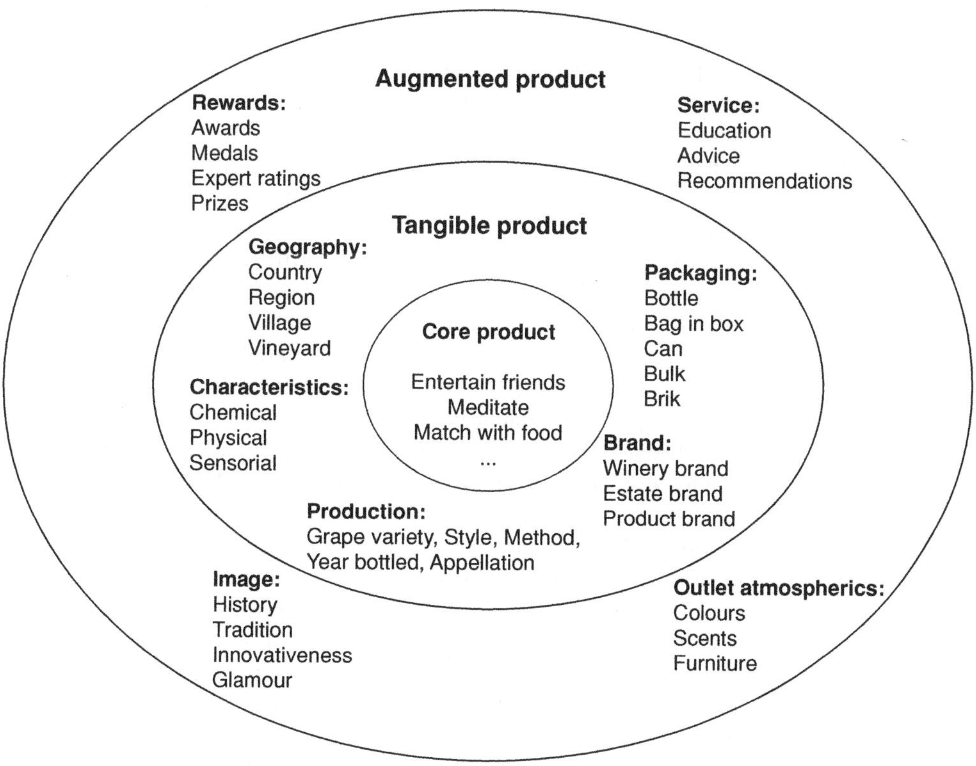

Figure 25.2 The product concept of wine. Adapted from Kotler (1983).

why a customer buys a good or a service. In the case of wine, the need that the consumer satisfies by purchasing the product could be to entertain friends, find a beverage that matches with food, meditate, and so on. Given that consumers buy products to satisfy their needs, producers have to track down these different needs and transform them into a suitable solution. This solution is called the physical, or *tangible*, product (Bergius and Kragell, 2001). It is the product consumers can touch; in the case of wine: look at, smell, taste. It includes packaging, label design and all the information that potential buyers might gain from observing and experiencing the wine. The third level includes intangible features that exceed normal standards and go beyond customers' expectations. It is called the *augmented product* and provides further benefits and value to consumers. It includes, for example, the image associated with the winery, the prestige of the wine provided by awards given by wine guides and magazines, the atmosphere created by the wine outlet as well as the education provided by the wine producer or seller. All these features can reinforce and enhance the wine-buyers' perception of the product (López Lluch *et al.*, 2011).

25.4.1 Sweet and fortified wines: some considerations

Considering both product features and contextual variables (demand, competition, etc.), we can try to point out the critical factors that should be taken into account in order to define the guidelines for a marketing strategy for sweet and fortified wines. Keeping in mind that each marketing lever has to be adapted, from time to time, to the different positioning of products (value, premium, luxury), some general considerations can be made.

- *Longevity*: The different production techniques (late harvest, use of dried grapes or grapes infected by *Botrytis cinerea*, addition of a spirit to stop the fermentation process, etc.) adopted to make sweet and fortified wines provide these beverages with a higher alcohol and/or sugar content which affect their chemical as well as their sensorial characteristics. Along with these qualities comes the longevity of the wine, that is, its capability to last in time without spoiling, and also once it is opened. It is important to note that this specific and differentiating characteristic of sweet, reinforced and fortified wines results from a tangible product feature related to the way the wines are produced.
- *Provenance*: Despite the multiple sources consumers can make use of in order to reduce the information asymmetry and moderate the perceived risk related to a possible mistake in the purchase of wine, a cognitive gap persists. This gap is particularly relevant in the case of experience and high-involvement goods such as quality wines. In this field, one of the most useful extrinsic cues to reduce the cognitive mismatch is the product's 'place of origin', again a feature related to tangible product attributes. Most fortified wines are named after the regions where they are produced, as each regional fortified wine has a distinct style (*Marsala* from the Sicilian port of Marsala, *Port* from the Douro Valley in the northern provinces of Portugal, *Sherry* that comes from the anglicization of the name of the Spanish town of Jerez, Malaga, etc.), as well as a distinct positioning in local habits and traditions (some examples of sweet and fortified wines are shown in Table 25.1).
- *High notoriety and awareness of single appellations*: The value embodied by the territory where wines are produced is witnessed by the fact that some of these wines are protected with an Appellation of Controlled Origin, meaning that only wines from that region may bear that name. Wines not made in that region can only be labelled as being in the 'style' of that particular area. Functioning as meta-brands (Zanni *et al.*, 2005), geographical

Table 25.1 Examples of sweet and fortified wines.

Nation	Product class	Product name	Brand
Italy, Trapani	Sweet wine	Passito di Pantelleria	Donnafugata
Canada, Ontario	Sweet wine	Ice wine	Inniskillin
Portugal, Douro	Fortified wine	Port	Quinta do Portal
Spain, Cádiz	Fortified wine	Sherry	Bodegas Hidalgo La Gitana
France, Sauternes	Sweet wine	Sauternes	Château d'Yquem

indications may exert positive effects on attitudes as well as on buyers' willingness to pay a premium price. In addition to playing an informative role (by letting consumers know about the wine's territorial provenance, grape variety used, adopted production techniques, etc.), in fact, these indications endow products with symbolic features that influence the decision-making process of consumers (Barham, 2000). Affective and emotional cues related to the beauty of the landscape, the history of the place where wine is made, local traditions and culture, and so on, represent powerful tools to consolidate over time the notoriety and reputation of wines.

- *Strong stories and long-lasting tradition*: Sweet and fortified wines boast very ancient and often fascinating stories, intangible attributes belonging to the augmented product, able to exert a powerful attraction in the minds of consumers. *Vinsanto* (the famous Tuscan 'holy wine') whose earliest references date back to the Renaissance; *Port*, popular already in the early 1700s, fortified by the Portuguese to preserve the wine during the long journeys needed to ship it abroad; *Marsala*, originally produced in 1772 by an English merchant as an inexpensive substitute for *Sherry* and *Port*; these are just a few examples of the variety and richness of traditions of sweet and fortified wines.

While the factors illustrated above constitute important elements in enhancing the potential of special wines, some additional issues should be considered.

- *Constrained market penetration and consumption frequency*: Due to the higher sugar and alcohol content, sweet and fortified wines have been 'relegated' to specific and limited consumption occasions: they are often used in cooking, or they are served as aperitifs or dessert wines. From a marketing point of view, consumption occasion represents the reason for buying a wine and is related to core product attributes. At the market level, 'other than' food-matching opportunities for consuming sweet and fortified wines have caused a constrained penetration and a limited consumption frequency hindering their wide diffusion.[2]
- *Low awareness of product category*: While other wine categories, such as red, white, sparkling and still wines, are well known among consumers, who would be able to describe them and recall some product or winery brands, especially in the so-called 'New World' countries,[3] there is a lack of understanding and perception of the distinctiveness

[2] That consumption occasion represents a discriminant factor between wines is witnessed also by some classifications that group wines into Table wines and Sweet and Fortified wines. These taxonomies emphasize the fact that Fortified wines are *not* Table wines and therefore their consumption occasion refers to other than food-matching opportunities. See CLP Regulation Service, 2012.

[3] *Australian Fortified Wines: The Dawning of a New Era*, 2009. Available at: http://www.wfa.org.au/resources/1/Reports/Fortfied%20Strategy/Fortified_Wines_Strategy.pdf (last accessed 8 January 2013).

of sweet and fortified wines as a category. As previously stated, product knowledge and awareness are much higher with relation to single appellations and names such as *Port*, *Marsala*, *Madeira*, *Ice wine*, *Sauternes*, and so on.

- *Dated and older user associations*: Probably led by the strong reputation of Port and Sherry, again in the New World countries,[4] fortified wines seem to be associated with a dated and older consumer. This could discourage consumption on the part of younger individuals, leading to reduced consumption frequency, and should spur producers and sellers to modernize their image.
- *Relevant quality differentials within appellations/regions of production*: As in other geographical areas (Nosi and Zanni, 2004) where wine is produced under the same appellation, quality – and price – differentials between products may be significant. We can buy, in fact, the same – only nominally – 0.75 L bottle of Tawny Port from as little as €4 to almost €2000. One of the most important functions of appellations consists in assuring quality standards and product homogeneity within a given territory. However, due to the qualitative divergence between wineries, neither the first nor the second condition seems to be granted. This matter ends up confusing potential consumers who find it hard to discriminate between wines and start to have misgivings about the power of the denomination to provide the customer with information about the product and the social environment from which it comes.
- *Imitation by non-traditional producing countries*: A further critical factor is represented by the threat of products provided by those countries that, not boasting of a tradition in the production of sweet and fortified wines, commercialize imitation brands (or 'fakes'). It is the case in New Zealand and some Asian countries, for example, where Ice wines are made from artificially frozen grapes (O'Dell, 2009); or in Australia where up until 2008, wines produced on the continent were being labelled as Tokay or Sherry. In that year, in order to stop this illegal commercial strategy, based on the Australia–European Community Agreement on Trade in Wine, Australia agreed to relinquish the use of these traditional European expressions.[5]

25.5 CONCLUSIONS, LIMITATIONS AND SUGGESTIONS FOR FURTHER RESEARCH

Due to the reasons introduced at the beginning of this chapter, both marketing management theory and practice have to adapt to the portraying features of wine. Specific traits of sweet, reinforced and fortified wines make the application of these concepts and techniques even more particular, compared to the already fairly specific but wide-ranging category of 'wine'.

Adopting a top-down analysis approach, two aspects are to be considered:

1. discrimination of marketing policies based on *value band*
2. product innovation limited to peripheral product features.

Discrimination of marketing policies based on value band. Nowadays, the wine market is fragmented into three major value bands (Mattiacci, 2003). The discriminant factor dividing

[4] *Australian Fortified Wines: The Dawning of a New Era*, 2009.

[5] *Australian Fortified Wines: The Dawning of a New Era*, 2009.

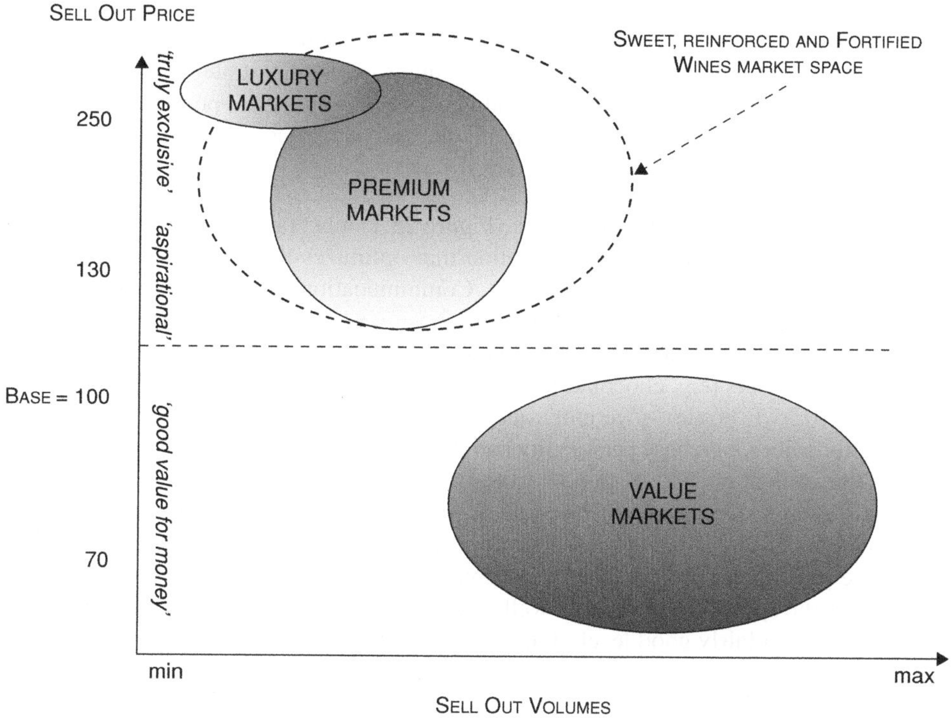

Figure 25.3 Market polarization: value bands.

up the different market spaces is the threshold level of the final price. Below this level, we find the *value* segment, above this level, the *premium* segment, and further above, is the *luxury* segment. Figure 25.3 shows this market structure. On the vertical axis, final prices valid in a specific geographical area with reference to a certain product type are displayed, on the horizontal axis, the relative volumes sold are exhibited.

Value markets are positioned in the medium-low and low portion of the product category, *premium markets* in the medium-high and high portion, *luxury markets* are positioned in the area that is above the *premium* area.

The highest sale volumes of the category are usually to be found in *value markets*, due to the apparent fact that there are positioned brands whose price, on average low (and anyway lower than a certain threshold), can be accessed by the majority of buyers in the considered market. Value offerings are of a *good-value-for-money* type. These wines are of a fairly acceptable technical and organoleptic quality, boast of basic performance focused on a *core benefit*, and are able to provide elementary utilities expected by consumers. They are offered at an averagely low price and are easy to find. The situation change totally as the price increases.

Beyond a certain threshold, in fact, we enter *premium markets* or 'value-added' markets. Price accounts for a certain amount and above (virtually limitless), and category sale volumes are lower, mirroring the fact that fewer people can afford to pay higher prices for the product. Purchasers looking for brands in these market bands have an interest in products boasting of high experiential performance (both of a technical and of an image-related kind). Finally, at the superior extreme, we find *luxury markets*. They represent a market space related to real luxury, inaccessible for most people, and basically non-conforming to common market rules.

This stated, it is meaningless to talk about a single marketing strategy. It is necessary to adjust the strategy based on the corresponding market. As it is not possible to consider volume-based policies due to the limited commercialized quantities of sweet, reinforced and fortified wines, these wines can be classified as follows:

- Up to €10 (0.75 L bottle) – producers (see Figure 25.1) must take special care to manage channel efficiency, through wide distribution that optimizes the presence in the points-of-sale where the category moves the most. Communication strategies can play a major role supporting both brand notoriety and predisposition of new consumption habits, with a moderate as well as accurate use of promotional levers and initiatives of co-marketing with related products (e.g. chocolate).
- Up to €50 (0.75 L bottle) – we enter another market, completely different in terms of sought-after values, purchase personality and (both expressed and unexpressed) consumer needs. Dominant attention has to be devoted to the harmonious combination of selected distribution, targeted communication and at-point-of-sale movement decisions (i.e. in-store merchandising). This can be achieved by accurately selecting points of sale compliant with price positioning and product/brand image, guaranteeing a discrete but continuous presence in the specialized press, as well as organizing special events. These initiatives are able to keep a fairly good level of attention and brand notoriety.

Product innovation limited to peripheral product features. Probably more than other product classes, sweet and fortified wines suffer from a major constraint hindering the possibility to deeply modify sensorial characteristics to follow the market. The relevant heritage – therefore the weight of tradition – and the strict linkage between heritage and product/brand reputation, impose fairly conservative product strategies. However, the model of innovation of peripheral product features shown in Figure 25.3 still applies.

This obviously represents an ideal mapping of the structuring and functioning of sweet and fortified wine consumer markets, based on the desk analysis and adaptation of the more comprehensive wine marketing literature. It would be worthwhile to test the interpretative framework through an empirical survey to further investigate both consumer and entrepreneurial behaviour. Interesting proposals for further research could be: (i) identify different strategic behaviours of wineries belonging to different socio-cultural markets, distinguishing between producing (therefore traditionally consuming) and non-producing countries; (ii) identify the best practices in product innovation pursued by wine sector players in order to increase the consumer needs to be met by the product offering; (iii) analyse brand communication initiatives, working out successful case histories representing useful benchmarks; (iv) define the role that these references play in the retailers' commercial portfolio.

Consumer demand for sweet and fortified wine classes can be significantly enhanced. It is in the hands of business players to innovate consumers' attitudes of mind by adopting more market-driven approaches to product development and commercialization, rather than the product-focused approaches traditionally used.

REFERENCES

Arnould, E.J. and C.J. Thompson. 2007. Consumer culture theory: twenty years of research. *Journal of Consumer Research*, 31:868–882.

Bagozzi, R. 1975. Marketing as exchange. *Journal of Marketing*, 39(4):32–39.

Balestrini, P. and P. Gamble. 2006. Country-of-origin effects on Chinese wine consumers. *British Food Journal*, 108(5):396–412.

Barham, E. 2000. Translating terroir: the global challenge of French AOC labeling. *Journal of Rural Studies*, 19:127–138.

Bergius, A. and Kragell, J. 2001. In Search of an Integrating Product Model. Integrating Consumer Behaviour and Customer Value. Available at http://www.ep.liu.se/exjobb/eki/2001/iep/018/ (last accessed 16 January 2013).

Beverland, M. 2004. Uncovering 'theory-in-use': building luxury wine brands. *European Journal of Marketing*, 38(3/4):446–466.

Beverland, M. 2005. The 'real thing': branding authenticity in the luxury wine trade. *Journal of Business Research*, 59:251–258.

Borden, N.H. 1964. The concept of the marketing mix. *Journal of Advertising Research*, 24(4):7–12.

Carù, A. and B. Cova. 2011. *Marketing e competenze dei consumatori*, Etas, Milano.

Chaney, I.M. 2000. External search effort for wine. *International Journal of Wine Marketing*, 12(2):5–21.

Charters S., L. Lockshin, and T. Unwin. 2000. Consumer responses to wine bottles back labels. *The Australian & New Zealand Wine Industry Journal of Oenology, Viticulture, Finance & Marketing*, 15(3):94–101.

CLP Regulation Service. Wine classifications available at http://www.rackwine.com/WineClassification/ (last accessed 8 January 2013).

De Magistris, T. 2004. Le determinanti del comportamento del consumatore: analisi teorica e verifica empirica per I prodotti biologici. Working paper no. 2, Istituto di Studi Economici, Università degli Studi di Napoli 'Parthenope'.

Fotopoulos, C. and A. Krystallis. 2003. Quality labels as a marketing advantage. The case of the 'PDO Zagora' apples in the Greek market. *European Journal of Marketing*, 37(10):1350–1374.

Gardner, B.B. and S.J. Levy. 1955. The product and the brand. *Harvard Business Review*, 33(March-April):33–39.

Gluckman, R.L. 1986. A consumer approach to branded wines. *International Journal of Wine Marketing*, 2(1):27–46.

Grunert, K.G. 1995. Food quality: a means-end perspective. *Food Quality and Preference*, 6:171– 176.

Grunert, K.G., T. Bech-Larsen, and L. Bredahl. 2000. Three issues in consumer quality perception and acceptance of dairy products. *International Dairy Journal*, 10:575–584.

Grunert, K.G., H.H. Larsen, T.K. Madsen, *et al.* 1996. *Market Orientation in Food and Agriculture*, Kluwer Academic Publishers, Boston.

Hauck, R. 1991. Buying behaviour and attitudes towards wine-findings of a field survey among younger consumers. *Acta Hoticulturae*, 295:127–132.

d'Hauteville, F., M. Fornerino, and J-P. Perrouty. 2006. Disconfirmation of taste as a measure of Country of Origin strength. An experimental approach on French wines. Paper presented at the 3rd International Wine Business and Marketing Research Conference, Montpellier, France, 6–8 July.

Jenster, P. and L. Jenster. 1993. The European wine industry. *International Journal of Wine Marketing*, 5(1):30–74.

Koewn, C. and M. Casey. 1995. Purchasing behaviour in the Northern Ireland wine market. *British Food Journal*, 97(11):17–20.

Kohli, A.K. and B.J. Jarowski. 1990. Market orientation: the construct, research proposition and managerial implications. *Journal of Marketing*, 54:467–477.

Kotler, P. 1967. *Marketing Management*, Prentice Hall, Englewood Cliffs, NJ.

Kotler, P. 1983. *Principles of Marketing*, Prentice Hall, Englewood Cliffs, NJ.

Kotler, P. 1984. *Marketing Management*, Prentice Hall, London.

Levitt, T. 1969. *The Marketing Mode*, McGraw-Hill, New York.

Lockshin, L. and J. Hall. 2003. Consumer purchasing behaviour for wine: what we know and where we are going. Paper presented at the International Colloquium in Wine Marketing, Adelaide, Australia, 26–27 July.

Lockshin, L., A. Spawton, and G. Macintosh. 1997. Using product, brand and purchasing involvement for retail segmentation. *Journal of Retailing and Consumer Services*, 4(3):171–183.

López Lluch, D.B., F. Vidal Jiménez, and F.J. Del Campo Gomis. 2011. Analysis of the importance of several aspects before the purchase of a wine bottle for Comunidad Valenciana (Spain) consumers, available at http://www.vdqs.net/2011Angers/documents/publication/Text/LOPEZLLUCH_VIDALJIMENEZ_DELCAMPOGOMIS.pdf (last accessed 9 January 2013).

Mattiacci, A. 2003. *Il Marketing Consumer-based*, Cedam, Padova.
Mattiacci, A., F. Ceccotti, and V. De Martino. 2006. Il vino come prodotto cognitivo: indagine esplorativa sui comportamenti giovanili. Paper presented at the Quinto Congresso Internazionale sulle Tendenze di Marketing, Venice, Italy.
McCarthy, E.J. 1960. *Basic Marketing: A Managerial Approach*, Richard D. Irwin, Homewood, IL.
McKitterick, J.B. 1957. What Is the Marketing Management Concept? In *The Frontiers of Marketing Thought and Science* (ed. F.M. Bass), AMA, Chicago, pp. 71–81.
Nelson, P. 1970. Information and consumer behavior. *Journal of Political Economy*, 78:311–329.
Nosi, C. and L. Zanni. 2004. From 'typical food products' to 'food-related services'. The Slow Food case as a new business paradigm. *British Food Journal*, 106(10/11):779–792.
Novak, L., L. Thach, and J.E. Olsen. 2006. Wowing the millennials: creating brand equity in the wine industry. *Journal of Product & Brand Management*, 15(5):316–323.
O'Dell, A. 2009. *A National Export Strategy for Canadian Wines*. Available at http://www.docstoc.com/docs/78902740/Go-to-Market-Strategy-Plan (last accessed 9 January 2013).
O'Neil, P. and S. Whatmore. 2000. The business of place: networks of property, partnership and produce. *Geoforum*, 31:121–136.
Odorici, V. and R. Corrado. 2004. Between supply and demand: intermediaries, social networks and the construction of quality in the Italian wine industry. *Journal of Management and Governance*, 8:149–171.
Olson, J.C. and J. Jacoby. 1972. Cue utilization in the quality perception process. *Advances in Consumer Research*, 3:167–179.
Orth, U.R. and P. Krška. 2002. Quality signals in wine marketing: the role of exhibition awards. *International Food and Agribusiness Management Review*, 4:385–397.
Orth, U.R. and K. Malkevitz. 2006. Packaging design as resource for the construction of brand identity. Paper presented at the 3rd International Wine Business and Marketing Research Conference, Montpellier, France, 6–7 July.
Orth, U.R., M. McGarry Wolf, and T.H. Dodd. 2005. Dimensions of wine region equity and their impact on consumer preferences. *Journal of Product & Brand Management*, 14(2):88–97.
Parasuraman, A., V.A. Zeithaml, and L.L. Berry. 1988. SERVQUAL: a multiple-item scale for measuring consumer perceptions of service quality. *Journal of Retailing*, 64(1):12–40.
Rao, A.R. and K.B. Monroe. 1989. The effect of price, brand name, and store name on buyers' perceptions of product quality: An integrative review. *Journal of Marketing Research*, 26:351–357.
Santini, C., A. Cavicchi, and B. Rocchi. 2006. Small wineries and strategic options: the role of premium bag-in-box. Paper presented at the 3rd International Wine Business and Marketing Research Conference, Montpellier, France, 6–7 July.
Smith, W.R. 1956. Product differentiation and marketing segmentation as alternative marketing strategies. *American Marketing Association. Journal of Marketing*, 21:3–8.
Thode, F.S. and J.M. Maskulka. 1998. Place-based marketing strategies, brand equity and vineyard evaluation. *Journal of Product & Brand Management*, 7(5):379–399.
Wilkie, W.L. and E.S. Moore. 2012. Expanding our understanding of marketing in society. *Journal of the Academy of Marketing Science*, 40(1):53–73.
Zanni, L., A. Mattiacci, and C. Nosi. 2005. Strategie emergenti nel business vitivinicolo: un'analisi comparata tra Australia e Italia. In *La valorizzazione economica delle tipicità rurali tra localismo e globalizzazione* (ed. C. Ciappei), Firenze University Press, Firenze, Italy, pp. 117–187.
Zeithaml, V.A. 1988. Consumer perceptions of price, quality and value: a means-end model and synthesis of evidence. *Journal of Marketing*, 52:2–22.

Plate 1.1 An image of a symposium from a Greek Attic red-figure cup attributed to the painter Douris (*ca* 500 to 460 BC). © The Trustees of the British Museum. All rights reserved.

Plate 3.6 Excess of sulfur bentonite on grapes cv. Sagrantino placed over a metallic net for dehydration.

Sweet, Reinforced and Fortified Wines: Grape Biochemistry, Technology and Vinification, First Edition.
Edited by Fabio Mencarelli and Pietro Tonutti.

Plate 3.10 Fruttaia of Avignonesi winery in Tuscany for the production of Vinsanto. Photograph by Mencarelli with permission of the owner (2005).

Plate 3.11 Erbaluce (left) and Trebbiano and Malvasia (right) bunches dried for Passito in Piedmont (photograph by Mencarelli, with permission of the owner, Ciek winery, 2010) and Tuscany (photograph by Mencarelli with permission of the owner, Frescobaldi Winery, Pomino, 2006), respectively, hung on wires or on a wooden frame.

Plate 3.12 Muscat of Alexandria (Zibibbo) bunches dried in the sun in a yard on Pantelleria island.

Plate 3.13 Malvasia bunches drying in the sun on the Aeolian Islands.

Plate 3.14 Zibibbo grapes drying in plastic tunnel on Pantelleria island.

Plate 3.15 New technology for air ventilation in a grapes dehydration facility.

Plate 3.16 Dehydration rooms for grapes with accurate control of the environmental parameters.

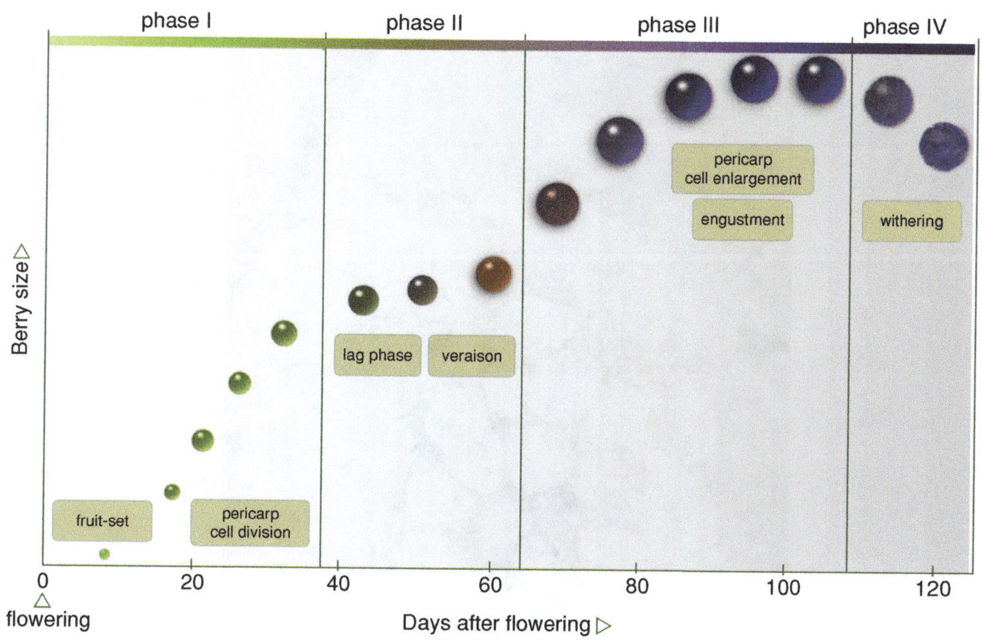

Plate 5.1 Diagram showing the four phases of berry development (D'Onofrio, unpublished).

Plate 9.1 Bunches of grapes untreated (left) and treated with copper sulfate (right).

Plate 11.1 Natural development of noble rot on vine.

Plate 11.2 Trays with bunches in the VIMACO withering chamber dedicated to postharvest botrytization and noble rot development.

Plate 11.3 Different phases of noble rot development on berry after *Botrytis* infection under controlled conditions in the VIMACO withering chamber.

Plate 14.1 White Muscat grape cluster before the harvest for Passito wine.

Plate 16.1 Grapes subjected to an off-vine drying process (soleo) in the so-called 'pasera' site. (a) Pasera located in Montemayor (Montilla–Moriles zone). (b) Pasera in the Axarquia (Malaga zone).

Plate 16.3 Musts obtained from Pedro Ximénez grapes in three different degrees of off-wine grape drying (soleo) process.

Plate 17.1 Different types of botrytized grains for Tokaji wine.

Plate 18.1 Grapes on straw mats before being pressed.

Plate 18.2 Grapes hanging up before being pressed.

Plate 19.1 Stages of noble rot development in berries.

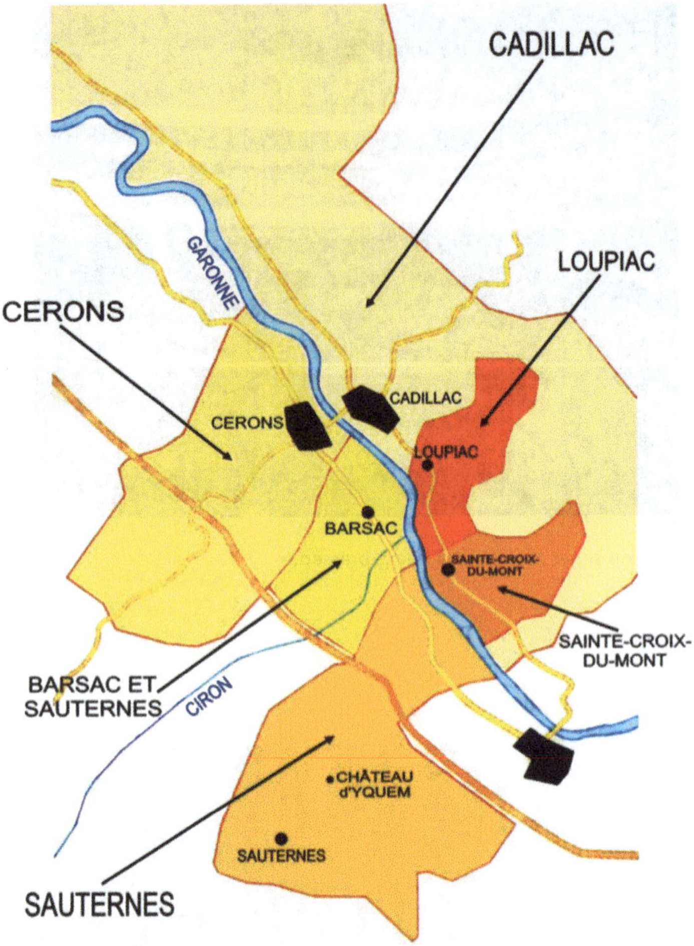

Plate 19.2 Map of the winegrowing regions of Sauternes.

Plate 19.3 Semillon grape variety botrytized (noble rot).

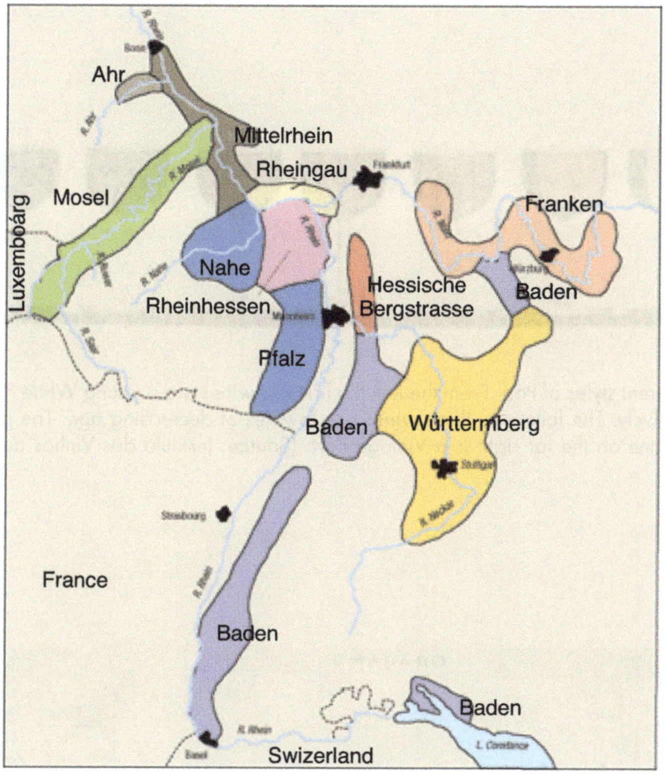

Plate 19.4 Map of the winegrowing regions in Germany.

Plate 19.5 Riesling variety botrytized grapes.

Plate 21.1 Different styles of Port. From the left, the first two wines are a young White Port and an aged White Port, respectively. The following three wines are Tawnies of decreasing age. The penultimate wine is a Ruby and the one on the far right is a Vintage Port. (Source: Instituto dos Vinhos do Douro e Porto, 2012.)

Plate 21.2 Robotic lagares in series, the gantry will work one ferment and then pass over to the adjacent one by means of the rails running under the tops of the lagares. Image courtesy of Symington Family Estates.

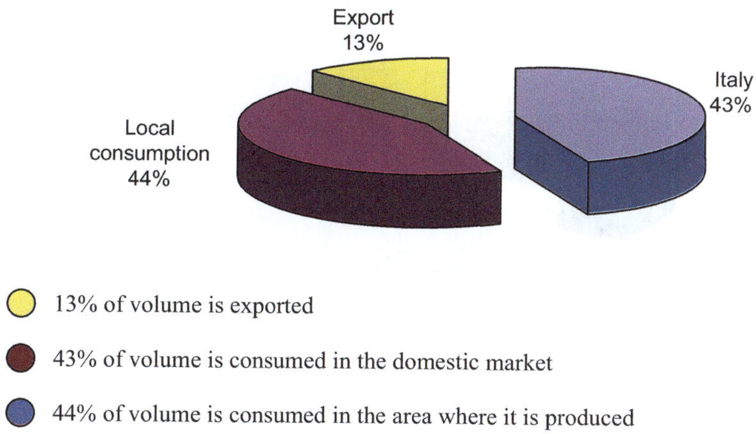

Plate 24.1 Italy's export and domestic consumption share by volume.

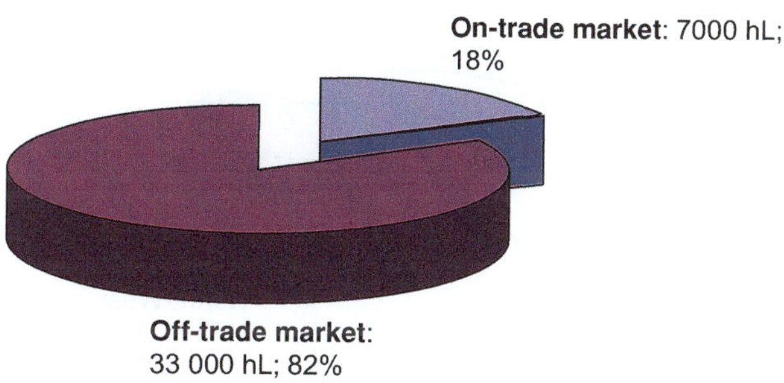

Plate 24.2 Italy's quantity and market share by channel.

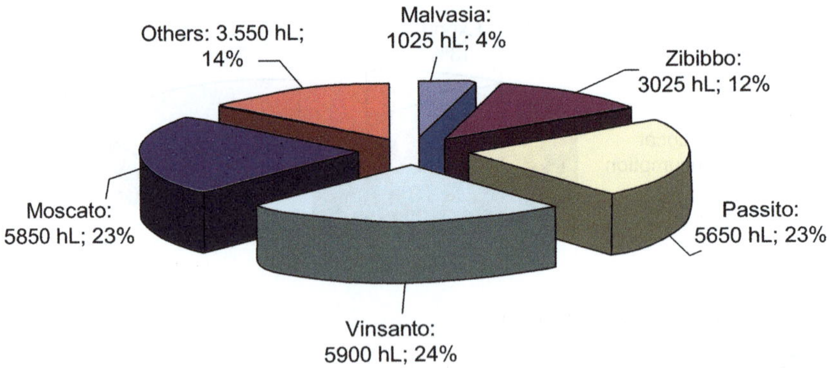

Plate 24.3 Major multiples off-trade sales and market share per wine.

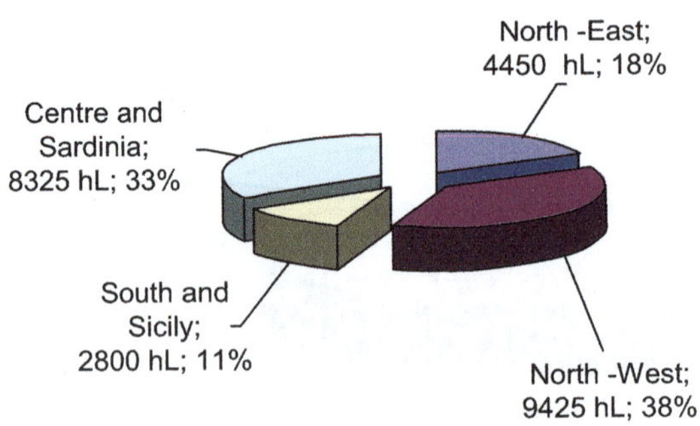

Plate 24.4 Major multiples off-trade sales and market share per region.

Analytical Index

Sweet, Reinforced and Fortified Wines: Grape Biochemistry, Technology and Vinification, First Edition.
Edited by Fabio Mencarelli and Pietro Tonutti.

Grapes and Wine Index

Sweet, Reinforced and Fortified Wines: Grape Biochemistry, Technology and Vinification, First Edition.
Edited by Fabio Mencarelli and Pietro Tonutti.